Endicott
College

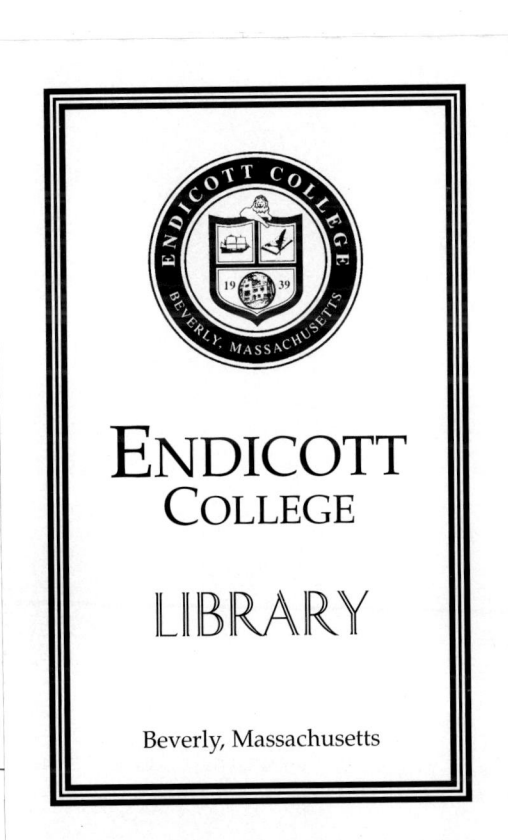

Library

Beverly, Massachusetts

African State and Society in the 1990s

African State and Society in the 1990s

Cameroon's Political Crossroads

Joseph Takougang
and Milton Krieger

WestviewPress
A Division of HarperCollins*Publishers*

All rights reserved. Printed in the United States of America. No part of this publication may be reproduced or transmitted in any form or by any means, electronic or mechanical, including photocopy, recording, or any information storage and retrieval system, without permission in writing from the publisher.

Copyright © 1998 by Westview Press, A Division of HarperCollins Publishers, Inc.

Published in 1998 in the United States of America by Westview Press, 5500 Central Avenue, Boulder, Colorado 80301-2877, and in the United Kingdom by Westview Press, 12 Hid's Copse Road, Cumnor Hill, Oxford OX2 9JJ

Library of Congress Cataloging-in-Publication Data
Takougang, Joseph.
  African state and society in the 1990s : Cameroon's political crossroads / by Joseph Takougang and Milton Krieger.
    p. cm.
  Includes bibliographical references and index.
  ISBN 0-8133-3428-4 (hc)
  1. Cameroon—Politics and government—1981–   2. Democracy—Cameroon.   I. Krieger, Milton.   II. Title.
JQ3525T35   1998
320.96711—dc21   97-43832
                                                                                                 CIP

The paper used in this publication meets the requirements of the American National Standard for Permanence of Paper for Printed Library Materials Z39.48-1984.

10   9   8   7   6   5   4   3   2   1

To the Takougang and Krieger families

Contents

Illustrations   ix
Acronyms   xi
Acknowledgments   xiii
Preface: The Research Setting, Methodology, and Theory   xv

Introduction: Cameroon's State and Society in African Context   1

A Metaphor for Africa in the 1990s, 1
Cameroon in Africa, 1960s–1990s: A Synopsis, 3

1 Scholarship on the African and Cameroon State   13

State-Civil Society Scholarship in the 1990s, 16
The Cameroon Scholarship, 22
The Questions Ahead, 27

2 Ahidjo and the Single-Party State, 1958–1982   35

The Emergence of Ahmadou Ahidjo, 36
The Years of Trial, 1958–1962, 37
Creating a Single-Party State, 1962–1966, 44
From Single Party to United Republic, 1966–1972, 47
Ahidjo's Presidential Monarchy, 1972–1982, 50

3 Biya's Early Presidency, 1982–1986   63

From Ahidjo to Biya: The Transfer of Power, 1982, 63
Disputed Succession and Legacy:
   Ahidjo and Biya, 1983–1984, 66
The New Deal: Aspiration, Compromise, and Reality, 76

4 The State Under Pressure, 1986–1990   89

Failed Promises and the Illusion of Reforms, 90
Cameroon's Declining Economy, 97
Global Politics and Domestic Demands for Reform, 102

5 Crisis Years, 1991–1992   115

The Press, 117
Opposition Challenge and Regime
  Response to Mid-1991, 123
Villes Mortes/Ghost Towns, Mid-Late 1991, 126
Political Parties and Mounting Tensions, 131
The Balance of Forces, Late 1991, 139
Legislative Electoral Test, Early 1992, 142
Presidential Electoral Test, Late 1992, 146

6 Impasse, 1993–1997   159

Party and Institutional Politics: The Early 1993 Profile, 160
Regional and Ethnic Politics, 1993–1997: Anglophones, 161
Regional and Ethnic Politics, 1993–1997: The North, 169
Regional and Ethnic Politics Miscellany, 1993–1997, 180
Constitutional Debate and Action, 182
Institutional, Electoral, and Party Politics, 1995–1997, 194

7 Cameroon and the Prospects for Democratization   217

Disarray in the Political Class:
  Principals and Auxiliaries, 219
An Alternative Political Class? 228
Emergent Social Experience, Formations,
  and Alignments, 231
Foreign Factors, 241
Conclusion, 243

Appendix   251
Bibliography   255
Index   271

Illustrations

Map of Contemporary Cameroon.	xix
Newspaper Kiosk, Bamenda, 1995.	122
Douala, June 27, 1991: The Response to President Paul Biya's "Sans Objet" Speech.	126
John Fru Ndi, Chairman, Social Democratic Front.	129
A Bamenda Rally Begins, July 1991.	130
Maïgari Bello Bouba, President, National Union for Democracy and Progress.	133
Adamou Ndam Njoya, President, Cameroon Democratic Union.	134
Display of the Federation Flag, Bamenda, 1992.	144
Blockade Against Security Forces During Bamenda's State of Emergency, November 1992.	151
Attack on John Fru Ndi, Yaounde, November 3, 1993.	196
John Fru Ndi Mediating Village Boundary Dispute, North West Province, mid-1995.	202
The SDF Cavalcade, Maroua Party Congress, May 1995.	205
John Fru Ndi with Commonwealth Delegation, Yaounde, 1995.	220

Acronyms

AAC	All Anglophone Conference
ALUCAM	Société Aluminium du Cameroun
ARC-SNC	Alliance for the Reconstruction of Cameroon-Sovereign National Conference
BCD	Banque Camerounaise de Développement
BMM	Brigades Mixtes Mobiles
CAM	Cameroon Anglophone Movement
CAMAIR	Cameroon Airlines
CBA	Cameroon Bar Association
CDC	Cameroon Development Corporation
CDF	Cameroon Democratic Front
CDP	Cameroon Democratic Party
CDU	Cameroon Democratic Union
CENER	Centre National des Etudes et des Recherches
CFA	Communauté Financière Africaine
CFDT	Compagnie Française pour le Développement des Fibres Textiles
CND	Centre National de Documentation
CPDM	Cameroon People's Democratic Movement
CPNC	Cameroon Peoples National Convention
CRATRE	Cercle de Réflexion et d'Action pour le Triomphe du Renouveau
CRTV	Cameroon Radio and Television
CUC	Cameroon United Congress
DC	Démocrates Camerounais
DIRDOC	Direction Générale des Etudes et de la Documentation
FAC	Front of Allies for Change
FPUP	Front Populaire Pour l'Unité et la Paix
GATT	General Agreement on Tariffs and Trade
IMF	International Monetary Fund
ISH	l'Institut des Sciences Humaines
KNDP	Kamerun National Democratic Party
LAVANET	National Veterinary Laboratory

MANC	Mouvement d'Action Nationale Camerounaise
MDR	Movement for the Defence of the Republic
MINAT	Ministry of Territorial Administration
MP	Mouvement Progressiste
NCDM	National Coordination for Democracy and a Multiparty System
NCOPA	National Coordination of Opposition Parties and Association
NDI	National Democratic Institute for International Affairs
NUDP	National Union for Democracy and Progress
NUCW	National Union of Cameroon Workers
OAU	Organization of African Unity
OCALIP	L'Organization Camerounaise pour la Liberté de la Presse
PDC	Parti des Démocrates Camerounais
PI	Paysans Indépendants
SAP	Structural Adjustment Program
SCB	Societé Camerounaise de Banque
SCNC	Southern Cameroons National Council
SCPC	Southern Cameroons Peoples Conference
SDECE	Service de Documentation Extérieure et de Contre-Espionage
SODECOTON	Société de Développement du Coton
SEDOC	Service des Etudes et de la Documentation
SDF	Social Democratic Front
SMIC	Société Mobilière d'Investissement du Cameroun
SNEC	Société Nationale des Eaux du Cameroun
SNC	Sovereign National Conference
SNI	National Investment Society
SONARA	Société Nationale de Raffinage
SONEL	Société Nationale d'Electricité
SOPECAM	Cameroon Press and Publishing Company
SOTUC	Société de Transports Urbains du Cameroun
UC	Union Camerounaise
UFC	Union for Change
UPC	Union des Populations du Cameroun
USAID	United States Agency for International Development

Acknowledgments

We owe many people many thanks for our collaboration. Victor Le Vine introduced us in 1993 and Mette Shayne of the Herskovits Africana Library at Northwestern University has consistently helped us locate and use materials. Before we met, and since, many others have shared documents and guidance; we thank in particular E. M. (Sally) Chilver, John Cinnamon, Cyprian Fisiy, Ambroise Kom, Célestin Monga, and anonymous readers of various drafts of the manuscript. Assistance from those who are part of our story is cited where appropriate in the text; Cameroonians we think may prefer anonymity will recognize our enormous debt.

Both our universities have been generous with funding and technical assistance. At the University of Cincinnati, the Charles Phelps Taft Memorial Fund provided a Summer Grant and publication support, and the Office of the Vice-President for Research and Advanced Studies provided a Summer Faculty Research Fellowship. Western Washington University provided research leave for time in Cameroon, publication support, and technical preparation; Geri Walker and Kevin Short were especially helpful. Thanks are extended also to the American Historical Association for a Bernadotte E. Schmitt Grant.

Work on Cameroon entails some complexities we have compounded by writing in considerable detail. Not everyone will like our naming conventions (the sequence of given and family names in Cameroon varies), our choices of what to translate from the French into English (translations are ours; we omit them where we expect our readers not to need them), and our choices of whether to use French or English names for political parties and the like (English is used, except where a French title is more familiar in the English language world, or is used by anglophones in Cameroon). More serious issues arise where we risk distortions, and giving offence, when calling people who are above all Cameroonians "an anglophone," "a Beti," "a Catholic," and so forth. We trust they will understand our needs and bear with choices made in good faith, so as to provide the first comprehensive study of Cameroon's recent history and politics written in English for nearly a decade.

We two, none of those above, share responsibility for any defects in the text. We also share loving and indispensable support over *many* years from Prudentia Takougang and Judy Krieger, and from immediate and extended family members in two countries, to all of whom we dedicate this book.

Joseph Takougang
Milton Krieger

Preface: The Research Setting, Methodology, and Theory

We are historians, concerned here with recent and contemporary politics in Cameroon, which is the site of one of Africa's least known but most tenacious, complex democratization struggles of the 1990s. The book starts with a survey of the research environment and pertinent literature, and then develops in three, increasingly detailed stages. The first is preliminary, about Ahmadou Ahidjo's presidency from independence in 1960 to his retirement in 1982. The second covers the period 1982–1990 and is transitional, recounting his successor Paul Biya's effort to define his own presidency and respond to new pressures while maintaining Ahidjo's legacy. The last stage, which is our primary focus, deals with the unprecedented challenges to Biya's regime since 1990. They show Cameroon, despite its relative anonymity and current political impasse, to be a pivotal arena for current African politics. As a sub-theme throughout, Cameroon is related as a country case study to the broader sub-Saharan experience.

Writing the book has required our attention to narrative lines, people, and institutions that are unfamiliar beyond Cameroon specialists, and its detail thickens as we move from 1960 to the present. The risks facing scholarship of such density, particularly for contemporary events that challenge us to separate essential history from ephemeral reportage, realities from representations, seem worth taking, to get the story told, and make it accessible to English language readers.

The material before 1990 is nationally focussed, with a monolithic state and party as the key agencies, using standard published documentation. Thereafter, reflecting key shifts in Cameroon's experience, sub-national coverage increases, using less conventional sources based in civil society, particularly newly issued print genres including an independent press, and a variety of oral material ranging from formal interviews with party leaders to bus, taxi, and street talk. Such evidence immediately raises questions about its character, and therefore the book's value. In order to ground the text in the conditions, procedures and protocols of our research, and to authenticate the work, we begin with a survey of

Cameroon's scholarly environment, emphasizing key changes in the 1990s which influence the book.

Research in Cameroon until 1989, the year Mark DeLancey published the most recent English language general study, was not as logistically demanding or politically sensitive as thereafter. Yaounde, the capital, gave scholars access to ministry data of the kind used systematically by agencies like the United Nations as well as domestically. It also provided a welcoming network of faculty at the University of Yaounde, a United States Agency for International Development (USAID) library with archive quality governmental and non-governmental material kept current, a United Nations document center, and the head office of l'Institut des Sciences Humaines (ISH). Elsewhere, ISH operated provincial capital research sites that on our limited evidence varied in quality but not in purpose. Pius Soh's Bamenda site, typical of North West Province marginality, had little room and few resources, but his own local studies and the help he extended made work there productive. In Garoua, handsomely endowed during Ahidjo's presidency, Eldridge Mohammadou directed a model bureau, well stocked and maintained, staffed with associates and apprentices. There was also the gem of a library at Bambui's Catholic seminary (North West Province) and a National Archive site in Buea (South West Province) which was more useful than Yaounde's. Conditions to 1989, then, favored research; a scholar's conventional diligence was rewarded.

The setting soon changed. Challenges to the regime began in 1990, persist to the present, and significantly alter scholarly activity in Cameroon. In mid-1991, for instance, when we were both there at a particularly turbulent time, travel and access were severely restricted in and between southern cities by a week-day general strike in virtually all public and private service sectors, except in and near Yaounde where the state's writ still held its force. But to settle there meant basic ignorance of conditions elsewhere; more was to be learned in cities like Douala, Bafoussam, Bamenda, and their hinterlands. Excepting Douala, however, travel from these areas to Yaounde meant weekend journeys and week long sojourns, unless one had a private car, which was risky to use in the face of defiant strikers. A choice between two polar opposite research environments was required for political study. Yaounde with its records was a less productive place to be than the others, especially Bamenda, a venue emphasized in our post-1990 coverage as regional, local, and especially anglophone events developed. The choice may raise questions of selection in our coverage, but it appropriately realigns the balance of attention both in politics and scholarship between the long-privileged capital city with its state apparatus and so many other sites and dimensions of Cameroon's experience neglected since the 1960s.

Preface   xvii

Research limitations mounted after 1991. A decree terminated ISH, to save money and disperse dissidents; Mohammadou's Garoua facility, for example, was closed and then vandalized, and he left to work in Nigeria. What previously passed for "normalcy" at the university vanished. With classes often suspended and teachers away, time spent there was less useful than before, and the creation of new universities throughout Cameroon from 1992 broke up Yaounde's core of faculty, including dissidents, a short term logistical loss for visiting scholars even if (one hopes) a long term gain for Cameroonians. When USAID closed its Yaounde office in 1993, a political signal to the regime following electoral fraud, the library was dispersed. Surviving research facilities continued to deteriorate through the decade.

But there were factors since 1990 that compensated for such limits to our sources. We cite three examples, vital to our work. First was a proliferation of print materials. Some appeared outside Cameroon, in a paradoxically crucial way. With all the conflict there, access to official government sources faced more than logistical problems; it would have been naïve, probably foolish, for us to approach key ministries in Yaounde where data was politically sensitive. But a major factor offset such deficits in the "official" domain. Structural adjustment conditions moved data from key economic ministries into publications abroad, and also opened Cameroon to independent study, ultimately printed, like the World Bank's landmark six month poverty survey of 1994. We could not have easily gained such documentary access inside Cameroon, and are grateful for the chance to use such sources published elsewhere, especially in the Introduction. A vast array of advocacy literature from political parties and civic associations also appeared, not conventionally published but widely available in Cameroon. Such material, ranging from draft constitutions to one page handbills passed out on the streets, constitutes a research arena we substantially rely on.

Second, and the most important print genre, independent journalism grew dramatically, increasing field and investigative reportage and broadening analytic and editorial content. This made the public record so prolific and diverse as to give Cameroon research a new quantitative *and* qualitative base. We have consulted close to 100 newspapers. Care using them and sorting their biases is essential, as all historians and African scholars know, and as an example shows. A minor figure in our story, Alhaji Tita Fomukong, from Bali near Bamenda, created the Cameroon National Party in 1991, proclaiming it oppositional. But he urged that taxes be paid during the general strike and that National Day, May 20, 1992, be celebrated, not boycotted, making credible a charge that the party was a regime front. Late in 1992, he burned to death in his torched house while North West Province endured a state of emergency following a disputed presidential election. Whereas regime newspapers called him a martyr to opposition mob justice, the opposition press ran a widely circulated local story that he had hired village youths to fan the conflict, refused to pay them, and was their victim. Given shadowy party and personal records, which source is credible, in a story quite vital for its local flavor? Accuracy *is* an issue. But only the press tracks such stories, or key people like Cameroon's premier writer Mongo Beti since his permanent return from exile in 1993, in interviews, stories of continued efforts to silence him, his April 1995 *Génération* reportage on his birthplace Mbalmayo, and his candidacy for political office there in 1997. It is curious that scholars of current Africa cite journalism's contribution to civil society but discount or ignore what it reports, the basis of its value and appeal: could one report eastern and central Europe in the 1980s without using *samizdat* sources? The study of Cameroon in the 1990s requires the comprehensive sampling and cautious use of its newspapers, as we have done.

A third new research channel emerged. Just as the press in the 1990s is a new resource, the people who conducted strikes, boycotts, and demonstrations speak freely. As we will see, the press faces reprisals, but a state authority which has lost much of its capacity for systematic punishment now seldom sanctions private citizens. By mid-decade, one could "catch up" with their reflections on five years of crisis, and on the deeper past. Few were reticent, and what follows records formal interview and focus group efforts to relate what a range of Cameroonians were saying and thinking beyond the published record, and very frank talk in "off-licences," city taxis, and country busses.

We use rumor sparingly, and call it that where it does appear below. It is rampant now in Cameroon. It ranges from commonplace charges, like bribes to switch parties and newspaper stories paid for or planted, to the more spectacular kind which circulated in Yaounde about a 1994 coup plot, based on armed troop units parading National Day among others unarmed, and attributed to a very high regime politician who was soon thereafter demoted. Evidence which follows, as distinct from rumor, has at least the authority and citation of print, formal interview, or informal conversation.

We acknowledge geographical gaps in coverage. Cameroon is more than California-sized, with 240 ethnicities, three colonial legacies prior to 1960, and three constitutional frameworks since then. The past is complex and the current experience differs everywhere in character and intensity. We are most familiar with the four provinces which are oppositionist, at times insurrectionary since 1990: North West, South West, Littoral, West. One of us, born, raised and schooled in Cameroon, is from that "Great West" area. The other came professionally late to Cameroon, and anglophone areas color his contribution, especially his time in Bamenda since

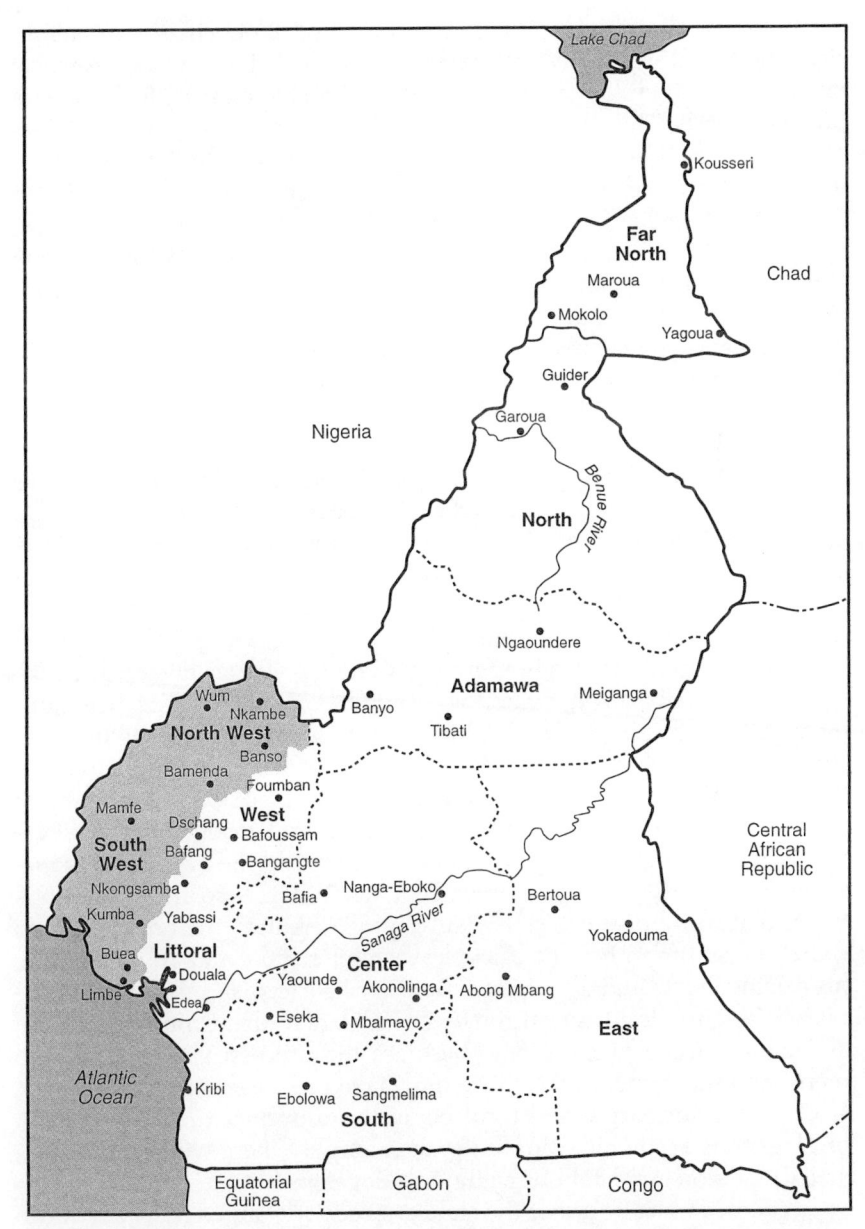

Map of Contemporary Cameroon. From Milton Krieger, "Cameroon's Democratic Crossroads, 1990–4," *The Journal of Modern African Studies* 32, 4 (1994), p. 604. Reprinted with the permission of Cambridge University Press.

1989 (adding up to two years, and addressing a deficit in scholarship since the 1960s, as noted above). There is nothing below on East Province; one has to go there to cover it, and (Peter Geschiere aside) few scholars do so regularly. There is not enough on the three provinces of the "Great North." Only *L'Harmattan* among newspapers has been to our knowledge its particular voice, although northern news and analysis is plentiful in the national press. Between us, recent travel there covered only two weeks, May 1995, although it was useful because it covered the Maroua Congress of the major opposition party, the Social Democratic Front (SDF). Center and South Provinces raise fewer problems of coverage. All national newspapers are prolific on life there, and *Le Temps, Le Patriote, La Caravane,* and *Le Courrier* support their regime-dominated politics, balancing hostility found in other sources.

Such is our documentary base; we believe that the value of our non-official sources far surpasses what comparable effort in conventional tracks would have provided for the 1990s. Cameroon data has engaged many scholars in eight nations since 1960: Cameroon itself, Canada, France, Germany, Great Britain, Japan, The Netherlands, and the United States of America. We have used their collective wisdom on politics, except where it appears in German and Japanese languages.

We rely on this varied, abundant record for Cameroon, and also on current Africana debates on methodology and theory. This Preface concludes by aligning our approach to this literature, both for Cameroon where our story unfolds and for Africa at large, so as to provide a comparative setting for this essentially one-country study. We place Cameroon among polities governed autocratically before 1990, and recount the achievements and vulnerabilities thus entailed. We explain the hegemonic structures of Ahmadou Ahidjo's presidency, 1960–1982, and their modifications to 1990 by his successor, Paul Biya. Thereafter, the book shifts to sharp, as yet unresolved conflict between his regime and an opposition, his calcified state and a fledgling, fluid civil society, and uses the vocabulary of regime transition through liberalization, democratic transition, and democratic consolidation. These are the processes the book documents and the terms it uses. Since they have precise scholarly meanings, globally and for Africa, and recur below, we introduce them here with a brief, generic academic scaffolding, then amplify them with more extensive discussion and citations as the text develops.

To characterize our use of the state-civil society language which dominates the recent study of Africa, we assume, first, the African state as the primary agency which assembles and allocates basic national resources, and will utilize Jean-François Bayart considerably when developing this theme below.[1] Cameroon's indigenous and colonial diversity surely brings "imagined community" language to mind; it took determined, often forcible state and nation building by political elites to secure the structure. The process transformed a federation into a unitary state by 1984, meeting Ahidjo's (and French) specifications; Richard Joseph called it Gaullist at the time and its foes called it Vichist by 1995.[2] It was a typically hasty African experience of post-colonial "late development," building a nation-state polity and starting a modern economy in ways which limited democratic initiatives. The state structure and a pervasive clientelism have been dominant from the start. By the time of Ahidjo's 1982 retirement or shortly thereafter, publications on the state apparatus, offices, and officeholders who mattered were well advanced. The key formulations were Bayart's "recherche hégémonique/hegemonic project" and Pierre Flambeau Ngayap's "classe dirigeante/ruling class."[3] The former covered Ahidjo's success moving Cameroon's governance from pacification of open rebellion in the 1960s to the achievement of acquiescence and the marginalization (but not the end) of dissent in the 1970s. The latter identified roughly 1,000 Cameroonians whose offices and connections to Ahidjo operated the system and transmitted its orders and values to the population at large. Simultaneously, however, as counterpoint, Joseph and Achille Mbembe documented the parallel force and legacy of the political party which resisted the regime until 1970, the Union des Populations du Cameroun (UPC), and its founder Ruben um Nyobe, murdered in 1958, the year Ahidjo emerged as the leader the French chose to continue their authority past independence.[4]

It is fundamental to understand that these two features of Cameroon's political landscape, hegemonic and counter-hegemonic, state-directed and civil society-oriented, remain in the field as legitimizing claims for the regime and opposition in the 1990s, and that they give the current struggle its indigenous texts and colors. The African state's capacity to dominate has, in fact, been generically at issue since Goran Hyden's work on escape and exit routes began to appear in 1980.[5] Richard Sklar's 1982 presidential African Studies Association address and Patrick Chabal's edited 1986 volume widened the path to democratization and civil society as key issues in Africa.[6] This literature has flourished since 1990.

Here, drawn from a mainstream, comparativist, global perspective and literature review, is John Hall's 1995 formulation of civil society as

> the self-organization of strong and autonomous groups that balance the state [where] membership of autonomous groups needs to be both voluntary and overlapping if society is to become civil.[7]

This text and his further stipulations about the state-civil society nexus, that "the coexistence of competing political traditions" is preferable to some dominant form and that "a country is only strong when an orderly civil society works with the state," represent a major paradigm for Africa and Cameroon. This view of civil society *empirically* covers many facts of life in the 1990s: critical expression, associational action, and (among their derivatives) new political parties which have emerged beyond the state's confines. It also *normatively* covers a two-fold insistence for a decade now from Michael Bratton, Naomi Chazan, and other key Africanists: for democratic outcomes to prevail, the public sphere needs both citizens with multiple identities and loyalties, and (recognizing the claims of both authority and liberty) some complementarity and convergence between state and civil society.[8] In Larry Diamond's phrase: "Authority must be questioned and challenged, but also supported."[9]

Two more facets of this literature need attention. Bratton in particular differentiates political society and civil society, so as to identify those actors, in the former, who contest for state power rather than keeping an autonomous distance from the directly political arena, as those in civil society do.[10] Theoretical debate considers whether political society occupies the state's or civil society's space, but the distinction matters to us more as a practical issue below, as we track the fluid acts and intentions of many of Cameroon's current public figures. And Thomas Callaghy, Robert Fatton, and René Lemarchand are skeptical about civil society's structures and actors, citing their potential for *in*civility, opportunism, cooptation, and fragmentation, a useful caution about the need to avoid demonizing the state and deifying civil society.[11]

A second vocabulary in this mainstream Africanist literature leads beyond the language of state and civil society to three phases of "regime transition" which *may* emerge when challenges to authoritarianism arise: liberalization, democratic transition, and democratic consolidation. Liberalization refers to events and processes visible in much of Africa by 1990, as a freer press, national conferences, political parties, elections and the like surfaced. Many reasons are cited and combined to explain their appearance, and the ground that state leaderships yielded to them, if grudgingly and only for a time. A standard version recognizes, in Chazan's words, that "hegemonic impulses rarely gained total ascendancy" for domestic "statist" interests during the first independence decades, and that dissident elites and (less prominent in much of this literature) broader community forces were at work.[12] There is also the look abroad for Africa's cues, to eastern and central Europe's changes in the late 1980s and to the politically conditioned demands of structural adjustment on Africa's state and market arrangements.

Whatever the specific historical framework or form of analysis for liberalization, it is now considered preliminary, necessary but not sufficient, for democratic transition and consolidation. Summarizing a mainstream literature he has helped shape, Diamond in 1997 outlines these

Preface xxiii

two phases.[13] He locates *transition* by contrasting merely "minimalist" forms of democracy and "pseudodemocracy," providing only periodic elections for which the regime controls the rules and procedures, with the movement to "a sufficiently fair arena of electoral competition to make it possible for the ruling party to be turned out of power" and for widespread personal freedoms, an effective rule of law, and other broadly democratic practices to prevail. *Consolidation* is the next step, predicated for Diamond on

> time to work with and become habituated to democratic institutions, to shape them to fit ... particular cultural and political circumstances, and to allow them to sink deep roots of commitment among all major players and the public at large.

These two conditions are precarious in Africa, although Diamond sees progress. He counts eighteen polities both authoritarian and pseudodemocratic out of fifty-three in 1997, leaving seventeen (mostly small) with some real democratization advances. Cameroon is not among the seventeen.

Some scholars, especially Africans, have raised different voices and issues in the 1990s, attacking the paradigms sketched above. Eghosa Osaghae finds that in the "democratic transition school ... a clearly African perspective is still far from crystallising."[14] That school's "strategic elite perspective" discounts struggle against regimes lodged deeper in African history, in social sectors disparaged as parochial but in fact autonomous and significant, than the "diffusing" account of democracy from abroad, and from above, recognizes. It also wrongly privileges elite cleavage over "ordinary peoples seeking better and more responsive governance" as the real oppositional force at work in the 1990s. The universal standards and comparative analysis of this "school" are not sufficiently aligned to African developmental needs. They use formal, institutional criteria derived from studies of Latin American and European democratic breakdown and renewal, and they reflect "the class character of liberal democracy" with its preference for an individualistic notion of citizenship and a reduced state presence. Dickson Eyoh echoes Osaghae in a 1996 text criticizing this form of "the grand narrative of modernization" and is more detailed about both the culpability of what Bratton called "political society" and the strong role of ethnic and kinship networks and other expressions of the "chaotic pluralism" from which "responses to the crises of state legitimacy" issue.[15]

They gain comprehensive support in a 1996 book by Célestin Monga, a Cameroonian whose direct contribution to democratization there made him a prime regime target before he left the country in 1992.[16] African states in his reading now have a short reach and an unsure grasp. Civil societies, which he roots deeper in the rural population than most scholars do, now test the "political market" with caution learned from harsh experience since they first surfaced around 1990. Democratization efforts may mislead conventional wisdom, by their lack of institutional definition the social science paradigms emphasize and their frequent look of nihilism more than civility. But even though stalled in the mid-1990s, they mask with "optimal anarchy" (recalling Eyoh's "chaotic pluralism") a patient, creative search for better leaders within the democratic movement itself, and ultimately for a legitimate polity not yet close to emerging in the ways the formulaic "transition" scholarship measures: "the threshold of effective action is particularly unstable. Theory is not linear in Africa..."

We find nothing mutually exclusive in these perspectives and use them both. The mainstream literature *does* draw much from publications based outside Africa decades ago, as its collective bibliography and its congruence with, for example, the first pages of a 1996 book by the pioneer comparativists Juan Linz and Alfred Stepan confirm.[17] But nothing in this fact denies the salient appeal for Africans of globally understood human and civil rights, institutionally crafted and formally protected against the oppressive states they have long endured. For its part, the critique by Africans correctly draws attention to people in social formations we know to be active, relevant, and potentially decisive in Cameroon, who are little known but who our book increasingly turns to in its latter stages.

As historians who wonder about the balance of priorities in the scholarship between basic evidence and intricate debate, and who are dealing with events as yet little known in the English language world, our ultimate concern is chronological, narrative and empirical, more about Cameroon than "discourse": how have Cameroonians conducted their lives since 1960, especially in their work to democratize the forms and conduct of their governance in the 1990s? But we hope also to contribute a case study to the broader social science Africanist scholarship we have introduced here and use below.

We complete our writing in 1997, with Congo (Kinshasa) in the background. Cameroon's local and national elections in 1996–1997 prolong *its* impasse and could lead to renewed crisis and chaos. Its nation state integrity is questionable. The familiar space on the map is still formally but in many ways not functionally intact, under conditions which rank it among collapsing (not yet collapsed) states, if one applies the criteria of William Zartman's 1995 study.[18] How our account of this historical experience and current condition emerges and whether it convinces, readers will judge. The Introduction expands the profile of Cameroon and Chapter 1 expands the literature review contained in this Preface. Chapter 2 surveys Ahidjo's leadership, 1958–1982. Chapters 3 and 4 cover Biya's presidency through 1990. Chapters 5–7 are about 1991–1997, what Cameroonians call (in their official bilingual usage) "la crise/the crisis" of those years, and how it may play out. Opening the book to scrutiny, we trust that our collaboration approaches comprehensive coverage and offers a fair balance between complex forces.

Notes

1. The study of Cameroon led to his fundamental work, Jean-François Bayart, *L'Etat en Afrique: La Politique du Ventre* (Paris: Fayard, 1989), translated by Mary Harper, Christopher and Elizabeth Harrison as *The State in Africa: The Politics of the Belly* (London: Longman, 1993).

2. Richard Joseph, ed., *Gaullist Africa: Cameroon under Ahmadu Ahidjo* (Enugu: Fourth Dimension, 1978). A francophone at a meeting of regime critics in Yaounde, February 3, 1995, referred to Biya's regime as Vichist, and was clearly understood and approved.

3. Jean-François Bayart, *L'Etat au Cameroun* (Paris: Presses de la Fondation Nationale des Sciences Politiques, 1979); Pierre Flambeau Ngayap, *Cameroun: Qui Gouverne? de Ahidjo à Biya, l'héritage et l'enjeu* (Paris: L'Harmattan, 1983).

4. Richard Joseph, *Radical Nationalism in Cameroon: Social Origins of the U.P.C. Rebellion* (Oxford: Clarendon Press, 1977); Achille Mbembe, ed., Ruben um Nyobe, *Le Problème National Kamerunais* (Paris: L'Harmattan, 1984), the first of his related texts we will use below.

5. Goran Hyden, *Beyond Ujamaa in Tanzania: Underdevelopment and an Uncaptured Peasantry* (Berkeley and Los Angeles: University of California Press, 1980).

6. Richard Sklar, "Democracy in Africa," *African Studies Review* 26, 3/4 (1983), pp. 11–24; Patrick Chabal, ed., *Political Domination in Africa: Reflections on the Limits of Power* (Cambridge: Cambridge University Press, 1986).

7. John Hall, ed., *Civil Society: Theory, History, Comparison* (Cambridge: Polity Press, 1995), p. 15 for this quotation, pp. 20, 23 for the passages which follows.

8. Michael Bratton, "Civil Society and Political Transitions in Africa," in John Harbeson, Donald Rothchild and Naomi Chazan, eds., *Civil Society and the State in Africa* (Boulder: Lynne Rienner Publishers, 1994), pp. 51–81; Naomi Chazan, "Between Liberalism and Statism: African Political Cultures and Democracy," in Larry Diamond, ed., *Political Culture and Democracy in Developing Countries* (Boulder: Lynne Rienner Publishers, 1993), pp. 67–105.

9. Larry Diamond, "Introduction: Political Culture and Democracy," in Diamond, *Political Culture and Democracy in Developing Countries*, p. 13.

10. Bratton, "Civil Society and Political Transitions," p. 56.

11. Thomas Callaghy, "Civil Society, Democracy, and Economic Change in Africa: A Dissenting Opinion About Resurgent Societies," in Harbeson, Rothchild and Chazan, *Civil Society and the State in Africa*, pp. 231–253, esp. p. 235; Robert Fatton Jr., "Africa in the Age of Democratization: The Civic Limitations of Civil Society," *African Studies Review* 38,2 (1995), pp. 67–99; René Lemarchand, "Uncivil

States and Civil Societies: How Illusion Became Reality," *The Journal of Modern African Studies* 30,2 (1992), pp. 177–191.

12. Chazan, "Between Liberalism and Statism," p. 85.

13. Larry Diamond, *Prospects for Democratic Development in Africa* (Palo Alto: Stanford University, Hoover Institution, 1997), pp. 3, 5 for the quoted passages which follow, pp. 54–55 for the tabulations on democratization.

14. Eghosa Osaghae, "The Study of Political Transitions in Africa," *Review of African Political Economy* 22, 64 (1995), pp. 183–197, using pp. 186, 191, 192 for the quoted passages.

15. Dickson Eyoh, "From Economic Crisis to Political Liberalization: Pitfalls of the New Political Sociology for Africa," *African Studies Review* 39,3 (1996), pp. 43–80, citing pp. 64, 65 directly.

16. Célestin Monga, *The Anthropology of Anger: Civil Society and Democracy in Africa* (Boulder: Lynne Rienner Publishers, 1996), pp. 117, 125 for quotations, the text throughout for recurrent passages on the larger themes summarized here and used later in this text.

17. Juan Linz and Alfred Stepan, *Problems of Democratic Transition and Consolidation: Southern Europe, South America, and Post-Communist Europe* (Baltimore: Johns Hopkins University Press, 1996), esp. pp. 3–9.

18. William Zartman, ed., *Collapsed States: The Disintegration and Restoration of Legitimate Authority* (Boulder: Lynne Rienner Publishers, 1995).

Introduction
Cameroon's State and Society in African Context: A Metaphor for Africa in the 1990s

The state in sub-Saharan Africa as parasite, as predator, even as vampire: such language emerged in article and book titles around 1990 and dots the current literature.[1] The French scholar Jean-François Bayart touched a common metaphoric nerve most deeply with his 1989 book about Africa's "politique du ventre/politics of the belly."[2] It aptly translated the idiom of the "body politic" in Chinua Achebe's sense from *A Man of the People:*

> a man could only be sure of what he had put away safely in his gut ... you chop, me self I chop, palaver finish

and linked the analytic and imaginative vocabularies about the state.[3] Their mainstreams reveal a good deal about the birth of the independent African state, and the mixed signals about its current health. There is a genre of commentary on how the state is nurtured, fed, clothed, and sheltered, then displayed by its adepts, those privy in its ways, to their followers, those who until recently have fallen passive and silent before its rituals.[4]

Conversely, registering challenges to the state, the body politic's constituents, Leviathan's extremities, command more attention as the 1990s advance. The "associational life" of "civil society" still seeks a recognized metaphoric voice, but let us propose the candidacy of another great novelist, Ngugi wa Thiong'o in *Petals of Blood*, where there is a counterpoise to Achebe's vision of the state's anomic and betrayed politics. Ngugi's Kenyans trek from the village of Ilmorog's drought and marginality to Nairobi, seeking relief. They begin with high hope. The maimed freedom struggle veteran, Abdullah, very much until this pilgrimage a character drawn from Africa's victims, awakes to the possibilities of renewing his lapsed forest skills in an urgent cause. He fells two antelopes with stones from a sling and provides the feast which becomes the celebratory part of Ilmorog's experience of the trek.[5]

But the closer the villagers get to Nairobi, ridicule and rejection mount, and they are turned away from homes where Kenya's parasites feast on artificial, adulterated, foreign food and drink. In Nairobi, they confront their parliamentarian Nderi wa Riera and recognize in him the medium of their systematic exploitation. He tries to deflect them, but a barrage from the street's refuse, food remnants included ("a hailstorm of orange peels, stones, sticks, anything") cuts him short.[6] Humiliated, his revenge takes the cynical form of a "Mass Tea Party" which the regime stages as a loyalty oath ceremony at the President's country estate and other venues, pitched to ethnic jealousies and meant to divert attention from or to distort any renewed consciousness of the earlier freedom struggle with *its* authentic feasts of communion and oaths. From tea, the colonists' crop and drink, Ngugi takes us to beer in its capitalist venture form, Theng'eta Brewery. It violates the name and spirit of the theng'eta flower which, ritually prepared and shared, had always distilled and transmitted Ilmorog's ancient wisdom.

This tension between what's eaten and drunk corruptly and purely, between alienation and fulfillment, shapes Ngugi's politics throughout the novel. It ends with a strike, and the visit to its jailed leader Karega by a young girl whose brewery job is to turn seed millet into Theng'eta beer. Representing all the workers, she assures him of their support in "bringing to an end the reign of the few over the many and the era of drinking blood and feasting on human flesh."[7]

Achebe's and Ngugi's alimentary guides to politics conclude quite differently. Achebe allows his novel's murdered regime critic only the posthumous, isolated revenge of his lover's bullets in his killer's, the regime leader's, chest: Max was avenged not by the people's collective will but by one solitary woman who loved him.[8] The book closes (so prophetically like the Nigerian reality just after its 1966 publication) with the populace given over to "unruly mobs and private armies having tasted blood and power."[9] By contrast, Ngugi's conclusion permits Karega's glimpse from behind bars into a collectively redeemed future. It foretells (not yet prophetically for Kenya) the restoration of community and of communion celebrated by sanctified food and drink.

Moving from these metaphors to the reality they frame, to understand the contours of state and civil society, and to orchestrate their gnarled relationship, is perhaps *the* major African scholarly and policy enterprise as the continent witnesses, at century's end, widespread confrontation between regimes and oppositions. Cameroon provides our case study.

Cameroon in Africa, 1960s–1990s: A Synopsis

This is not the likeliest study at first glance. Cameroon's recent experience remains by and large anonymous, at least in the English language world, despite significant regime-opposition struggle since 1990. More familiarity attends its neighbors like Nigeria, nations once or twice removed like Rwanda and Congo (Kinshasa), and distant South Africa. But Cameroon's lesser drama, as well as the African "crossroads" and "microcosm" character imparted by its diversity, may offer a more typical continental profile. Three colonizers, 240 ethnicities, all the intricacy of a three-stage movement since 1960–1961's independence and reunification from federal to unitary constitution, the slide from some prosperity to virtual bankruptcy: these are ingredients for a timely, representative analysis.

Africa's experience 1960–1990 was, at best, mixed. Optimism about the 1960s "development decade" yielded to falling standards of living or the more stark, worst case scenario impacts of hunger, disease, and violence. Where once a foreign aid package meant benefits without many immediate costs, foreign creditors began to demand a reckoning, and in some places foreign arms intervened. For the continent at large in the 1990s there are the shatter zones like Rwanda and, more within the range of "normalcy," the contrasts between, say, Botswana and Congo (Kinshasa). The median experience lists more toward Congo (Kinshasa), a central, gaping wound in 1997. As many soldiers as unambiguous, elected, civilian democrats still occupy presidencies. What it all means for Africa's people registers in the prediction one encounters that it will be the mid-21st century before quality of life standards recover to 1960 levels. Thus, for Cameroon itself, the headline in *La Nouvelle Expression*, February 24–27, 1995: "La Banque Mondiale Prévoit: Encore un démi-siècle de misère pour les Camerounais/The World Bank Forecasts: Another half-century of misery for Cameroonians."

Cameroon bypassed the worst of the trauma through the mid-1980s, although it is necessary to emphasize the late colonial and early national history of struggle waged from 1945 until 1970 between the UPC and, first France, then the regime of Ahmadou Ahidjo which France placed and kept in power. Seminal works by Achille Mbembe and Richard Joseph record that struggle's ferocity, ideological intensity, and legacy.[10] After this quarter century of substantially armed struggle, politics were by and large orderly. The UPC's eclipse through the 1960s and the folding of some UPCistes and all other functioning parties into the Cameroon National Union (CNU) in 1966 created a *de facto* although never a *de jure* single party politics. A referendum on May 20, 1972, staged and carried out within a month from its announcement to the decree which confirmed its result, transformed Cameroon from a Federal Republic reserving powers since 1961 to the minority anglophone population of the former British Trust Territory into a unitary state, officially the United Republic of Cameroon.

With domestic peace enforced and politics centralized, Ahidjo used state and party to integrate what looked from the dominant *étatiste* perspective like Cameroon's still diverse, potentially centrifugal forces, and achieved a measure of prosperity and reputation in the post-referendum decade until his resignation from the presidency in 1982. Revenue from the newly integrated anglophone zone's petroleum profited elites; kept out of the national budget, its discretionary use from the presidency fuelled parastatals, development projects, and a "high politics" of patronage sufficient to attract and hold the loyalty of its beneficiaries (known as "the barons") and their clients. The agriculture sector's sales abroad and domestic surplus protected commoners from shocks felt elsewhere in Africa. Ordinary Cameroonians aspired realistically to and even accumulated some of this lesser wealth: the people did eat. With most favored French client status also a factor, a privileged state leadership and apparatus managed patronage, resources, and repression well enough to keep the state's and civil society's composite interest base either content or at least sufficiently silent for Cameroon to make most lists of countries in Africa which "worked." The development of state and market while other countries faltered marked Cameroon as stable and "safe for investment." Ahidjo, if not quite a Senghor or Houphouët-Boigny, looked a cut above most heirs to Franco-African power. Cameroon was a steady performer, if not viewed with the same enthusiasm abroad as Côte d'Ivoire, Tanzania or Botswana, or considered the visionary beacon for African unity in diversity which its own hallmark intellectual Bernard Fonlon hoped for.

There was a succession crisis after Ahidjo turned the presidency of Cameroon over to Biya in 1982 but kept *party* leadership well into 1983, then left the country (dying exiled in 1989). Residual Ahidjo loyalists in the military mutinied in 1984; it was bloody but quickly put down, and soldiers have never yet governed Cameroon.[11] The formal adoption of unitary state features was completed that same year. The word "United" disappeared from the Republic of Cameroon's name and its flag discarded one of two stars; most of the remnant anglophone presence in the constitution and its symbolic traces thus disappeared (though memories lingered and, we will see, proved combustible in the 1990s). Another Ahidjo creation disappeared in 1985, when Biya replaced the CNU party with his own Cameroon People's Democratic Movement (CPDM).

With Biya apparently secure, the first reversals in terms of foreign trade and currency in the mid- and later 1980s set off few, or at most muted alarms. From 1990, however, circumstances worsened dramatically, a reality which must furthermore be measured against the brief perception of buoyant democratization hopes elsewhere.[12] Following Europe's profound changes of 1989, the 1990s opened with Nelson Mandela's release and Sam Nujoma's presidency in a free Namibia. A series of multiparty, electoral and National Conference initiatives then successfully retired presidents in countries like Zambia, Benin and Mali. Many other entrenched regimes with previously docile populations were challenged. Liberalization came to much of Africa and a democratic transition gained momentum.

Cameroon's new politics began in 1990. First, the prominent Douala lawyer Yondo Black was arrested and tried along with nine associates for the most vigorous attempt in many years to form an opposition party; five were convicted and imprisoned in April. The case brought regime and opposition militants into very public demonstrations, leading to a key episode in the North West Province capital, Bamenda, May 26. The SDF led by John Fru Ndi, a party which proclaimed its right to organize, sought official registration weeks earlier, but went unacknowledged, rallied at least 20,000 people that day for a defiant launching ceremony. By all accounts, the demonstration was peaceful. But as it dispersed, massively reinforced regime troops facing taunts and scattered rocks shot six young people to death. The killings transformed a country which had known no real political alternatives for twenty years. The victims became martyrs and Fru Ndi moved from obscurity to populist leadership prominence and is known ever since as "Ni John," a title conferring the honor due elders. Another prosecution early in 1991 following a letter by Célestin Monga, critical of Biya and published in *Le Messager*, intensified the demonstrations in francophone cities. Cameroon (like South Africa for so long after Soweto) has not had a tranquil day since. The opposition claimed 400 were dead by mid-decade as open conflict escalated, reached the level of a general strike in most southern cities the last half of 1991, and peaked during a 1992 presidential election which many participants at home and observers abroad believe Biya stole by fraud and manipulation from Fru Ndi.[13]

Crisis waxed and waned in the next years. The evidence of its impact and of Cameroon's condition revealed by 1995 a frayed politics at impasse and an economy in shambles, shattering any residual domestic sense of well-being and compromising a quarter century's credibility abroad. Nicolas van de Walle reported the state's 1991 domestic revenue generation at 15 percent of 1990's, an extractive deficit (tied to the politics of 1991, covered in more detail below) which it has not yet closed, so that minimal public functions, ever more oriented to state security and less to social services, came to depend on fewer, less friendly foreign sources.[14]

A 1995 General Agreement on Tariffs and Trade (GATT) text on Cameroon, with data current to late 1994, provided the economic context. Real Gross Domestic Product (GDP) per capita rose from US$500 in 1970 to a peak near US$1,200 in 1986, but fell back to US$500 in 1994. The ratio of external debt to GDP doubled, 1986–1992. Net foreign direct investment from all sources of US$300 million in 1985 became a US$80 million disinvestment in 1990, and there was a 40 percent drop in the value of petroleum exports, 1990–1993, revenue which financed parastatals pre-1990 but dwindled thereafter.[15] Simultaneously, a World Bank profile based on massive field studies (issued the November 1994 week of Biya's 12th presidential year celebration, provoking the regime's rage), measured the human consequences of systemic crisis. Cameroon's productivity was the worst among forty-one African countries from which statistics were available, 1988–1992, so steep down the graph's cliff at 5 percent per annum GDP decline from the next worst case (Togo, at –1.1 percent) as to look like a misprint.[16]

The public service sector was in shreds. Salaries, erratically paid except to security forces, dropped 60–70 percent from 1990 to mid-decade. Retirement was enforced at age 55 or before, with pensions as problematic as salaries. The 50 percent CFA franc devaluation in 1994 took its added toll. Evasion strategies and informal economy alternatives flourished; a sampling of sources reveals estimates of 40–60 percent clandestine petroleum and 80 percent smuggled textiles (mostly Nigerian) among imports, 60 percent evasion of customs duty from all import goods, and an increasing transfer of savings and credit activity from large banks to the informal sector *tontine* (francophone) and *njangi* (anglophone).[17]

Was a now traumatic experience the result of structural adjustment and world prices, external forces beyond the government's short term power to reverse but within its wisdom and capacity to ride out and repair, given time it has bought abroad, as Biya constantly argues? Or was it caused by corruption and fundamental misrule, requiring a moral and institutional reconstruction under new leadership, Fru Ndi's position? Or was a combination of such factors responsible? Such debate about the causes was academic for most Cameroonians as consequences rippled down the social order into commodity prices and income levels.

Consider Mendankwe, a village near Bamenda, and how prices rose for staple goods every pantry stocked, from late 1991 to early 1995 (in CFA francs, per basic unit of sale): flour from 150 to 250; bulk rice from 120 to 220; soap from 150 to 300; groundnuts from 25 to 75; local palm oil from 200 to 500; toilet roll from 100 to 200; cooking gas from 2,500 to 3,650. Mendankwe, while coping with this inflation, simultaneously faced two other consequences of the more general crisis, which further compromised its basic needs. Jobless indigenes returned from other domiciles, swelling the population and adding pressure on land which previously cushioned hardship by the food it produced. Urban Bamenda needed more firewood from the village as a cheap energy source. It yielded cash but wore down the children who hauled it and further degraded the environment. There was also strong anecdotal evidence of increased death rates because cash flow no longer paid for long accustomed courses of medical treatment, and clearly visible evidence of fewer children in school and a shift by those still enrolled to more crowded but less costly schools.[18]

As the cost of essentials rose, income levels fell. One of the authors conducted focus group research along Commercial Avenue, central Bamenda's thoroughfare, in May 1995. Seamstresses, market women, car mechanics, and male curbside vendors of goods ranging from clothing and jewelry to stationery and soap were asked their income in 1990 and 1994, so as to compare the last year of "normalcy" with the first year of devaluation. Thirty took part; of the twenty who worked in *both* years, only one (factoring in the new CFA rate) earned more in 1994 than 1990. The CFA475,000 average income of the twenty-one who worked in 1990 fell to CFA213,000 for the twenty-nine who worked in 1994. This 55 percent drop among middle to marginal income people (ranging from master mechanics running their own shops to apprentice seamstresses) was in line with the 60–70 percent salary and benefits drop or forced early retirement for civil servants noted above, cushioned for a time but then caught by structural adjustment and devaluation.

Many Cameroonians knew of the World Bank report through the independent press, and most shared some measure of Mendankwe's and Bamenda's experience. Long used to the envy of others in Central Africa, they also faced and feared pariah status as former satellites turned the tables. Smaller neighbors used to depend on Cameroon to buffer their employment and trade shortfalls, but Cameroonians were now deported from Gabon and incursions on Cameroon's border towns took place not just (as is well publicized) from Nigeria but from Chad and Central African Republic.[19] Forecasts for the future were bleak, as reports circulated about the end of most earnings from petroleum by the year 2000, and from timber by 2015.[20] The situation registered emotionally when 1990's football pride turned to shame; charges spread that the "Indomitable Lions" threw their 1994 World Cup Brazil match to gamblers, and the money publicly subscribed for their 1994 support was found to be unaccountable.[21] Intensifying this dismal profile early in 1995 was a public display on Yaounde and Douala streets of work crews in tunics marked PSU (Program Social d'Urgence) worn by day laborers clearing garbage on World Bank funds. The government's own *Cameroon Tribune*'s front page, February 27, read "Garbage Situation Draws Cabinet Attention" and the independent *Le Messager*'s March 27 satiric column asked how Cameroonians felt about the situation: "Toi tu trouves normal que dans un pays comme le nôtre on soit obligé de solliciter l'aide extérieure pour curer nos caniveaux et enlever nos ordures?/Do *you* find it normal that in a country like ours we are obliged to solicit foreign aid in order to clean out the gutters and take away the trash?"

Grinding stalemate in the polity, grinding poverty in the economy as opposition initiatives since 1990 face the regime's checks: it is not clear whether the term "moderation" one often used to hear for and in Cameroon will keep the forces contesting state and civil society in check, or whether a zero-sum game between "heroes" and "villains" will unfold, with a winner-take-all rather than peaceful resolution to the struggle.[22] It could be that Cameroon's recent and contemporary experience removes it, for now, from the ranks of "middling" Africa, toward the more endemically or epidemically morbid polities. A 1996 monograph makes (then) Zaire its chief point of comparison; even if, Philip Burnham writes, "the Cameroonian state has not descended to the level of the *Zairois* state," the word "yet" is close to the surface.[23]

Given this gap between the year 1990's hope and the mid-decade reality—is it to be Botswana or Congo (Kinshasa)?—the fact and depth of what Cameroonians have since 1990 called "la crise/the crisis" must be made clear for those who do not yet recognize its depth and duration. To do so, we revert to our earlier line of metaphor. Well into the 1980s, there was enough for most Cameroonians to eat, or for them to aspire to do so, both literally in a food-sufficient or surplus economy and figuratively in a carefully articulated polity. In 1986, Michael Schatzberg, drawing on Claude Meillassoux, likened Cameroon's state apparatus to the African domestic community. The "paternal premise of nurturance" by and large worked, satisfying "a complex and largely unarticulated moral matrix of legitimate governance."[24] But in 1993, expanding his analysis to the entire continent, he included Cameroon among countries whose "fathers" in the presidency and high offices fattened up on "sons" left with little nourishment now, or prospects in the future. Governmental legitimacy was gone, and would have to be restored before any democratization could take place. This was not imminent, and in the meantime Biya made Schatzberg's "politically endangered species" list.[25]

Few people as *this* study appears would recognize either legitimacy or democratization in Paul Biya's Cameroon. Against the force and reputation of Ahidjo as a provisioner, a sustainer, a "big man" in the presidency, Biya is ridiculed as a "small boy" in the independent and savagely critical press which (we will see) developed from 1990—a predator, not a provider. His CPDM party endures the popular acronym "Chop People Dem Moni" in Pidgin, loosely paraphrased, "They eat our money like it's

Introduction

their own."[26] The idiom persists; Bamenda's people, especially children, swarmed outdoors the night of February 10, 1995, protesting hunger nation-wide by beating pots and pans with spoons, the most common kitchen utensils, as Biya began his televised speech on the eve of Youth Day, a major national holiday. Two weeks later, reinforcing the idiom, recalling our central metaphor and Ngugi's barrage on Nderi in *Petals of Blood,* there appeared below the headline cited earlier from *La Nouvelle Expression* (on another 50 years of misery) a refuse dump photo, with an insert of Biya and this caption: "L'homme-ordure nous aura apporté ça/The garbage-man will have brought us that."

As we conclude writing this Introduction in mid-1997, the time is long past since Biya could credibly blame "structural adjustment" for the difficulties. The formal state institutions and informal networks of patronage and alliance inherited from Ahidjo have worn out. Cameroonians now witness stalled initiatives, a tense marking of time, a vacuous political space after disputed, inconclusive elections to local offices in January 1996 and the National Assembly in May 1997, and await presidential elections scheduled for October 1997 which are likely to be inconclusive and could be disruptive. There is some recent evidence of macro-economic renewal in franc zone Africa since devaluation and the GATT report, with a roughly 5 percent growth in productivity for Cameroon and the Central African region in 1996, and similar growth with lower inflation for West Africa.[27] Cameroon's budget allocation for fiscal 1996–97 surpassed the previous year's by 61 percent.[28] This on its face is good news for the regime, justifying its approach and arguing the case that a recomposition of its vital parts is at hand. It is, optimistically, conceivable that compromise managed by actors inside and outside the borders will alleviate, perhaps end, the very serious crisis we document here, before anything more happens to distort Cameroon beyond hope of a peaceful settlement. But will favorable macro-economic figures offset Cameroonians' micro-experience and political sense of the 1990s? Impasse could continue indefinitely. And there is also the pessimistic reading; force of arms could be brought to these issues, through leaders risking all to secure their stake in a diminished resource, or through a citizenry with only a few remnant crumbs and trickles left to chase, and thus disposed to violence unknown for three decades. The democratization vocabulary is not the only one current in Cameroon's region.

As conflict engages regime and opposition, state and civil society, as some of those engaged are rendered more volatile the longer a resolution is delayed, Cameroon's people, precariously situated, bend without breaking, endure, respond. As the book now proceeds with their story, we align ourselves with the kind of realism called for in a recent overview of the continent, seeking the differences between the discourse on Africa and what in fact Africans themselves have confronted and still have to confront in their daily political lives [and] a different order of realities from what can be described as the more intellectualized discourse of some contemporary theorists.[29]

We undertake to draw as historians on the sources from political science, anthropology and political economy discussed next, in Chapter 1, and then to review the national experience under Ahidjo and Biya until 1990, to present and analyze some less known national *and* local experience since then, and to account for the larger public forces which frame the current crisis.

Notes

1. Jonathan Frimpong-Ansah, *The Vampire State in Africa: The Political Economy of Decline* (Trenton: Africa World Press, 1992), on Ghana.

2. Jean-François Bayart, *L'Etat en Afrique: La politique du ventre* (Paris: Fayard, 1989), translated by Mary Harper, Christopher and Elizabeth Harrison as *The State in Africa: The Politics of the Belly* (London: Longman, 1993).

3. Chinua Achebe, *A Man of the People* (New York: Anchor, 1967), p. 141.

4. Essays in James Manor, ed., *Rethinking Third World Politics* (London: Longman, 1991) about Africa and elsewhere are exemplary, keyed (p. 4) to "the theatrical and imaginary dimensions of politics."

5. Ngugi wa Thiong'o, *Petals of Blood* (New York: Dutton, 1978), pp. 134–139.

6. Ngugi, *Petals of Blood*, p. 183.

7. Ngugi, *Petals of Blood*, p. 344.

8. Achebe, *A Man of the People*, p. 140.

9. Achebe, *A Man of the People*, p. 138.

10. The key works are Achille Mbembe, ed., Ruben Um Nyobé, *Le Problème national kamerunais* (Paris: L'Harmattan, 1984) and Richard Joseph, *Radical Nationalism in Cameroun: Social Origins of the U.P.C. Rebellion* (Oxford: Clarendon Press, 1977). It is as well to begin any English language study by insisting on the not always familiar brutality of that struggle, especially from the capture and execution of the UPC leader Ruben um Nyobe in 1958, through a number of his colleagues' murders abroad and a fierce domestic *maquisard*, to the 1970 trial leading to the execution of Ernest Ouandié and the jailing of Bishop Albert Ndongmo. There was French napalm, rebel corpses displayed, the frustration and revenge unleashed after Indo-China and Algeria, and initiatives and reprisals in kind from Cameroonians, against the French and each other. The intensity of conflict rivalled e.g., Kenya's, better known to those Anglo-African oriented for its 1950s rebellion and the later deaths of Pio Pinto, Tom Mboya and J. M. Kariuki, where standard accounts report 15,000 military and civilian deaths. Mark W. DeLancey, *Cameroon: Dependence and Independence* (Boulder: Westview Press, 1989), p. 41, records 15,000 civilian deaths in Cameroon. See Phyllis Martin and Patrick O'Meara (eds.), *Africa* (Bloomington: Indiana University Press,

3rd edition, 1995), p. 162, for a recent example of the Cameroonian violence and costs understated.

11. For general coverage of the 1980s, see DeLancey, *Cameroon: Dependence and Independence*.

12. For the economic decline and its political consequences from 1990, see Nantang Jua, "Cameroon: jump-starting an economic crisis," *Africa Insight* 21,3 (1991), pp. 162–170, and Nicolas van de Walle, "Neopatrimonialism and Democracy in Africa, with an illustration from Cameroon," in Jennifer Widner, ed., *Economic Change and Political Liberalization in Sub-Saharan Africa* (Baltimore: Johns Hopkins University Press, 1994), pp. 129–157.

13. For the casualty figure, see a prominent opposition newspaper, *The Herald*, January 16–18, 1995. For politics through 1994, see Luc Sindjoun, "Cameroun: Le système politique face aux enjeux de la transition démocratique (1990–1993)," in *L'Afrique Politique 1994* (Paris: Karthala, 1994), pp. 143–165; Milton Krieger, "Cameroon's Democratic Crossroads, 1990–4," *The Journal of Modern African Studies* 32,4 (1994), pp. 605–628; Jean-Germain Gros, "The Hard Lessons of Cameroon," *Journal of Democracy* 6, 3 (1995), pp. 112–127.

14. Van de Walle, "Neopatrimonialism" p. 146, used *Africa Confidential* (London) as source for the domestic fiscal gap since 1990. France by 1994 was the only reliable foreign source of finance and guarantor for loans; other former partners placed Cameroon on a short leash as its performance faltered.

15. General Agreement on Tariffs and Trade, *Trade Policy Review: Cameroon, 1995* (Geneva, May 1995), pp. 5, 7, 27, 76, 169.

16. *Cameroon: Diversity, Growth, and Poverty Reduction*. World Bank Report No. 13167-CM, April 4, 1995. This official version omitted the most damning Africa-wide data, which were publicized in Cameroon the previous November.

17. For the examples, see GATT, *Trade Policy Review*, p. 32, and David Blandford, et al., "Oil Boom and Bust: The Harsh Realities of Adjustment in Cameroon," in David Sahn, ed., *Adjusting to Policy Failure in African Economies* (Ithaca and London: Cornell University Press, 1994), pp. 147, 155, 157. For public finance, see, e.g., *Africa Research Bulletin: Economic, Financial and Technical Series* 33,3 (March 16th-April 15, 1996) p. 12515 ("Cameroon's credit risk rating is abysmal and worsening"), more recent work by van de Walle, and Philippe Hugon, "Sortir de la récession et préparer l'après-pétrole: le préalable politique," *Politique Africaine* 62 (1996), pp. 35–44, part of a special Cameroon issue.

18. Judith Krieger, "Women, Men, and Household Food in Cameroon," Ph.D Dissertation, Department of Anthropology, University of Kentucky (1994), and subsequent research.

19. *La Nouvelle Expression*, January 19–22, 1995 and February 24–27, 1995, citing the 1995 World Bank Atlas and other sources; *Jeune Afrique Economie* 190 (February 13, 1995), pp. 68–75.

20 *La Nouvelle Expression*, January 26–29, 1995.

21. These were "human interest" themes the independent press featured across its spectrum the first two months of 1995. "Hot line" government radio four afternoons a week was similar.

22. One basic index of danger, ethnic conflict, threatened to erupt, with the Beti considered the state's sole recent beneficiaries by those outside their ranks, and with Bamileke and anglophone opposition facing the Beti charge of disloyalty to "la patrie." These forces will be analyzed below.

23. Philip Burnham, *The Politics of Cultural Difference in Northern Cameroon* (Edinburgh: Edinburgh University Press, 1996), p. 160 (the emphasis is Burnham's).

24. Michael Schatzberg, "The Metaphors of Father and Family," in Michael Schatzberg and William Zartman, eds., *The Political Economy of Cameroon* (New York: Praeger, 1986), p. 14.

25. Michael Schatzberg, "Power, Legitimacy and 'Democratization' in Africa," *Africa* 63,4 (1993), pp. 445–461, p. 456 for the Biya reference.

26. The perception outside the country registered in Cameroon's ranking by international business people as the world's sixth most corrupt nation in a Berlin-based poll; *New York Times,* reprinted *Seattle Post-Intelligencer* June 2, 1996.

27. *Africa Confidential,* May 24, 1996, p. 6, citing higher crop prices abroad; *Le Monde,* September 22–23, 1996.

28. *Africa Research Bulletin: Economic, Financial and Technical Series* 33,6 (June 16–July 15, 1996), p. 12630.

29. This passage from the editors' introduction, David Apter and Carl Rosberg, eds., *Political Development and the New Realism in Sub-Saharan Africa* (Charlottesville: University Press of Virginia, 1994), p. 38, parallels our perspective as historians.

1

Scholarship on the African and Cameroon State

Recurrent motifs of food and drink, drawn from novels, scholarship and reality to distinguish Africans who are powerful from those who are less so, or powerless, shaped our Introduction. We move now from metaphor to more systematic analysis, and place our approach to Cameroon within the pertinent African historical and social science literature. We establish our scholarly bearings for Africa at large until 1990 in this chapter's first section, then for Africa in the 1990s in the second section, and for Cameroon itself, 1960 to the present, in the third. The last section poses questions which lead to our coverage of Cameroon's major historical processes of the 1990s.

Reduced to essentials, two issues have dominated recent scholarship on Africa: the character of the state; its relation to productive forces, classes, and in the broadest sense civil society. Even if it stands classical materialism on its head to discuss the nature of the state before mode of production and class issues, most Africanist writings in the early times of independence for most countries recognized the salience of the state as the crucial protagonist and arbiter in African affairs.Inheriting and expanding the colonial state, it accumulated and disposed of much of the continent's material and social wealth. Political science in the U.S.A., the dominant discipline and most optimistic genre early on, emphasized through the 1960s the creation and performance of the new states. Findings and projections ranged widely. Liberal capitalist sources, even where cautious, charted Africa's "progress" at least in pockets, and argued that more mature states and markets in course of formation, some of them extended experimentally by alliances across national borders, would mobilize resources and people more widely.[1]

But dissenting views emerged, from a range of Marxist and dependency theory sources. There were many lines of attack, most of them keyed to "modes of production" analysis about different forms of wealth:

the immature and fragile nature of markets and states; their shallow roots in material, productive wealth (although social wealth created by mobilizing kinship was deeper); the unequal patterns of distribution built into state structures and the opportunism of those directing their practices and policies, especially among domestic "gatekeepers" where foreign interests shaped and channelled Africa's wealth; and the challenge of entrenched "primordial" formations to the entire fabric of centralized economic activity and political authority where progress through class development and consciousness might otherwise take root.[2]

By 1980, with nation-state and market oriented political science less dominant, two distinctive, competing vocabularies shared the field. One was the Gramsci variant of the Marxist tradition with its focus on hegemony and ideological ascendancy, designating (in ways not always clearly distinguished) either the added acquiescence of those under the state's control, thus amplifying its firm rule, or the state's capacity through cunning to paper over its objective weakness.[3] The symbolic apparatus and trappings of leadership in Africa, the "big man" syndrome, made the notion of hegemony especially compelling in this reading of Africa. It tended to emphasize strong initiatives and orderly processes at least on the surface of governance and it reflected the 1970s realism rather than earlier, normatively higher hopes.[4] The other came from skeptics regarding the state's real or apparent power. Goran Hyden took up the language of exit and withdrawal from state and market structures, and made them look "softer" than before.[5] Clearly, most Africans in the time of Amin, Nguema, and Bokassa were *subjects,* marginalized even brutalized, but evidence emerged of the choice for avoidance, though it was made by exiles embracing anonymity, unproductive in that role, not yet by *citizens* waiting in the wings for the autocrats and tyrants to leave, or working to displace them.[6]

The next major scholarly source, the growing 1980s body of anthropological research on sub-national, "micro-level" Africa, challenged all previous readings. It validated rather than dismissed (as either irrelevant or reactionary) the local base of life, and helped steer African Studies away from the cul-de-sac or collision course limitations of work narrowly focused on whether or not the state and market were effective mobilizers of surplus, often with reductionist, mechanistic tendencies in its approach to African peoples on their own ground. This constituted both a critique of earlier social science literature and a window on the broader base of African experience, viewing people actively wherever they were, not as passive or mute to the degree they were distanced from "the center." Anthropology, for example, drew attention to African governance which long ago concentrated and stratified both wealth in things and wealth in people. It thus corrected political science and dependency theory views of African customary authority as static and historically benign, or awaiting the developmentalist breath of life from afar.[7]

The result was a more empirically dense and theoretically sophisticated study of wealth and power, disclosing loose ends, rough edges, sharp divisions. Conflict within and between state and civil society was sharply revealed, with class emerging as a major arena in anthropological analysis alongside previously focussed kinship and ethnicity, but not (in Peter Geschiere's term) as an apodictic (absolute) category.[8] Anthropology also expanded the study of gender and age as factors in Africa's experience.

Global political economy analysis also emerged, adding the vocabulary of Immanuel Wallerstein to Africanist theory and case study.[9] Scholars from Africa or based there like Walter Rodney, Samir Amin, and Claude Ake augmented this work. Rodney's *How Europe Underdeveloped Africa* (1972) drew from his time's dependency literature and world system variants of political economy a full-fledged account of African surplus as the basis of European capitalism, with its transfer arresting the prospects for indigenous development.[10] He was more attentive than some others of his persuasion to contradictions generated among Africans as well as between Europe and Africa, in a history which spanned slave trade and colonial times. Of all such writers more contemporary than Rodney in their coverage, Amin in *The Class Struggle in Africa* (1973) and Ake in *A Political Economy of Africa* (1981) most closely focussed the alignments within post-colonial state and class structures, and showed the now rapid push to make the state an agent of coercive accumulation.[11]

Africana thus became by 1990 a repository of important state and class scholarship, lowering epistemological and experiential barriers between stateless and state-like polities, lineage and broader modes of production, primordial and instrumental ethnicities, rural and urban life, the place of men, women and children, what was the state's and the society's domain, and what was foreign and indigenous. Many *pre*-colonial indigenous communities were recognized as accumulationist and significantly marked by inequalities. The *colonial* state and its *post*-colonial successors (or remnants), maintained for a time after independence in most settings, were revealed as shrewdly opportunistic alliances of class and ethnic fragments among (Basil Davidson's term) "inheritance elites" fronted by "big men" using the trappings of real and symbolic authority well known to their subjects.[12]

And so the questions are posed: how does one understand the state-society issues presented thus far, to prepare the coverage of Cameroon's experience in the body of this work? Was, and is, the state in Africa hard or soft? The adjectives might vary, but the noun holds its place. By either measure, hard or soft, whether its efficacy is affirmed, questioned or denied, and factoring in whatever direction it takes from interested parties beyond its borders, the state apparatus and its class core and their affiliates (now understood in very dense terms) are the key forces to reckon with. Africans who both control the state and contest that control have no doubt about its operational primacy in the short run. Life with, within or despite the state, not beyond it, is what our study must emphasize, even if its salience is challenged or reduced in ways we record below.

State-Civil Society Scholarship in the 1990s

We continue this survey of Africanist writings on state and society into the 1990s, first by considering Robert Fatton Jr. and Basil Davidson as representative sources for a wide stretch of sub-Saharan Africa, including its three major colonial arenas, British, French and Portuguese. We then return to Bayart in more detail, for his rebuttal or reformulation of virtually all the scholarly strands we have covered.

While Africa's optimism about 1989–1990 still lingered, Fatton's 1992 book *Predatory Rule: State and Civil Society in Africa* reminded anyone needing it of the state's hard side. Most striking was a truly apocalyptic subchapter "EXIT AND ANARCHY: THE DESCENT INTO HELL" about "anarchic armies of macoutes" and "Afro-narcotism," ferociously chasing funds from informal and illicit channels as others dried up. Central to his analysis was the leadership of predators who hold formal power in state offices. They face and interact with a second layer of political society, from those class formations perhaps inclined toward, even teetering on the edge of opposition. But the latter are still capable of (and are as likely to choose as to refuse) a strategy of "disarticulation" they pursue against a third level of political operatives, the actively resistant class fragments in the civil society where Fatton found things truly soft. His key operational issue for actors on a political stage much like Bratton's: in whose interests will class choice within defensive, even desperate elites be made? His answer: their own, on increasingly narrow grounds, with harsh consequences for the truly disempowered and (Hyden a decade earlier notwithstanding) few viable "exit options." The challenge to more optimistic democratization hopes, experience and literature, then and still abundant, could not have been clearer.[13]

Davidson's critique of state operations since independence, made most explicit by the phrase noted earlier, inheritance elites, was reinforced in *The Black Man's Burden* (1992, like Fatton). Central to *his* analysis, as for many years, was the argument that whereas early Africa's polities operated within "rules and structural restraints," Europe's slavers and colonial states and now their successors have destroyed those balances. The result in too many cases, described in a chapter ("Pirates in Power") rhetorically pitched like Fatton's text noted above, was some variety of "the typical pathology of the times" exemplified by a Doe or Mobutu. Davidson found in the debate about where to situate strength and weakness "the fearsome dilemma of the 1980s . . . which taught that a strong state had to mean dictatorship but a weak state must collapse into clientelism," and it was a long way from any favorable resolution. His book's last question, despite Davidson's hallmark good will and optimism: "If one or other pirate falls in Africa, what promises that another will not take his place?"[14]

Fatton and Davidson are important sources for the 1990s, but it is Bayart who now requires our full attention. *L'Etat au Cameroun* (1979) and *The State in Africa* (1989, trans. 1993) fuse our Cameroon case study and the continent through a rich blend of empirical mastery and of surgery on most of what was c.1990, and still is, current methodology and theory in Africana sketched above. The immense scholarly apparatus of the 1989 text has little in common with the literature on democratization emerging in the 1980s and leading to "regime transition" studies surveyed earlier, and stands against the post-1990 writings surveyed just above. Notions of "disarticulation" (a basic premise of Fatton's *Predatory Rule*) assumed for Bayart (as for Geschiere, noted above) a formulaic class structure not yet in place.[15] Bayart's view of Africa's past was less pristine than Davidson's, and he would disagree with Davidson's language about "pathology" quoted above. Here is Bayart: forms of power practiced today in Africa "should not be labelled as pathological"; they shape a "banal Africa" and are also known elsewhere, past and present.[16]

More serious attention was given Hyden, but there was ultimate skepticism, worth quoting because it helps to align Bayart himself:

> "an uncaptured peasantry" . . . is a suggestive but nonetheless unfortunate expression. Actors are found neither inside nor outside the State. Rather, all actors, depending on circumstance, sometimes participate within a statist dimension and sometimes turn away from it.[17]

Emphasizing Africa's still open demographic space, he subsumed what Hyden called the exit option under longer term migration processes, and argued that the state should be considered incomplete, not soft.[18]

Bayart drew on the anthropology noted above. Against *any* approach applying binary thought (to state and society, realms of the institutional and the personal, foreign and indigenous, subject and citizen), or in other ways fixing (for instance, and above all) class analysis too firmly to account for the fluidity he perceived in the continental experience, Bayart issued a powerful injunction to respect the trial and error, *longue durée* process under way. Thus, another characteristic challenge emerged, to but also beyond Hyden:

> The private and State paths to accumulation share, broadly speaking, the same ethos of personal enrichment and munificence . . . [in] . . . [T]he maelstrom of ever-shifting disagreements and alliances sweeping them along. . . . State and private operators act complementarily when they are not just confused, changing hats to fit the circumstances.[19]

Again the word *act:* what emerged from Bayart was a sweeping insistence that sub-Saharan Africans have acted in accord with the logics of their own local community structures during their long and often painful encounters with each other and with *all* (not just European) interventions from abroad.

They have shrewdly rolled with the punches, absorbed and turned to advantage what was unavoidable, and fashioned distinctive historical structures (note the plural). These are flawed, like all peoples' untidy histories in the pursuit of wealth and power among their near and distant partners and rivals in "la recherche hégémonique/the hegemonic project" through the state, the human lot which Bayart considers decisive. Here is his classic definition of "la recherche hégémonique" from *L'Etat au Cameroun*, which he brought forward for use in *The State in Africa:*

> un système de hiérarchie social cohérent [qui] reposer sur un processus d'assimilation réciproque et de fusion des groupes dominants anciens et de nouvelles élites nées de la colonisation et de la décolonisation. Les lignes contemporaines d'inégalité et de domination semble ainsi s'inscrire dans le prolongement direct des structures sociales précoloniales, les dominés d'hier constituer la masse des dominés d'aujourd'hui/a system of coherent social hierarchy [which] rests on a process of reciprocal assimilation and fusion of the former dominant groups and of new elites born from colonization and decolonization. The contemporary lines of inequality and domination thus seem to register in the direct extension of the precolonial social structures, the dominated of yesterday constitute the mass of the dominated of today.[20]

Bayart, again, emphasized Africans as *actors*, far less acted upon than others perceive, bringing customary experience from the intimate level of small domestic communities keyed to kinship, age, gender, and other grounds of unequal identity and exchange that have survived in familiar ways, to bear upon the newer global reality they still grapple with. He drew on Georges Balandier to fuse Africa's past and present self-construction, restating the hegemonic quest:

> The creation of a clan in the anthropological sense of the term in the context of a lineage society is, in Georges Balandier's words, 'a global enterprise which challenges kinship rights over women, wealth and genealogical conventions' ... the same could be said of the 'politicking' within the postcolonial State.[21]

This action, as elsewhere, was and remains the state-building *process* (not the condition, too static a word) Bayart called Africans' "politics of the belly" which, high and low, they practise, understand, and share—we are back with Achebe and Ngugi—pulling together to feast, feeding their complex network of clients, pushing others away to wait their turn. He bared Africa's ancient disequilibria and accounted for their recent modifications, as the state's "space" has responded to pressures from abroad and enlarged, filled in by the work of each community's adaptive practice ("by chameleon's footsteps") to processes not yet easily controlled.[22]

The keys to understanding this state-building process in most of the continent are the "reciprocal assimilation of elites" and their "straddling" strategy.[23] These should not be mistaken as firmly rooted in specific modes of production or class alignments, which are not fixed or even primary roots of interest and motivation—thus, the image of the "rhizome" (resembling the banyan, not the oak) for the pragmatic, constantly adjusted hegemonic quest by those who operate the state.[24] They include and exclude, open and shut its doors, and alongside the straddlers (Fatton's macoutes) are the stragglers (Hyden's exiles). There are resemblances between them forged in "second" clandestine economies like Congo (Kinshasa)'s, where "corruption and predatoriness" are common to the patron-client networks at the center *and* the villagers who cannibalize the world's longest overhead power line for parts.[25] All seek and may take some place at Bayart's table, the State (he generically capitalizes the word), where the "national cake" is divided. Though it serves them differentially, even idiosyncratically, since its workings are far from fixed and predictable, Africans recognize it as the venue where their activity most matters; they enact and transact simultaneously diverse roles across its multiple settings.

In one respect at least, Bayart agreed with most writers whom he criticized for the overly categorical terms they apply to the actors in and the process of the quest for hegemony. Africans' struggle for mastery, as their domestic and global circumstances unfold, is crude and volatile, and will remain so. Despite his different typologies, Bayart perceives the *conduct* of politics in terms we have seen:

> the combats are ruthless ... hence the unbridled predatoriness and violence of political entrepreneurs ... the social struggles which make up the quest

for hegemony and the production of the State bear the hallmarks of the rush for spoils in which all actors—rich and poor—participate in the world of networks ... [it] is a zero sum game where the only prize is the accumulation of power.[26]

But this, he insists, is not a pathology, or exotic; the experience of Africans is consistent with peoples' history subject to the same pressures elsewhere. The viewpoint rings true in most of Africa today, and surely applies to Cameroon.

This primacy of the state and its "rhizome" density are Bayart's anchors and have become common elements in most cognate literature of the 1990s, but what now can we make of the "civil society" Fatton in particular found so fragile? As the Preface showed, the term designated alternative approaches emerging in the 1980s to dominant *étatisme* in scholarship around the world, reflecting change in both superpowers and satellites. It then became a descriptive marker for experience on *African* soil which from 1990 challenged long entrenched regimes. Scholarship and reality converged. The literature about civil society and regime transition we surveyed earlier applied theoretical and historical precedents to the explosion of African events and inititiatives in 1990–1991: Mandela's and Nujoma's leadership; Benin's Sovereign National Conference leading to others (whether or not "Sovereign"); multipartism and a cycle of elections like Zambia's replacing or weakening some heads of state; the massive rallies and strikes to support these movements, like Cameroon's.

There were advances for civil society on the ground, but also setbacks, and the scholarship responded to both. With all the pluralist and democratic hopes raised, René Lemarchand noted in 1992 the need for caution in

> the persistent tendency ... to locate state and society in separate conceptual niches: one inhabited by a potentially predatory species and the other by a defenceless and fully domesticated pigeon which could lead, further, to privilege the pigeon's search for its own wings in misleading ways.[27]

He warned the flourishing genre of civil society analysis not to overlook the gap between the search for democracy and the reality of Africa's "universe of rural societies." Such care *was* taken in the best of that writing about new political practice in the 1990s. When Naomi Chazan specified indices for "democratic experience" and civil society's advance (specific notions of authority, distributive justice and conflict resolution, and respect for the rule of law), and Michael Bratton and Nicolas van de Walle developed the criterion of an alternative ruling coalition with a sustainable multiclass social base and a coherent platform for governing, they insisted on the gap thus far between democratization as ideal and reality.[28]

Recognizing in Albert Hirschman's terms that there was a "voice" as well as "exit" option emerging in Africa, Lemarchand called it "extremely problematic," for it was "politically meaningful" in just a handful of countries (Cameroon not among them). The long reliance on "exit" has been costly:

> It has shaped perceptions and praxis in ways that make it difficult if not impossible for the sundry components of the civil society to make their voices heard in coherent terms. Different dialogues are unfolding in different arenas, and more often than not the interlocutors are talking past each other.[29]

Fatton in 1995 added a new critical apparatus to the debate on civil society by noting *its* multiple components, parallel to those he found in the state—predatory, quasi-bourgeois, and popular—and the convoluted, often deceptive, or even consciously disguised forms of alliance they assume under the fragile conditions of Africa's democratization process.[30]

Still, civil society's recent emergence was acknowledged. Anthropologists, as we saw, made the case that those outside the state's ranks can not simply be dismissed as subjects, merely acted upon, or as escape artists. Their voices and actions matter, clearly so after 1990, as the presence and impact of civil society, and the very notion of a *citizen*, with all the attendant claims, augment and renew Africa's public life and its scholarly lexicon—all this is pivotal in Célestin Monga's work.[31] Latter parts of our book will reflect this change, which we argue is of fundamental importance in Cameroon and needs to be introduced as evidence against a certain reductionism of politics not just to the belly but to the feedbag and trough which pervades Bayart's work.

More than incidentally for our purposes, note how the historical and social science framework of analysis offered here parallels the continent's imaginative literature our Introduction cited. Achebe and Ngugi, as we have seen, Wole Soyinka, Mongo Beti for Cameroon (more on him follows), and most creative writers of consequence shaped their work significantly or dominantly around the encounter between the state apparatus and African people—elites, commoners, marginals. They took sides with the powerless against the powerful, spoke for citizenship against subjection, before most scholars did so. They may earlier have read like writers' voices in a wilderness: not so since 1990, although all of them have at some point been voluntary or forced exiles, or prisoners, two of them (Ngugi and Soyinka) remain so, and there is the ultimate malevolence of the Ken Saro-Wiwa case always to remember.

Writings which establish Africa's vulnerable governance condition in the 1990s thus recounted—whether state and market structures shape people and conduct affairs by brute force and intimidation, hegemonically with or without mirrors, rhizomatically à la Bayart, or not at all; whether civil society yet makes a difference—we now turn to key works on Cameroon itself.

The Cameroon Scholarship

Writing on Cameroon's experience since independence generally fits scholarly rhythms traced above (it was, after all, where Bayart began as a scholar, and led him to "politics of the belly"). Victor Le Vine's *The Cameroon Federal Republic* (1963) was the first major political study to include both French and British trust zones and to recount independence and reunification (1960–1961). One of the models from its generation, the book demonstrated state-building for diverse settings like Cameroon's as the art of the jigsaw puzzle, on the architect's scale. It typified a positivist and programmatic American vein of political science, aligned to views of the state and its ruling party as agents of modernization and progress. It was reissued and updated to 1971, the 10th anniversary of independence; Cameroon had weathered a good deal, spawned critics and created victims, but it clearly endured and in some ways thrived, and Le Vine judged that "praise is largely merited... pride and confidence mostly justified." Ahidjo's defeat of the UPC and the maquis "under firm but insistent pressures" was never "vindictive": "The style of the regime appears to have been actively reconciliationist, pragmatic and tactically consistent." Given the good management of domestic politics, "military intervention seems remote and unlikely" and "the situation looks hopeful" on the economic front. Noting that "north-south ethnic antagonisms... the expansion of the dynamic Bamileke... [and] unemployment and poverty in the main towns" were sources of tension, and that leadership succession could prove vexing, Le Vine's epilogue chapter still summarized the prospects as "relatively good."[32]

Another key study echoed Le Vine, although Willard Johnson's *The Cameroon Federation: Political Integration in a Fragmentary Society* (1970) made the French (East)-British (West) divide a more salient focus (as the sub-title signified). He recognized the daunting task and original insecurity which faced Ahidjo's integrative effort. Still, Johnson acknowledged the president's bargaining skills and inclusive style of governance, containing and absorbing all the challenges, and concluded (like Le Vine) optimistically. The regime embodied Cameroon's hopes for justice, peace, and progress:

Cameroon has been remarkably successful in consolidating the state. By doing so it has given itself an even chance for some day becoming the 'One Cameroon Nation.'[33]

Between them, Le Vine and Johnson reflected the consensus of work on Cameroon in the 1960s. They perceived an enterprise worthy not of a mere politician but a statesman (the reputation Ahidjo cultivated). Comparing Cameroon with neighbors and Africa at large, they found its people rallied to a party and government where civilians remained in office, where competence prevailed over corruption. In our terms, state and civil society looked capable of converging, with a viable balance between authority and liberty.

But Ahidjo's governance was challenged by critics at home and abroad, including exiles like Mongo Beti and Abel Eyinga.[34] Both scholarly and imaginative critiques surfaced. A key early study appeared at the end of the first independence decade, just after Le Vine's and Johnson's books, just before the 1972 referendum dismantled the federal constitution. This was Jacques Benjamin's *Les Camerounais occidentaux: La minorité dans un Etat bicommunitaire*, scholarship based not in the U.S.A., but in Québec, influenced by De Gaulle's "Vive le Québec libre!" and the Front de la Libération du Québec (FLQ) and state of emergency. This experience included significant numbers of people who around 1970 considered themselves marginalized, and some in open rebellion. With this background, Benjamin made Cameroon's language cleavage, one which Johnson (recall his sub-title, "Fragmentary Society") recognized but finessed, central to its scholarship at the same time the referendum did so in terms of reality, and detected serious tensions elsewhere.[35] Favorable studies like Le Vine's and Johnson's stopped appearing, except from official or apologist circles.[36]

A linkage developed between the pre- and post-1972 victims and enemies made, and memories forged, by Cameroon's regime. They scored the surface of consensus in writings full of disclaimers and dissent published during Ahidjo's most settled period of rule, 1972–1982, which, after a pause when Biya took the presidency in 1982, resumed in the latter 1980s and dominates the 1990s. Foremost among the critics has been Alexandre Biyidi, who in the 1950s under the pen name Mongo Beti first drew attention to Cameroon's literature abroad (alongside Ferdinand Oyono, whose later path to cabinet posts was quite different). Beti's early anti-colonial novel sequence ended in 1958, when he turned against Ahidjo, was silenced at home, and went into exile in France. But in 1972 (that year again), he broke a period of silence with *Main basse sur le Cameroun: autopsie d'une décolonisation*.[37] The subtitle is self-explanatory for our purposes. The tract was a detailed indictment of the postcolonial ruination of Cameroon's society by its state, and the French neocolonial arch over that domestic structure. It was fierce enough to be banned not just at home but in France, which tried for a time to deport Beti. A new five novel sequence followed, along with his journal *Peuples Noirs-Peuples Africains*, constituting by 1980 a systematic assault and rallying point for other writers and critics against the regime.

Most directly in the title of the novel *Remember Ruben*, one of Beti's motifs was the UPC leader Ruben um Nyobe, killed by the French in 1958 and the martyr for those who believe that Ahidjo betrayed authentic Cameroonian nationalism. So primordial was his force both in Cameroon's literary and political terms that Richard Bjornson's study of the country's writing in 1991 made the confrontation between um Nyobe's ghost and the Ahidjo-Biya state a central paradigm.[38] The parallel evidence from Achille Mbembe on the texts of politics conveyed orally rather than in print, which keep um Nyobe alive in popular imagination decades after his death despite the state's effort to bury his life and memory, is equally striking proof of an enduring oppositional legacy in Cameroon's literature and life.[39]

These written, spoken and sung protests from Cameroonians were joined by the domestic and foreign scholarship. Richard Joseph contributed heavily, publishing *Radical Nationalism in Cameroon* (1977) and the edited volume *Gaullist Africa: Cameroon under Ahmadu Ahidjo* (1978). The first of these texts documented the UPC and created alongside Eyinga's French language work from exile the first systematic scholarly rebuttal to the Ahidjo regime. The second stretched back to the French Revolution for the Jacobins as precursor critics of an old regime, then used De Gaulle as the immediate model for Ahidjo's creation and management of Cameroon's state and its impact on society. His own opening essay disclosed fault lines, not stability, in the presidential monarchy and the CNU's one party rule, and etched a very different profile for Cameroon than did political science writing of the 1960s:

> The period of instability during the 1960's . . . derived in large part from the search for the particular structural "mix" that would enable the ex-colonies to continue to play their role in generating international surplus, but also be in a position to deal with the internal challenges that such a process was certain to provoke.[40]

Joseph argued from the emerging global "left" perspective that Ahidjo's version of the state, though civilian in form, shared essential features with praetorian rule elsewhere, and was far more the creation of foreign capital than had yet been revealed. Bayart contributed *Gaullist Africa*'s second chapter. Details from the 1970s abounded, supporting Joseph's model: offices and *apparatchiki* added to the presidency; its enlarged powers of appointment throughout government; continuation of the 1960s state of emergency by means of a tight internal security apparatus. Here was Bayart's 1978 summation:

> Since any crisis at any level in this monolithic and autocratic system could directly bring into question the authority of President Ahidjo, there is nothing therefore, which can be said to be "outside politics."[41]

The rest of Joseph's book offered more than a scholarly critique of Ahidjo; its appendices translated the first widely available excerpts for English readers from Beti's *Main Basse* and Eyinga's account of regime torture.

Bayart's own *L'Etat au Cameroun* a year later indelibly stamped the term "l'état hégémonique" on Cameroon studies, signifying

> un procès d'autonomisation de l'Etat qui constituerait une réponse globale et cohérente de facture bonapartiste, à une crise structurelle/a process of autonomization of the State which would constitute a total and consistent response, Bonapartist in technique, to a structural crisis

which a century of upheaval since European intervention had produced.[42] A seamless ideology of national unity now, he stated, framed all evidence of pluralism as "idées de division, d'intrigue, de tribalisme." Cameroon, from the standpoint of this critical scholarship, thus experienced

> l'émergence progressive d'une vaste alliance régroupant les différents segments régionaux, politiques, économiques et culturels de l'élite sociale ... les principaux groupes dirigeants du pays, hérités du passé ou nés des mutations contemporaines ... travaillent de concert sous le couvert des institutions du régime présidentiel de parti unique/the progressive emergence of a vast alliance comprising the different regional, political, economic and cultural segments of the social elite ... the principal ruling groups of the country, heirs of the past or born from contemporary mutations ... work together under the cover of the single party presidential regime.

This classic 1979 text told an additionally nuanced story when its last chapter moved from the apparatus of the state to the fabric of society. A prescient analysis cited drugs, workplace absenteeism, and energies displaced from politics to football among the symptoms of a civil society, especially its youth, deemed "sans importance" and victimized by the "recherche hégémonique/hegemonic project" of the state and dominant class. This was evidence of incivility, anomie and morbidity, tattered montage trappings, not the substance of a consensual hegemony.

Mbembe in particular pursued critical enquiry through the 1980s, alongside his detailed work on um Nyobe. What emerged (Bayart's later texts on Africa at large notwithstanding) was the pathology of both Cameroon's state and society, and the causes and symptoms of what was happening beneath the opaque surface of Ahidjo's, then Biya's regime.[43] The cumulative effect of his work was to document the emerging stalemate in Cameroon which would take overt political form after 1990, with (to repeat our earlier phrase) neither a state hegemony nor an opposition counterhegemony yet prevailing.

The fissures which would open in 1990 were there to see in simultaneous 1989 publications, Mark DeLancey's *Cameroon: Dependence and Independence* and a Leiden University symposium. One a macro-level synthesis, the other a series of micro-level analyses, they both covered Cameroon since Joseph, Bayart and Mbembe broke the critical scholarly ground.[44] Here was DeLancey on the CPDM party-state Biya formed to succeed the CNU in 1985:

> What we have so far seen in the Biya administration is a continuation both of the pattern and process of rule established by Ahidjo and a reliance on personnel drawn from very much the same strata of society.[45]

The Leiden symposium offered the composite evidence of thirty-six country specialists, from Cameroon and abroad, who tracked the emergence of forces in society confronting each other and the state apparatus. Although neither DeLancey's nor this collective work predicted the crisis which started in 1990, they certainly documented its origins.[46]

Mbembe reviewed the national condition in a 1993 essay, steeped in the crisis. His trajectory helps summarize our text thus far and points our way ahead. Parts read much like Bayart; the essay was laced with the intricacies of collusion in belly politics, and insisted on the multiform rather than class-defined and driven character of Cameroon's networks of power and wealth. There was also, however, more clear attention to some prevailing trends, amounting to causal patterns which are hard to find in Bayart's less differentiated, often reductionist swarm of details and variables. Mbembe specified two poles in Cameroon's politics for almost half a century: the soft nationalism accommodating France leading from Louis-Paul Aujoulat's Bloc Démocratique Camerounais through Ahidjo to Biya; the UPC's authentic, endogenous resistance politics. Another variation was Mbembe's more concrete sense than Bayart's of distinct regional blocs as the bases of domestic political alliances, especially Ahidjo's North-South axis, which (we will see) Biya has compromised.[47]

The last segment of Mbembe's 1993 text spelled out a close to worst case scenario for his own country.[48] Phrases like "Enkystement et tonton-macoutisation" and "un système définitivement ankylosé" translated into a "barnacled" and "sclerotic" condition, and verified Fatton's Africa-wide profile for Cameroon. Mbembe went on to make what was still Zaire into Cameroon's reference point:

> une vénalité et un degré de prévarication, dont on ne trouve l'équivalent que dans des pays comme le Zaïre, se sont progressivement institutionalisées/a venality and a degree of evasion, which one finds equalled only in countries like Zaïre, has become progressively institutionalized.

This was how Cameroon looked in Mbembe's last lines:

> le Cameroun poursuit lentement sa course folle vers des ruptures qui, parce qu'elles ne cessent d'être différées, risquent à la fin d'être particulièrement sanglantes/Cameroon slowly pursues its mad course toward ruptures which, because they continue to be put off, risk in the end being particularly bloody.

Relieved only by speculation that the SDF might draw on the UPC's earlier legacy and become a pluralist and democratic rallying point, a call our last chapter returns to, Mbembe's gloomy diagnosis rang true in the mid-1990s.[49]

The Questions Ahead

Joseph in fact anticipated a reckoning to come, but also a democratization potential, twenty years ago. Here is the last sentence of Gaullist Africa:

> The proper question, for which this study can perforce supply no answer, is: when will circumstances permit the popular forces of Cameroon to renew the struggle for their own liberation?[50]

There is little doubt that Bayart's constantly self-renewing *étatisme* accounts for most of Cameroon's basic experience from independence until 1990. But popular forces came forward that year, and remain prominent. Thus, the effort to discover whether Cameroon in the 1990s offers an experience of and model for civil society in process of mobilization, and whether Cameroon may be the source of any new and better African politics, ranks high among our tasks ahead. Was 1990 a watershed and do the years since then mark a democratic advance of some consequence, which Joseph looked for, Mbembe at least glimpses, and Monga believes is brewing, there and elsewhere? Or do we now witness merely episodic turbulence in Cameroon, likely to dissipate and contribute, at best, to the history of Africa's missed opportunities?

These are, we think, open questions, with an optimistic resolution to consider. There is nothing transient about Cameroon's crisis described in the Introduction and viewed in terms of the scholarship this chapter surveys. It manifests fundamental economic and social tensions. Viewed as fault lines and complications in modes of production and class structures, the crisis softens the state's hard profile, however accurate that is for Africa at large, as economic hardship pushes more Cameroonians to the margins or completely out of market arrangements and, in some regions, in many ways, beyond the state's effective command. There is, for instance, the reversion to customary, often subsistence oriented practices of cultivation and marketing as cash crops like coffee and cocoa are cut back or abandoned, which makes women ever more crucial to livelihoods as retrenchment sets in. There emerge, in bold relief, strategic choices to withdraw goods and services from centralized state and market control, to confine them to private channels where they can be bartered, smuggled or secreted, or even not made into productive forces, "capital" in a true sense, at all.[51]

In these circumstances, it becomes necessary to revamp some of the familiar scholarly approaches to the state in African experience. It makes ever decreasing sense to locate decisive power in the alliance assembling a national bourgeoisie from its private and public sector fractions—business, the professions, civilian and military operatives using the state—as its enterprise shrinks and simultaneously recedes from this elite's control, even accepting Fatton's litany of coercion and violence. To register this change in the most relevant Cameroonian terms for the 1990s, consider Ahidjo's elite alliance, which Ngayap analyzed in ways both the Preface and this chapter use: 1,000 strong at its core; constantly recruited and finely tuned; its sectors rewarded or punished according to a presidential will which (part of its strength?) was unpredictable but was accepted; the tight articulation of the capital city and the central state apparatus with provincial capitals, and of both with the rural sector.[52] Then compare the 1990s: nothing so orderly obtains.

One relevant way to understand the Fru Ndi-SDF phenomenon in these terms is the defection it represents from the consolidated national bourgeoisie. He was very much its product, recognizable from Bayart. Before he surfaced as the opposition's probe in 1990 and pivot thereafter, he was a CPDM legislative candidate as recently as 1988, his book store business in the North West Province capital, Bamenda, held a government contract to supply North West Province offices and schools, and he owned property which he planned to develop (and later did so, on a large scale) on Commercial Avenue, Bamenda's main street. May 26, 1990, was his political catapult, and rallied popular forces beyond anything known for years or anticipated at the time. The SDF wedge constantly widened and deepened the regime's fault lines as the 1990s advanced and other significant desertions accumulated. National politics became less and less manageable by the direction of its reduced band of loyalists left with the state apparatus in Yaounde. Provincial and local affairs in much of Cameroon have gravitated toward the control of opposition elites like the Fru Ndi-SDF network in North West Province but regionally different in class, religion and ethnicity.[53] But they are not nationally consolidated, their connection to commoners is uncertain, and the latter's initiatives are not easy to articulate. In a crucial sense already stated, Cameroon's mid-1990s impasse reflects the weakness of both the state's hegemony, now declining and perhaps lost, and of the opposition's, or civil society's, counterhegemony, yet to emerge.

There are very few tangible guides to these processes, which the text ahead examines. One of the best is the SDF's vote tally of nearly 40 percent in the disputed presidential election of 1992, based primarily on its 70–90 percent success in southern cities, and the similar pattern for local elections early in 1996. These provide some evidence that Cameroon's formal politics has broken through Ahidjo's elite ranks and become more inclusive in pockets, more ideologically driven by class alignment which might dissolve long established distinctions of language, ethnicity, religion, and the like, in other words, that a "civil society" most would recognize as such is forming.

The precise disposition of power between the remnant state, dissenting elites including Fru Ndi's circle among many others, and newly active and articulate commoners, remains unclear, but there are creative and democratic as well as dead-end possibilities.[54] Tracing these processes, and their consequences for the character and interaction of Cameroon's state and civil society, provides the remainder of this study one of its major challenges. Keyed by generic questions touching the viability of Cameroon's state and by Joseph's pointed, Mbembe's tentative, and Monga's sweeping questions about its civil society and democratic possibilities, guided as historians by Bayart's insistence on actors and specificities while seeking patterns and contours like Mbembe's, we can now move to the account of what since 1960 has transpired in Cameroon.

Notes

1. Representative edited collections from the U.S.A. at that time can be usefully cited. Rupert Emerson and Martin Kilson, eds., *The Political Awakening of Africa* (Englewood Cliffs: Prentice-Hall, 1965) read African developments posi-

tively, variously stated pp. 2, 42, 106, 145. At the conclusion of a three volume series, 1962–1966, with 19 country studies in print, its primary editor Gwendolen Carter found moderate success in Africa's quest for national unity, social reconstruction and economic growth; Gwendolen Carter, ed., *National Unity and Regionalism in Eight African States* (Ithaca: Cornell University Press, 1966), pp. 539, 548, 549, 552.

2. René Dumont, *L'Afrique noire est mal partie* (Paris: Editions du Seuil, 1962), translated as *False Start in Africa* (London: André Deutsch, 1966), was perhaps the opening volley, on economic grounds. There rapidly developed distinct but overlapping theoretical and case study work in political economy. This literature is too vast to survey here, but a sense of how such analyses altered the U.S.A.'s political science surveyed in the previous paragraph emerges if one compares the authors, themes and bibliographies in cognate edited volumes 15 years apart: William Friedland and Carl Rosberg Jr., eds., *African Socialism* (Stanford: Stanford University Press, 1964); Carl Rosberg Jr. and Thomas Callaghy, eds., *Socialism in sub-Saharan Africa: A New Assessment* (Berkeley and Los Angeles: University of California Press, 1979). A more comparative analysis than we attempt would cite here the literature on "state-led" polities and economies in Latin America and Asia.

3. Joseph Femia, *Gramsci's Political Thought* (Oxford: Clarendon Press, 1987), is a useful introduction. Jean-François Bayart, *L'Etat au Cameroun* (Paris: Presses de la Fondation Nationale des Sciences Politiques, 1979), on which more follows, exemplifies the force of "hegemony" in African and Cameroon scholarship, as does, more recently, Peter Geschiere, "Hegemonic Regimes and Popular Protest—Bayart, Gramsci and the State in Cameroon (1)," in Wim van Binsbergen, Filip Reyntjens and Gerti Hesseling, eds., *State and Local Community in Africa* (Brussels: Centre d'Etude et de Documentation Africaines, Cahier 2–4, 1986), pp. 309–347. For a very clear statement of these issues in a specific setting, see Toyin Falola and Julius Ihonvbere, *The Rise & Fall of Nigeria's Second Republic: 1979–84* (London: Zed Books, 1985), pp. 238–243.

4. On the "big man" in various guises, see Robert Jackson and Carl Rosberg Jr., *Personal Rule in Black Africa: Prince, Autocrat, Prophet, Tyrant* (Berkeley and Los Angeles: University of California Press, 1982).

5. Goran Hyden, *No Shortcuts to Progress: African Development Management in Perspective* (Berkeley and Los Angeles: University of California Press, 1983), with its "economy of affection" (introduced p. 8), its "state with no structural roots in society . . . [like] a balloon suspended in mid-air" (p. 19), its denial of a hegemonic ruling class (p. 39), and its "uncaptured peasantry" (introduced p. 52). He recognized, pp. 39–40, the debt to Peter Ekeh, "Colonialism and the Two Publics in Africa: A Theoretical Statement," *Comparative Studies in Society and History* 17,1 (1975), pp. 91–112, and the "primordial public realm" discussion there. Note also the skepticism in Richard Sandbrook, "Hobbled Leviathans: Constraints on State Formation in Africa," *International Journal* XLI,4 (1986), pp. 707–733. Bayart draws attention on this front below.

6. Hyden, *No Shortcuts to Progress*, p. 209, stated that living with capitalism and returning to the state and market were at some point going to be necessary.

7. Many anthropologists, especially Europeans, became interlocutors, working out from their own discipline (and often from both their countries of

origin and sites of first field research) to enlarge social science perspectives. Peter Gutkind's contribution is notable for his cosmopolitan range of study, editing of *Labour, Capital and Society*, and the collaborative Peter Gutkind and Immanuel Wallerstein, eds., *The Political Economy of Contemporary Africa* (Beverley Hills: Sage, 1976), publishing multi-disciplinary scholars from eight countries.

8. Geschiere, "Hegemonic Regimes," p. 320. Jan Vansina, "Mwasi's Trials," *Daedalus* 111,2 (1982), pp. 49–72, esp. p. 58, also reflected this shift, for (then) Zaire.

9. Following his first major studies, based on Africa, Immanuel Wallerstein, *The Modern World System* (New York: Academic Press, 1974) began his three volume analysis.

10. Walter Rodney, *How Europe Underdeveloped Africa* (Dar es Salaam: Tanzania Publishing House, 1974). Quite remarkably for our purposes, the book's last paragraph cited Cameroon as a place where that history has created an arena of highly intense struggle, one of the few such examples in an English language Africa-wide work.

11. Samir Amin, *Neo-Colonialism in West Africa* (New York: Monthly Review Press, 1973), translated from *L'Afrique de L'Ouest Bloquée* (Paris: Les Editions de Minuit, 1971); Claude Ake, *A Political Economy of Africa* (Harlow: Longman, 1981), especially pp. 126, 129, 182. Bayart's critiques of this work as well as much else we have previously surveyed are considered below.

12. Did Davidson borrow or create the term "inheritance elites"? It first informed his post-1960 coverage in Basil Davidson, *Let Freedom Come: Africa in Modern History* (Boston: Little, Brown and Company, 1978), from p. 300.

13. Robert Fatton Jr., *Predatory Rule: State and Civil Society in Africa* (Boulder: Lynne Rienner Publishers, 1992), especially pp. 7–9 for the theoretical premise, p. 86 for these striking quotations.

14. Basil Davidson, *The Black Man's Burden: Africa and the Curse of the Nation-State* (New York: Random House, 1992), pp. 247–248, 294, 320.

15. Jean-François Bayart, *The State in Africa: The Politics of the Belly* (London: Longman, 1993), p. 218; there is a later swipe, p. 331, at Fatton's use of Gramsci in his Senegal book as "theoretically dubious."

16. Bayart, *The State in Africa*, pp. 262, 268–269.

17. Bayart, *The State in Africa*, p. 254.

18. Bayart, *The State in Africa*, pp. 258, 261. For Cameroon in this sense, see Georges Courade and Luc Sindjoun, "Le Cameroun dans l'entre-deux," *Politique Africaine* 62 (June 1996), pp. 8–10.

19. Bayart, *The State in Africa*, p. 95.

20. Bayart, *L'Etat au Cameroun*, p. 19.

21. Bayart, *The State in Africa*, p. 228.

22. Bayart, *The State in Africa*, p. 254 for the chameleon metaphor from Mali.

23. Bayart, *The State in Africa*, pp. 150 ff. for the first of these key terms (which was also prominent in the 1979 Cameroon text), *inter alia* for the second.

24. Bayart, *The State in Africa*, pp. 220ff.

25. Bayart, *The State in Africa*, pp. 237–238 for the power line example.

26. Bayart, *The State in Africa*, pp. 234–235, 239. There was the slightest hint of respite in his conclusion, pp. 263, 267, in passages like the "orbit of politics . . . is

capable of change ... the long-term prison sentence is more like probation" and "the possible invention of a democratic culture."

27. René Lemarchand, "Uncivil States and Civil Societies: How Illusion Became Reality," *The Journal of Modern African Studies* 30,2 (1992), p.177, introducing a distinctively comprehensive literature review.

28. Naomi Chazan, "Africa's Democratic Challenge," *World Policy Journal* 9,2 (1992), p. 289; Michael Bratton and Nicolas van de Walle, "Popular Protests and Political Reform in Africa," *Comparative Politics* 24,6 (1992), p. 440.

29. Lemarchand, "Uncivil States and Civil Societies," pp. 186–187.

30. Robert Fatton Jr., "Africa in the Age of Democratization: The Civic Limitations of Civil Society," *African Studies Review* 38,2 (1995), pp. 67–99, especially the "Class and Civil Society in the Plural" section, pp. 77–78.

31. Recall his 1996 book's title: *The Anthropology of Anger*. This change in perspective first registered clearly in the landmark title and chapters of Patrick Chabal, ed., *Political Domination in Africa: Reflections on the limits of power* (Cambridge: Cambridge University, 1986), where even Bayart, for instance, wrote cautiously of a "democratic problematic [with] a texture which has usually been neglected ... we shall not underestimate a society's capacity to 'invent democracy,'" p. 124.

32. Victor Le Vine, *The Cameroon Federal Republic* (Ithaca: Cornell University Press, 1971), pp. 179–184.

33. Willard R. Johnson, *The Cameroon Federation: Political Integration in a Fragmentary Society* (Princeton: Princeton University Press, 1970), p. 374.

34. Beti, surveyed below, is well known. Eyinga, less recognized but the leading UPC political analyst, issued a series of critiques from France, most comprehensively in Abel Eyinga, *Introduction à la politique camerounaise* (Paris: L'Harmattan, 1984). The anglophone Ndiva Kofele-Kale added similar work from North America.

35. Jacques Benjamin, *Les Camerounais occidentaux: La Minorité dans un Etat bi-communitaire* (Montréal: Les Presses de l'Université de Montréal, 1972), p. 28 ("Le Cameroun est né en pleine guerre civile"), p. 149 for reference to federalism's potential for thwarting rather than satisfying anglophones, p. 192, n. 43 for his differences with Johnson.

36. It must be said, before moving to the critical literature, that they might have swelled it had they remained fully active in Cameroon studies, especially Le Vine, judging by a study he soon published on corruption in Ghana.

37. Mongo Beti, *Main basse sur le Cameroun: autopsie d'une décolonisation* (Paris: Maspero, 1972).

38. Richard Bjornson, *The African Quest for Freedom and Identity: Cameroonian Writing and the National Experience* (Bloomington: Indiana University Press, 1991), pp. xv, 50–51, 282, 327–332, and 349–376 for detail on the full print legacy in prose, drama, poetry, philosophy and religion.

39. Achille Mbembe, "Pouvoir des morts et langage des vivants: les errances de la mémoire nationaliste au Cameroun," *Politique Africaine* 22 (1986), pp. 37–72.

40. Richard A. Joseph, ed., *Gaullist Africa: Cameroon under Ahmadu Ahidjo* (Enugu: Fourth Dimension, 1978), p. 34. We noted the 1995 transposition to "Vichist" language in the Preface.

41. Joseph, *Gaullist Africa*, p. 81.

42. Bayart, *L'Etat au Cameroun*, p. 52; for passages quoted or summarized directly below, see pp. 53, 138, 233ff. His base definition of "la recherche hégémonique" in Cameroon was quoted and translated above.

43. Achille Mbembe, "Pouvoir, violence et accumulation," *Politique Africaine* 32 (1990), pp. 7–25, and "Provisional Notes on the Postcolony," *Africa* 62,1 (1992), pp. 3–37.

44. Mark DeLancey, *Cameroon: Dependence and Independence* (Boulder: Westview Press, 1989), and Peter Geschiere and Piet Konings, eds., *Proceedings of the Conference on the Political Economy of Cameroon—Historical Perspectives/Colloque sur l'économie politique du Cameroun—perspectives historiques* (University of Leiden: African Studies Centre, 1989), two volumes.

45. DeLancey, *Cameroon*, p. 78.

46. Papers by Jua and van de Walle, who were cited in the Introduction for works following the Leiden session, gave blunt economic analyses of the emerging crisis.

47. Achille Mbembe, "Crise de légitimité, restauration autoritaire et déliquescence de l'Etat," in Peter Geschiere and Piet Konings, *Pathways to Accumulation in Cameroon* (Paris: Karthala and Leiden: Afrika-Studiecentrum, 1993), pp. 345–374. See p. 364 for the first point, *inter alia* for the second.

48. Mbembe, "Crise de légitimité," pp. 370–374 for passages in this paragraph. A year earlier he noted the limited gains of political struggle since 1990 in a passage which resonates: "la démocratisation est une illusion, un multipartisme administratif"; Achille Mbembe, "La violence derrière le multipartisme," *Afrique Magazine* 97 (November, 1992), p. 56.

49. Mbembe, "Crise de légitimité," p. 366 for the SDF's potential as "la voie à une recomposition de la famille nationaliste et sa recristallisation," drawing like Bayart, *The State in Africa*, p. 251, on the UPC's "collective historical memory" and "symbolic capital."

50. Joseph, *Gaullist Africa*, p. 199.

51. The study of coping strategies sketched here is best developed for Congo (Kinshasa). For very pointed items from a rich scholarship including work by Crawford Young, Thomas Callaghy and Georges Nzongola-Ntalaja, see Vansina, "Mwasi's Trials," and Janet MacGaffey, *The Real Economy of Zaire* (Philadelphia: University of Pennsylvania Press, 1991), especially the conclusion, pp. 154–157.

52. Pierre Flambeau Ngayap, *Cameroun: Qui Gouverne? de Ahidjo à Biya, l'héritage et l'enjeu* (Paris: L'Harmattan, 1983), pp. 8ff. for "la Classe dirigeante" defined, *inter alia* for its characteristics.

53. Although this was not its overt point, Michael Rowlands, "Accumulation and the Cultural Politics of Identity in the Grassfields," in Peter Geschiere and Piet Konings, *Pathways to Accumulation in Cameroon* (Paris: Karthala, 1993), pp. 71–97, takes soundings on these issues in the core opposition territory where, he argues, the state was never considered or used as the primary agent of accumulation. Jean-Pierre Warnier, *L'esprit d'entreprise au Cameroun* (Paris: Karthala, 1993) covers the topic far more thoroughly, with similar conclusions, esp. pp. 274ff.

54. A great virtue of Fatton's work is his unremitting probe of these themes, and recognition of the fluidity of these actors.

2

Ahidjo and the Single-Party State, 1958–1982

Like many former colonies in Africa, Cameroon entered the postcolonial period under difficult political circumstances, complicated by the divided legacy of United Nations Trust Territory status under French and British administrations and the paths leading not simply to independence but to reunification of the earlier German-ruled terrain. Because of the preponderance of people and land in the French zone and the earlier initiatives and subsequent domination by Cameroonian politicians from that side of the border, our coverage of the transitional period since 1960 emphasizes conditions in the territory emerging from French rule.

Among the problems French and Cameroonian authorities faced as they organized the transfer of sovereignty was armed resistance by the radical UPC following its proscription by French authorities in July 1955, particularly in the Littoral and Western regions.[1] The UPC resistance did not attain the level and ferocity of the struggles for independence in Vietnam, culminating in France's humiliating defeat at the battle of Dien Bien Phu in 1954, or in Algeria with its National Liberation Front. But it was among Africa's fiercest, and a cause for concern to French authorities and later to the postindependence government. Less dangerous but politically challenging was the proliferation of some ninety ethnic and regionally based political parties and groups, some of which had been formed with French support in the late 1940s and the 1950s in an effort to counter the popularity of the UPC.[2]

It was Ahmadou Ahidjo, a Northern Moslem of Fulbe background, who became Cameroon's first president, and then managed the integration of the two colonial heritages following the reunification of the former French Cameroon and the British Southern Cameroons on October 1, 1961. For the next two decades Ahidjo was obsessed with how to stabilize this delicate political patchwork and how to transform the various particularist (ethnic, regional, religious, and linguistic) interests or what he called "separate fatherlands" into a single nation. As the president indicated: "The project of the Nation is to assemble the fatherlands in order to transcend them, and thus build a new fatherland founded on their realities, their virtues, their values and their emotional impact in order to universalize them."[3] How Ahidjo constructed and maintained that nation and the problems he faced in doing so are the focus of this chapter.

The Emergence of Ahmadou Ahidjo

Despite initial objections to the concept of independence for its African territories, French authorities and a majority of Frenchmen toward 1960 abandoned the romantic idea of creating a greater *France d'Outre-mer*[4] and reconciled themselves to the fact that independence for her colonies was inevitable.[5] With that in mind, the main concern for French authorities was to identify and promote into leadership positions moderate Africans such as Léopold Senghor in Senegal and Félix Houphouët-Boigny in Côte d'Ivoire who, despite minor disagreements among themselves and with French leaders on the process of decolonization, still favored France's continued presence in Africa.[6]

In French Cameroon, that opportunity initially fell in the hands of André-Marie Mbida, a Roman Catholic from the South-Central region of the country and leader of the Parti des Démocrates Camerounais, who was appointed Cameroon's first pre-independence prime minister in May 1957. However, Mbida's tenure as prime minister was shortlived. In February 1958, he was pressured to resign following disagreements between himself and Jean Ramadier, the French high commissioner, and also because of a series of censures passed by the Cameroon legislature against his government.

The collapse of Mbida's government was partly orchestrated by French authorities who objected to his resistance to early independence and reunification for Cameroon, and because of an imperious leadership style which had alienated both his political friends and foes.[7] For instance, Mbida failed to recognize widespread support for independence among Cameroonians. Instead, he proposed a ten year waiting period before Cameroon could attain independence on grounds that the territory lacked the necessary capacity, especially in skilled personnel to administer itself after independence.[8] He was also unwilling to negotiate a political solution with the UPC insurrection or grant amnesty to members of the party who had committed acts of violence since its ban in 1955.[9]

Although one might have expected France to support Mbida's objection to immediate independence for Cameroon since that would mean retaining its colonial relationship with Cameroon, that was not the case. In fact, by 1958, France had recognized that immediate independence and reunification were popular themes among various political groups in the territory, not just the UPC. Therefore, it was politically expedient to support those objectives, if France expected to prevent the rise to power of the radical UPC and to enhance her position in a postindependence Cameroon.

The man selected to replace Mbida by Xavier Torre, Ramadier's successor in January 1958 as the French high commissioner to Cameroon, was Ahmadou Ahidjo, vice-prime minister and minister of the interior in Mbida's government. The selection of Ahidjo as Mbida's successor was influenced by three factors. First, French authorities saw him as a compromise candidate whose moderate views were necessary in bridging the gap between the more conservative and economically less advanced northern region and the more progressive southern section that threatened national unity.[10] Second, unlike Mbida, Ahidjo supported the UPC's demands for immediate independence and reunification, albeit reluctantly and for his own political objectives. Finally, he favored continued French influence in Cameroon following independence.

The Years of Trial, 1958–1962

Perhaps the most difficult period in Ahidjo's tenure was from 1958 to 1962. First, he had to gain the confidence of his French benefactors as a leader who was not only capable of reconciling the various political factions, but was willing to protect French interests in the territory. Second, he had to establish himself in charge in order to carry out his domestic political agenda.

Although French authorities may have perceived Ahidjo as a "second-rank political figure"[11] whose appointment as prime minister was seen as a temporary measure until a more suitable candidate capable of protecting French interests was recruited from the ranks of aspiring Cameroonian politicians, he proved to be skilled and astute. He quickly realized that he had to play the tune of his colonial benefactors if he wished to remain in power. In a speech in May 1958, for example, Ahidjo alluded to the historical connection between Cameroon and France that he hoped to continue:

> In a world in which seclusion is harmful to individuals and nations, we cannot remain isolated and in these conditions, how can we conceive having any other partner than this country we know and love? How can we forget its accomplishment all these years that we have learned to understand and appreciate her, how can we ignore the cultural education that is leading us today in the course whose direction was determined by her? It is with France that Cameroon, once emancipated, wish to bind its destiny

and in concert, sail freely along with her through the turbulent seas of today's world.[12]

This text was activated in a series of Franco-Cameroonian agreements Ahidjo signed with France soon after independence, in November 1960, guaranteeing closer political, social, military, and economic ties between the two states. Like similar treaties with its other former African colonies, the agreement with Cameroon not only guaranteed the safety of French investments in Cameroon, but also granted France extensive influence in the territories. For example, these cooperation agreements obliged the various African states to consult with France before establishing any higher educational institutions in their countries. Additionally, France was given priority in the recruitment of expatriate personnel to work in them.[13]

As part of his goal of attaining power, Ahidjo also pursued a well calculated domestic strategy that involved creating a broadly based government including representatives from other parties, not just members of his Northern-based Union Camerounaise (UC). That may explain why his first cabinet in 1958 included two members from his UC party, one from the Paysans Indépendants, two from the Mouvement d'Action Nationale Camerounaise (MANC), one independent, and three others selected from outside the National Assembly.[14] The composition of the cabinet not only provided political and regional balance—the first version of the skilfully crafted North-South axis he came to rely on—but also broadened the legitimacy and support Ahidjo needed to carry out his political agenda.

Moreover, as indicated earlier, Ahidjo also favored, at least for political expediency, the UPC's agenda of immediate independence and reunification. And unlike his predecessor, he was willing to grant amnesty to UPC rebels who laid down their arms. These overtures had some success, especially after the September 13, 1958, assassination of Ruben um Nyobe, secretary general of the UPC, and the surrender of his chief lieutenant, Théodore Mayi-Matip. In fact, by the end of 1958, the violence engendered by UPC resistance, especially in Sanaga-Maritime Division, had subsided substantially following the surrender of more than 2,500 of its fighters.[15]

Another indication that Ahidjo was willing to accommodate former UPC rebels was his decision early in 1959 not only to allow Mayi-Matip to organize his followers into a coherent political faction known as the *rallié upecistes*, but for them to participate in the April 12 by-elections. The latter were necessary to fill the seats of six members of the Legislative Assembly from the Sanaga-Maritime, Nyong-et-Sanaga and Bamileke regions who had either resigned, died, or been assassinated by

UPC terrorists after the December 1956 legislative elections.[16] In fact, to make sure that the seats were won by UPC candidates, the UC did not field any serious candidates to compete against the UPC in these regions.[17] However, because the UPC had not been legalized, the victorious candidates, all former UPC rebels, were admitted into the legislative assembly as independents.[18]

Independence came to the French territory on January 1, 1960. Perhaps as a last gesture of good faith to the UPC, Ahidjo on February 25, 1960, signed a decree repealing the French authority's law of July 13, 1955, banning the UPC and all its associated organs, including the Jeunesse Démocratique du Cameroun and the Union Démocratiques des Femmes Camerounaises.[19] With the repeal of the July 1955 law, the UPC could again be considered a legal party, even if it was perceived to be collaborating with the administration. By making sure that the *rallié upecistes* was represented in the legislature, Ahidjo hoped to weaken the UPC by widening the rift between the collaborators and the external wing of the party which was still opposed to any participation by UPC members in Ahidjo's government.[20] Clearly, "la classe dirigeante" which Pierre Flambeau Ngayap documented in 1983 was on its way to state dominance.[21]

While Ahidjo was prepared to accommodate the UPC and its leadership, he was equally tough in dealing with members of the party who were unwilling to compromise. In fact, because of the intensification of UPC violence in the Bamileke region on the eve of independence, Ahidjo in October 1959 had requested emergency powers (pleins pouvoirs) from the legislative assembly, designed for rule by decree until new legislative elections were held in March 1960. Despite acrimonious debate in the legislature and serious objections by members of the opposition parties who were afraid that such authorization could lead to dictatorship and the suppression of individual liberties, the request was passed by a vote of fifty in favor and twelve against, with one abstention.[22]

The passage of the emergency power was important for several reasons, including the fact that it gave Ahidjo sole authority in requesting additional French troops who were critical in the regime's suppression of UPC rebellion in the 1960s. Apart from its role in suppressing UPC resistance, however, France's military involvement in Cameroon also included the building of the country's national army and one of the most effective intelligence services in sub-Saharan Africa, the Service des Etudes et de la Documentation (SEDOC).[23] Both institutions were crucial in Ahidjo's control of the state for over two decades. As Richard Joseph argues:

> The actual task of destroying the revolutionary maquis was only one side of French military activities in Cameroon up to the mid-1960s. The other side

was building and giving practical experience to a Cameroon national army, while making that army wholly subservient to the political will of the Head of State.[24]

French military support was not restricted to eliminating Ahidjo's political enemies at home. The November 1960 death in Switzerland of Félix-Roland Moumié, one of the key leaders of the UPC in exile, was attributed to William Betchel, an agent of the French secret service. In fact, without actually confirming France's direct orders in the murder of both Moumié and um Nyobe, Jacques Foccart, the veteran advisor to various French presidents on African affairs since the 1950s, in a recent interview before his death in 1997 left little doubt as to France's involvement in the death of both men.[25]

The emergency powers Ahidjo obtained in 1959 also allowed him to unilaterally postpone the legislative election scheduled for March 1960. Had the election occurred as planned, the newly elected members would have been responsible for drafting the constitution for the new republic. But with its postponement because of the unstable political environment caused by the UPC rebellion, a special forty-two member, largely pro-Ahidjo Constitutional Committee was formed to draft a new constitution. The result was the emergence of a Gaullist-style constitution with a strong executive and a weak legislature. Despite strong objections from some members of the committee, the Catholic Church and members of opposition parties, the constitution was ultimately approved by the electorate by 60 percent (797,498 for and 531,175 against) of those who cast their votes in a special referendum on February 22, 1960. Ahidjo's authority in this highly centralized form was later reaffirmed at the Foumban Constitutional Convention of July 17–21, 1961, when the Southern Cameroons delegation responsible for that territory's transition from British-supervised United Nations Trust status and representatives from the now independent Republic of Cameroon (République du Cameroun) convened to draw up a new constitution for the union between their polities. Attempts by the Southern Cameroons delegation to create a loose and decentralized federation that would have given greater authority to the different states in the federation were unsuccessful. Instead, what emerged from the constitutional negotiations joining the two states called East and West Cameroon after unification took effect on October 1, 1961, was a highly centralized federal structure and an executive President, whose authority, Victor Le Vine argued, combined "attributes of a British-style governor-general, a Fifth Republic president and an American chief executive."[26] In short, the naive, ill-prepared, bungling, confused, and perhaps "over-trusting" Southern Cameroons delegation was simply outmaneuvered in negotiations by a well prepared delegation from Yaounde, which also benefitted from the expertise of their French advisors.

Johnson notes that even the minor revisions that the Southern Cameroons delegation was able to make in Ahidjo's original constitutional proposal, aimed at diluting the power of the president, had a boomerang effect because they strengthened rather than weakened both the federal government and the federal president.[27] For instance, Article 5 of the constitution deprived the states of control over various institutions such as national defence, foreign affairs, higher education, and scientific research that would have given each of the federated states considerable power. As for the few institutions (managing human rights, public health and secondary and technical education, for example) that were under the temporary control of the states (Article 6), the states could legislate on such matters only after consultation with a Federal Coordinating Committee. Additionally, Article 15 of the Federal Constitution noted that the president *may* after consultation with the prime minister of the federated states declare a state of emergency and rule by decree "in the event of grave peril threatening the nation's territorial integrity or its existence, independence or institutions."

However, the constitution did not define what events or circumstances might be considered "grave peril" nor did it require the president to consult the prime ministers in determining if an event or circumstance was of grave peril to the state or nation. Therefore only the president's discretion led him to consult with the prime ministers or define what he perceived as "great peril."[28] This broad authority and vaguely worded provision cleared the way for the passage of the March 12, 1962, antisubversion law and other similar repressive laws that Ahidjo used in depriving Cameroonians of their freedoms for the next two decades. The framework of the Gaullist state, what Ngayap later identified as "la classe dirigeante" and Bayart as "la recherche hégémonique," began to emerge. For even before Ahidjo's formal presidential authority was legitimized and expanded into British Cameroon at the Foumban constitutional convention, he had already started the process of establishing himself as a national leader by expanding support for himself and his party from northern into southern regions of the former French territory. The first opportunity to achieve that goal occurred in the first postindependence legislative election of April 1960 for the election of members into the new 100 member Legislative Assembly.

In an effort to win a majority standing in the legislature, the electoral laws were tailored to favor UC candidates. For instance, because the UC was sure of winning most of the seats in the North (Ahidjo's region of origin), the electoral laws allowed for the creation of single-list electoral constituencies in the northern electoral districts. Consequently, any party that won the majority votes in any particular electoral district in the North got all the seats in that district. By contrast, in the southern regions where the UC had little chance of winning, the election was conducted on a single-member basis. Under this system, victory was based on the performance of each candidate in a district rather than the performance of a party. In other words, a party that won the plurality of votes in a particular district did not necessarily win all the seats in the district. This system of voting in the South not only allowed candidates from the same party to run against each other in the same district, but was also intended to diminish the chances of any of the southern-based parties from winning the majority seats in the legislature and as a result challenging UC control of the legislature. In fact, as a result of this dual electoral system, the UC was able to win fifty-one seats in the new 100 seat legislature. This included all of the forty-four seats in the North and in Bamoun Division where the UC was dominant and seven seats in the South-East.[29]

Following its victory in the legislative elections, the UC tried to organize local party cells in the South. But when the effort failed, Ahidjo saw the creation of a single party as the best means to achieve the political domination of his party throughout East Cameroon. In order to create such a party, Ahidjo either co-opted members of various opposition parties into his administration or used force and intimidation to silence those who were reluctant to join the UC. For example, his first fourteen member postindependence cabinet included six representatives from his party, the UC, three from the Front Populaire pour L'Unité et la Paix (FPUP) led by Pierre Kamdem Ninyim, a former leader of one of the guerrilla wings of the UPC, two from the Démocrates Camerounais (DC), two from the Parti Progressistes and one independent. The large number of cabinet positions, according to David Gardinier, was not designed to make the task of government more efficient, but to satisfy the various parties that were prepared to form a coalition government with Ahidjo.[30] Other important members of the various opposition parties who were not given cabinet positions but who were willing to join Ahidjo's government were given other administrative posts.

It is apparent from the aforementioned strategy that Ahidjo's objective was not only to weaken and eventually neutralize the opposition parties by rewarding those who were prepared to join his administration with government posts, but also to punish those who refused to follow his agenda. Richard Joseph points out that Ahidjo's tactic in this case was

> to use Southern politicians and their coteries of supporters as tools for beating the others into submission, but not in the mere sense of divide and conquer; it was a case of absorbing one's enemies to gain just that much addi-

tional power to proceed with the increasingly extralegal means needed to deal with the next group of recalcitrants.[31]

The tactic appears to have been successful. Soon after the legislative elections the MANC, led by Charles Assale, dissolved itself and declared its support for the UC. By June 1961, the FPUP had taken a similar measure after most of the party's leadership had agreed in April 1961 to collaborate with Ahidjo's government. Not surprisingly, these moves occurred soon after many of the party leaders had been co-opted into Ahidjo's government. For instance, Assale became the prime minister of East Cameroon, while the FPUP's Pierre Ninyim Kamdem and Victor Kamga were appointed minister of health and minister of justice respectively.

Although Ahidjo had succeeding in co-opting many opposition politicians into his administration and in reducing the number of opposition parties in East Cameroon by the end of 1961, he had still not achieved his goal of creating a unified party or what he later characterized as a "Great National Party." According to Ahidjo, this "Great National Party" to be realized through persuasion not coercion would include Cameroonians of all tendencies. In addition, democracy, freedom of speech and debate would prevail within the party.[32]

However, the president's intimation of this hegemonic strategy at a press conference in Yaounde on November 11, 1961, faced a quick, negative response from opposition leaders. In a letter to Ahidjo and the Central Committee of the UC, four opposition party leaders, including André-Marie Mbida of the DC, Theodore Mayi-Matip of the legalized wing of the UPC (rallié upecistes), Charles Okala of the Parti Socialistes Camerounais, and Dr. Marcel Bebey-Eyidi of the Parti des Travailliste Camerounaise expressed their willingness to work towards national unity. But they rejected Ahidjo's proposal for a Unified Party, arguing that such a party would lead to an end to democracy and create a "fascist-type dictatorship."[33] Instead the four leaders who had banded together on June 15, 1962, to form a United National Front (Front National Unifié) proposed the creation of a Unified Front (Front Unifié), a coalition of parties that would allow each party to maintain its identity and independence.

Ahidjo and the executive committee of the UC rejected the concept on grounds that the Unified Front would simply encourage the existence of a multiplicity of parties, which "was a source of inefficiency and a negation of all action."[34] In response to this retort, the four opposition leaders reiterated their earlier points in a lengthy June 23, 1962, memorandum titled *Manifeste du Front National Unifié* in which they called on Ahidjo to "conduct himself as an impartial arbiter and not as a partisan."[35]

Their action was not only an affront to the president but also an opportunity for Ahidjo to eliminate the last vestiges of what he perceived as resistance to his administration. On June 29, the four men were arrested under a relatively new ordinance (Law No. 62/OF/18 of March 12, 1962) authorizing the arrest and trial of anyone guilty of subversion. They were charged with threatening national security and spreading news liable to be harmful to public authority. After a brief trial, the four men were convicted on July 11, sentenced to thirty months in prison, fined CFA 250,000 francs, and lost their political rights. The prison sentences were extended to three years in December after they lost their appeal. Like much else recounted here from these early years of Ahidjo's power, this ordinance remained effective as a regime mainstay, not repealed until Law No. 90/046 late in 1990.

The imprisonment of such prominent politicians, together with the expansion of the UC in the southern region of the country forced many other members of the opposition parties to either dissolve their parties or join the UC. Those who remained in opposition were increasingly marginalized and rendered politically ineffective as Ahidjo consolidated his power and authority throughout the country. By the time of the UC's Fourth Congress held in the southern city of Ebolowa in July 1962, Ahidjo was confident enough to declare the UC as "the only political party in East Cameroon with a truly national character."[36] The DC and a few other parties that had managed to hobble along during this period of attrition and co-optation were so weakened and sapped of their support that they did not even consider it worthwhile fielding candidates in the 1965 elections to the East Cameroon legislative assembly.[37]

After successfully creating a *de facto* single-party state in East Cameroon, Ahidjo's next step in consolidating the Gaullist strategy was to convince the various political parties in West Cameroon to join with the UC in forming the single national party which he perceived as central to national construction.[38]

Creating a Single-Party State, 1962–1966

It was easier for Ahidjo to deal with West Cameroon's political leaders and to convince them of the benefits of a single-party state than had been the case in East Cameroon. Perhaps through ignorance, and certainly because of the greed and the need for personal aggrandizement, they not only competed for Ahidjo's favor, but promoted (perhaps inadvertently) the idea of a single national party. For example, during Ahidjo's trips to West Cameroon in 1961 and 1962, Dr. Emmanuel Endeley, leader of the Cameroon People's National Convention (CPNC), West Cameroon's minority party, expressed his willingness to support Ahidjo's vision of national unity and a centralized federation, including the formation of a single party.[39] Perhaps Endeley's support for a single-party in Cameroon was not surprising. As a leader of West Cameroon's minority party, Endeley not only saw Ahidjo's plan as an opportunity for him to play a role in national politics, but also as a means of preventing his party from being dominated by the majority Kamerun National Democratic Party (KNDP).

His scheme was not the only one in West Cameroon. The fear that an Ahidjo-Endeley agreement could undermine Foncha's authority as prime minister of West Cameroon and leader of the KNDP may also have prompted Foncha on April 27, 1962, to form a National Coordinating Committee between his party and the UC. Among other things, the committee was designed to explore ways of merging the two parties into a single national party. But until that goal was achieved, the agreement called on both the KNDP and UC to restrict their political activities to their own states.

Such a restriction was important to both Ahidjo and Foncha because it eliminated the chances of an alliance by either man with any political parties or groups in either East or West Cameroon that might undermine their authority. As Willard Johnson reminds us: "Foncha never seemed to fear losing his power and prerogatives to the federal government as much as losing it to his own opposition within the state." By the same token, Ahidjo was afraid of an alliance between the KNDP and any of the southern-based parties he had brought into the UC in East Cameroon, which could threaten his position as president.[40]

Rivalry between West Cameroon politicians was not limited to inter-party competition.[41] As early as August 1963, a rift occurred within the KNDP following the unsuccessful bid by Solomon Muna and Emmanuel Egbe, both federal cabinet members and supporters of a strong central state, for the positions of vice-president and secretary general of the KNDP, respectively, at the party's Ninth Congress in Bamenda. With the impending resignation of Foncha as prime minister of West Cameroon to become the vice-president of Cameroon in accordance with Article 3 (Section 3) of the constitution separating those posts, it was expected that the vice president of the KNDP would in all probability be nominated to fill the post of prime minister of West Cameroon.

Although Foncha had initially expressed support for Muna's and Egbe's higher party aspirations, he later changed his mind when the KNDP leadership overwhelmingly elected Augustin Jua and Nzo Ekah Nghaky to fill the posts. Disenchanted by their failure to win the positions, Muna, Egbe, and six of their supporters who were ultimately expelled from the KNDP for insubordination formed a new party, the Cameroon United Congress (CUC).

The intraparty rivalry between the pro-Muna and pro-Jua forces within the KNDP came to a head in May 1965. Jua as expected was appointed prime minister of West Cameroon when Foncha took over the federal vice-presidency. Although Ahidjo would have preferred Muna as the prime minister because of the latter's support for a more centralized government, he refrained from imposing his choice after extensive consultation with members of the West Cameroon legislative assembly, important traditional chiefs, and administrative authorities. In fact, Ahidjo was not only concerned that going against the wishes of the KNDP leaders could be politically disruptive, but that it would have been virtually impossible for Muna's appointment to be approved by the KNDP-dominated legislature as required by the constitution. At the same time, however, Ahidjo retained Muna and Egbe as federal ministers despite pleas from Foncha that both men be expelled from the cabinet.

The rivalry among West Cameroon political leaders provided an excellent opportunity for Ahidjo to call for the dissolution of West Cameroon's parties and the formation of a single national party. In making that case, the president argued that multipartism was not only a hindrance to effective execution of government policies, but could also lead to "anarchy, inefficiency and ultimately to the suicide of the Cameroon nation."[42]

Because they were afraid of being upstaged by members of the CUC and the CPNC who had expressed support for Ahidjo's call for a single party, some leaders of the KNDP were forced to embrace the idea even though they had initially been opposed to it. According to Ndiva Kofele-Kale, this change of heart among anglophone politicians had much to do with their perception of changing power relationships and forces in the nation, and the extent to which the changes would be profitably exploited to serve their narrow class interest.[43] The perception fits the scholarship surveyed in Chapter 1.

Having apparently convinced West Cameroon politicians of the benefits of single-party rule as the secure base of their political fortunes, Ahidjo swiftly formed the new single party. In fact, the entire process of dissolving old parties (three from West Cameroon and Ahidjo's UC in East Cameroon) and creating the new Cameroon National Union (CNU) took less than three months. The new party was formerly established on September 1, 1966. Ahidjo became its president, Foncha and Dr. Simon Tchoungui, the prime minister of East Cameroon, were elected as its two vice-presidents.

For Ahidjo, the birth of the CNU represented an important milestone in the kind of "democracy" he hoped to create in Cameroon,[44] and a critical step in maximizing his power and authority.[45] After all, he was both the head of state and president of the country's only functioning party. Despite his successes, however, especially in creating a single-party system in Cameroon, one step remained in Ahidjo's political agenda: the dismantling of Cameroon's 1961 federal constitutional structure.

From Single Party to United Republic, 1966–1972

From the beginning of the reunification process, Ahidjo and many East Cameroon politicians were more interested in creating a strong centralized unitary state rather than a loosely organized federation.[46] However, they conceded the latter in 1961 as a temporary measure to appease Foncha and other West Cameroon politicians. Even so, it could be argued that the foundation for the union Easterners hoped to create was established under Articles 5 and 6 of the federal constitution. As indicated earlier, Article 5 of the constitution gave the federal government jurisdiction in important responsibilities such as national defense, foreign affairs, immigration and emigration, higher education, and scientific research, which if controlled by the states could have given West Cameroon considerable political freedom to define its own policy in those areas. Additionally, activities such as public health, secondary and technical education, and prison administration, over which the states had been given temporary control under Article 6, were transferred to the federal government within two years of the federation.[47] The speed of this transition demonstrated both the enormous authority of the president and his desire to neutralize the states by denying them any independent authority.

Therefore, from the birth of the federation, the federal government and the president exercised their constitutional authority to weaken the federal structure by making the states, particularly West Cameroon, politically and economically dependent on the federal government. In fact, soon after reunification, Ahidjo issued Decree No. 61/DF/15 dividing the federation into six administrative regions. While East Cameroon was divided into five regions, West Cameroon was considered a single administrative unit. This administrative setup not only ignored the federal structure provided for in the constitution, but gave Ahidjo the opportunity to exercise inordinate control over the states. For example, each of the six units was under the authority of a federal inspector of administration appointed by the president and each of the inspectors was directly responsible to the president and not to the prime ministers. This in effect made the inspectors the local representative of the president in their jurisdictions. Each inspector had the authority to institute rules and regulations in the provinces without consulting the prime minister so long as the rules and regulations were under the purview of the president.[48] Frank Stark and Willard Johnson have discussed the effect of this apparent dualism of authority between the prime minister and the federal inspector in West Cameroon.[49] Suffice it to surmise here that the system clearly undermined the authority of the prime minister of West Cameroon. For example, Johnson argues that this administration

> brought Yaounde too close not only to Buea, the state capital, but equally close to Kumba, Mamfe, Bamenda and other principal administrative centers. The dangers in this were obvious to those interested in preserving a pattern of life in West Cameroon much as it had been before reunification, as Foncha had promised. As it turned out, not only was the state apparatus left porous and pitted, leached of its life substance like laterite in the dry season, but old, indigenous structures were paralleled by new foreign ones. This interpretation of the system was enhanced by the conduct of the inspector who often mentioned in his speeches that he was the "direct representative of the Federal President" in West Cameroon and that local administrative officers were directly responsible to him.[50]

Ahidjo also went about replacing West Cameroon's particularist politicians who advocated greater state authority with procentralist supporters of a unitary state. The decision by the president to move quickly in that direction became even more urgent after the December 1967 legislative election in West Cameroon which saw the election of thirty-one particularist candidates in the thirty-seven seat state Legislative Assembly.[51] Soon thereafter, on January 11, 1968, Ahidjo appointed the procentralist Muna, the federal minister of transport, to replace Jua as prime minister of West Cameroon. Although the appointment was made without the constitutionally stipulated prior consultation with members of the state legislature, the decision went unchallenged because of bickering within the KNDP, financial scandals, and charges of nepotism that had plagued Jua's administration since 1965. Ahidjo then removed the constitutional hazard by a November 1969 amendment abolishing the right under Article 39 for each state legislature to override the president's choice of a prime minister by a simple majority vote. This gave him sole authority to appoint the prime ministers of each federated state.

The next move to weaken the particularist or profederalist politicians in West Cameroon occurred in January 1970. Rather than reappointing Foncha, a proponent of state authority as his vice-presidential running mate in the March presidential elections, Ahidjo nominated Muna. They were elected by over 93 percent of the votes cast. Muna became both prime minister of West Cameroon and vice-president of the Federal Republic of Cameroon. To enable him to hold both offices, Article 3 (Section 3) of the constitution prohibiting the president or the vice-president from simultaneously holding any other office was then amended by Law No. 70/LF/1 of May 4, 1970, restricting the president and vice-president only from holding any other "elective office." With the federal prime minister now appointive from the presidency rather than elective, Muna assumed both posts, as Cameroon's vice-president and West Cameroon's prime minister.

Thus, West Cameroon's governance authority diminished further. We have seen how Ahidjo used the power granted him under Articles 5 and 6 of the constitution to reduce the state's control of most of its important institutions, including the police, civil service, prisons, and secondary education by 1963[52] and then removed profederalists from leadership positions. Additional fiscal and economic reverses followed. Between 1961 and 1972, West Cameroon lost an important source of revenue when the responsibility for collecting import and export tariffs in West Cameroon was moved from the state to the federal government. At the same time, most imports and exports to and from West Cameroon were diverted from the two West Cameroon ports in Victoria (Limbe) and Tiko to the port of Douala in East Cameroon. To further facilitate the transfer of West Cameroon products through Douala, the Tiko-Douala road and the extension of the Douala-Mbanga railway line to Kumba were opened in the mid-1960s. Both infrastructures allowed the transfer of West Cameroon's exports through Douala.

Furthermore, the closure of the West Cameroon Electric Power (POWERCAM), the company that supplied electricity throughout West Cameroon, and its replacement with the Société National d'Electricité (SONEL) from East Cameroon were seen by some anglophone Cameroonians as another attempt to make West Cameroon economically dependent on East Cameroon. Most West Cameroonians also blamed the federal government for the failure to negotiate an extension of its Commonwealth preference with Britain that would have allowed West Cameroon to continue exporting its bananas to Britain at a price that was 15 percent higher than in the world market.[53]

These moves, together with an already fragile economy dating back to colonial days,[54] not only deprived Anglophone political leaders of the resources that were needed to sustain a patronage system in the state, but made West Cameroon even more dependent on the financial largesse of the federal government.[55] And as Kofele-Kale points out, by the time Ahidjo suggested abolishing the federation in 1972, the idea "met with a collective sigh of relief because at that point the federal system was no longer able to improve on the accrued psychic and material gains made by the anglophone bourgeoisie."[56]

President Ahidjo proved to be as persuasive and effective in replacing the federal with the unitary state in 1972 as when he created Cameroon's single party in the mid-1960s. He argued that the federal structure was costly since it required financial support for four separate legislatures: a fifty member federal legislature; a 100 member legislature in East Cameroon; and a thirty-seven member Legislative Assembly and twenty-six member House of Chiefs in West Cameroon. Pointing to the wage differential between workers in East and West Cameroon, the lower standard of living, the poor infrastructure and the relatively low level of industrial development in West Cameroon,[57] Ahidjo also argued that the federal system was inefficient and a hindrance to economic development. Most forceful of all was the argument that federalism undermined national unity and fostered cleavages and conflicts between francophones and anglophones who considered themselves as members of separate states.[58]

With all possible channels of opposition eliminated, the actual process of creating the United Republic was very brief and typical of Ahidjo's authoritarian leadership style. On May 6, 1972, just after conferring with members of the Political Bureau and the Central Committee of the CNU party, the president informed members of the National Assembly of his intention to replace the federal structure with a unitary state, and of his decision to hold a referendum to allow the Cameroonian people to voice their opinion on the plan.

Despite the fact that the president had taken the decision unilaterally, the announcement to both political bodies met with the traditional "blind" enthusiasm that had greeted almost all past presidential decisions. Three days later, in a radio broadcast to the nation, Ahidjo scheduled the referendum for May 20; Cameroonians would be asked to vote on a draft constitution establishing the United Republic of Cameroon. In order to allay any skepticism, especially among anglophones, Ahidjo added that the introduction of a unitary constitution would not compromise the principle of bilingualism and polyculturalism.[59]

The referendum was approved by 99.99 percent (3,177,846 for and 176 against) of the votes cast. The United Republic was formerly ushered in on June 2, 1972, by Decree No. 72/270. According to Bayart, the creation of the unitary constitution was "the logical crowning of the twin processes of harmonising the administration of the two federal states and the maximising of presidential powers."[60] It was also the culmination of the goal of much of the francophone elite, who from the beginning preferred a more centralized unitary state to the loose federation created by the Foumban constitution.[61] Together with the creation of the United Republic, a constitutional amendment in July 1972 eliminated the vice-presidency and the post of prime minister of East and West Cameroon.

With the establishment of a unitary state and constitution finally realized, the fundamental task for Ahidjo until his resignation on November 6, 1982, was twofold: how to stay in power and how to maintain the union he had created.

Ahidjo's Presidential Monarchy, 1972–1982

In order to achieve both objectives, Ahidjo employed and perfected the various strategies that had been instrumental in the formation of a single

-party state and the creation of the United Republic. They included coalition building, repression, and the establishment of a highly centralized administration which vested most decisions with the president. All administrative decision-making processes were centralized in Yaounde and concentrated in the office of the presidency.

Soon after the introduction of the unitary constitution, Ahidjo restructured the country by Decree No. 72/349 of July 24, 1972, creating seven instead of the original six provinces of the federation. West Cameroon was divided into two provinces, South West and North West, while the five provinces in the former East Cameroon were retained. Each of the seven provinces was further divided into divisions, subdivisions and districts. While each province remained under the jurisdiction of a governor (the former federal inspector of administration), the divisions, subdivisions and districts were under a senior prefect, prefect, and district head, respectively, all subject to the governor's authority. Although the governors were under the jurisdiction of the minister of territorial administration, they were nevertheless the direct representative of the president in the province, and were directly responsible to him. In fact, on most important issues the governors could deal directly with the presidency rather than through the minister of territorial administration.

Centralization also gave the president tremendous authority over most aspects of politics, both at the international and domestic levels. Under Article 8 of the unitary constitution, for instance, the president had the sole authority to appoint and terminate his ministers and vice-ministers without the approval of the legislature. These officials were directly responsible to the president:

> He did not need to seek legislative approval for his appointments: He appointed his ministers, his governors, his judges alone, and they in turn were entirely dependent upon him and his favor if they were to remain in office. The National Assembly had no role to play in the process and could exert no pressure on it.[62]

Whereas in the former federal structure some administrative functions were carried out by the prime ministers and the various state legislatures, almost all decisions now emanated from the presidency in Yaounde. Even cabinet members and their senior staff required approval from the presidency before making the most fundamental decision. In fact, Mbu Etonga notes that Ahidjo's control of all aspects of government was so complete that most members of parliament were not even aware that they had the right to initiate legislation.[63] By contrast, the president was not accountable to anyone.

This centralized political structure gave Ahidjo complete control of all the levers of power. But it was his additional skills in using that power and authority to create supportive networks and coalitions reflecting the nation's regional, linguistic, and ethnic diversity that were critical in his ability to maintain political stability in Cameroon for over two decades.[64] Ahidjo's ability to create such supportive networks was based on a patronage system that allowed him to control and distribute the resources of the state.[65] Those who received important positions within the party or in the administration were not only able to exploit those positions for their personal gains, but were also able to build clienteles for themselves and for the president through employment opportunities and by providing other social and economic benefits for their clients. The president was thus able to create an extensive patron-client network that transcended various social, ethnic, and regional groups. Mark DeLancey cited the case of Egbe Tabi, who as minister of post and telecommunications for many years was able to establish an extensive patron-client network among anglophones, especially those from his ethnic group, by providing them with jobs in his ministry.[66]

Elections were also excellent opportunities for the president and members of the CNU's Political Bureau to extend patronage by rewarding supporters with seats in the legislature and other important party offices. That is because elections under the single-party system were simply occasions for Cameroonians to either approve candidates for the legislature and other party offices or put their seal of approval on important political decisions already made by the president and the Political Bureau. Although several candidates could seek nominations for legislative seats, electoral lists were drawn up by the president and the Political Bureau. Selection was usually based on a candidate's loyalty to the regime and the party rather than on political skills or levels of education.[67] And because these individuals benefitted from the system, they not only had a profound interest in making sure that Ahidjo remained in power, but also in guaranteeing the survival of the political structure he had created.

Meanwhile, in an effort to prevent any of his clients from either becoming too powerful or posing a threat to his authority, Ahidjo frequently brought new members into his cabinet or transferred those already in office to new cabinet positions. In his study of Cameroon's ruling class (classe dirigeante), Ngayap noted that once appointed to office, a cabinet member had less than a 10 percent chance of serving between four and six years in the same ministry. The percentage dropped to 2 percent probability of serving a period of more than eight years. The leading exceptions were Ayissi Mvodo and Egbe Tabi who kept territorial administration and post and telecommunications, respectively, for more than a decade each, and Sadou Daoudou who served as minister of armed forces for two decades (June 20, 1960–July 17, 1980).[68]

Such frequent cabinet changes were not only occasions for the president to preserve the all-important regional and ethnic equilibrium in government; they were also opportunities to bring in "rising stars" who, because of their expertise and the need for new initiatives, deserved a tryout in the "major league."[69] Perhaps the need to bring in these "new stars" accounts for the dismissal from his cabinet in July 1980 of longtime cabinet members such as Charles Onana Owana, Jean Keutcha, Paul Fokam Kamga and Enoch Kwayeb.[70]

Frequent cabinet changes also provided Ahidjo with the opportunity to replace those who may have consciously or inadvertently tried to upstage or to undermine the authority of the president. That may explain the dismissal and subsequent arrest in November 1966 of Victor Kanga, minister of information and tourism, for publishing information detrimental to the regime. He was tried, found guilty and sentenced to four years in prison.[71] Bernard Fonlon, one of the most educated members of Ahidjo's cabinet and a strong proponent of bilingualism, was also dismissed just before the United Republic's creation because he was considered to be too "independent" and outspoken. Similarly, in 1978, Vroumsia Tchinaye, minister of public service and one of Ahidjo's closest associates from the 1960s, was dropped from the cabinet after he reportedly criticized the president for "consulting Paris" before making decisions and because he objected to the large number of French citizens (14,000) working in the country.[72]

Governors, high-ranking military and police officers, and administrators frequently moved between command sites for the same reason.[73] Meanwhile, in order to prevent any threat from the military, Ahidjo compartmentalized the various units (army, navy, air force, police, gendarmerie, and the republican guard) under separate leadership. The different branches were coordinated from the presidency through the ministry of armed forces. As indicated earlier, the latter was under Sadou Daoudou, a northern Muslim and one of Ahidjo's closest allies, for nearly two decades. The close attention given military appointments and promotions made the security force's loyalty to the similarly managed civilian authority a distinctive feature of Cameroon's governance under Ahidjo.

Perhaps Ahidjo's most effective channel to maintain the system he had created was in the domain of legislation and enforcement of repressive measures to eliminate all dissenting voices or more direct threats to his authority. For instance, although Article 3 of the constitution allowed for the formation of opposition parties, a meeting by some anglophone leaders including Foncha and Jua in June 1972 to discuss the formation of a new party was broken up by security police; some of the participants were briefly detained.[74] Also, in April 1970, the Jehovah's Witnesses sect was banned in Cameroon because its members had boycotted the presidential elections. In both instances, the administration's actions could be justified under Article 4 of law No. 67/LF/19 of June 1967 prohibiting the formation of groups or associations having an exclusively tribal or clanic character or those set up to promote causes likely to endanger the national territory or the government.

Similarly, the March 12, 1962, anti-subversion decree prohibited Cameroonians from making any statement, political or otherwise, that could be construed as critical of the regime and the party. Articles 2 and 3 of the 1962 decree called for a fine ranging from CFA 200,000 to 2 million francs, imprisonment for a period of one to five years, or both fine and imprisonment, for anyone who "publishes or reproduces any false statement, rumors or reports or any tendentious comment or any statement or report which is likely to bring hatred, contempt or ridicule any public authority."[75] The decree was often so broadly applied by the authorities that any senior member of the administration could accuse any citizen of making "tendentious" comments or statements. That was the case in September 1970, when two issues of the Catholic weekly *L'Effort Camerounais* were seized, the first because it carried the picture of Colonel Ojukwu, leader of the breakaway republic of Biafra during the Nigerian civil war, and the second because it contained an article denouncing the seizure of the previous issue.[76] Even foreign newspapers and magazines were scrutinized to make sure that they did not contain articles that were critical of the regime.

Meanwhile, the actual task of maintaining the state of fear was largely in the hands of two well-organized and tightly controlled units: SEDOC, reconstituted in 1975 as the Centre National de Documentation (CND), and the BMM. While the former served as the political police responsible for spying on potential enemies of the regime, the BMM maintained sites of torture where physical punishment was used to extract confessions from suspects. Agents of both units were rewarded handsomely through promotions and financial remunerations for uncovering subversive elements or eliciting incriminating evidence from the population.[77] Unfortunately, this system often led to false accusations by overzealous agents who were more interested in attaining higher offices or more financial gains.

The state of fear and uncertainty created by the network of informers was so pervasive that many Cameroonians were even afraid to criticize the regime in front of their family members.[78] Charles Assale, former prime minister of East Cameroon, compared Cameroon under Ahidjo to the Soviet Union under Stalin.[79] Perhaps this detailed description by a correspondent for the weekly magazine *West Africa* best describes the repressive nature of the state under Ahidjo:

> Anyone who knew Cameroon well in Ahidjo's time can testify to the effectiveness of the state there. It avoided the unnecessary and headline-catching excesses of Amin and Macias, but succeeded in keeping a normally lively and articulate people under an iron grip for over 20 years. The gendarmerie, modelled on France's but more powerful, patrolled cities and roads to enforce possession of identity cards, tax receipts and, until 1975, laissez-passer obligatory in theory for travelling from one town to another, their brutality was notorious.... Everywhere the National Security (Sûreté Nationale), an all-pervading and very public body far removed from the French Sûreté, watched over every sort of activity to see the slightest signs of dissent from the one-party state and its rules. Every newspaper's copy was submitted in advance to a government official as part of a rigid censorship, under which, in addition, foreign journals ... were often seized, because of an article, a news item, or even a reader's letter offensive to the regime. And then there was the highly sinister SEDUC (Service de Documentation) and BMM (Brigades Mixtes Mobiles), the secret police, the torturers with their dreaded cells in Douala (the camp Mbopi), Yaounde and elsewhere.[80]

All the aforementioned mechanisms (centralization, patronage and the use of force and intimidation) contributed to Cameroon's political stability at a time when other countries on the continent were rocked by civil wars and other forms of political instability. Accordingly, Ahidjo deserves some credit for maintaining a great measure of political stability in Cameroon (albeit at the cost of severe human rights abuses), which could also be attributed to the visible measure of social and economic progress experienced during his tenure.[81] Tremendous strides were made in the areas of health, transport and education. For example, the number of students at various levels of secondary and technical institutions increased from 42,774 pupils in 1966–67 to 462,225 in 1980–81.[82] In just three years, the number of students enrolled at the University of Yaounde more than doubled, from 2,572 in 1970–71 academic year to more than 6,000 in 1973–74. By 1982, branches of the university had been established in Douala and Dschang, while two more campuses were planned for Buea and Ngaoundere.

Advances were also made in the areas of health and transportation. Although we noted bitterness among anglophones about the latter arena, its growth included the expansion of the Douala seaport, the extension of the Douala-Mbanga railroad to Kumba and the Douala-Tiko road linking the former West and East Cameroon, and the extension of the Douala-Yaounde railroad to Ngaoundere. Altogether, the railroad system was expanded from 517 km in 1965 to 1,172 km in 1977. Similarly, the total road system had increased from 23,250 km (bitumenized and non-bitumenized) in 1974 to 40,000 km in 1982. In the health field, there was an increase in the number of hospitals and clinics. The number of hospital beds had more than doubled, from 10,000 beds in 1962 to 24,541 in 1980. Meanwhile, the medical school at the University of Yaounde that was designed to address the nation's health problems graduated its first class in 1974.

The Cameroonian economy received a boost beginning in 1978 as a result of additional revenue obtained from the exploitation and sale of oil. According to government figures (considered by many analysts to be very conservative) oil earnings increased from US$425 million in 1981 to US$469 million in 1982, an increase from 39 percent to 44 percent of the country's total export earnings during the same period.[83] Although Ahidjo downplayed the importance of oil on the nation's economy and was very careful in using the oil revenue, part of the revenue was nevertheless used to offset the high cost of industrialization, and also to initiate new economic projects.[84] In fact, the 1982–83 budget of CFA 410 billion francs, the last presented by the Ahidjo administration, was financed entirely from Cameroonian resources without loans from any foreign country.[85]

By the time he resigned in November 1982, Cameroon, which once had the dubious reputation as one of the poorest and least developed economies of Africa, was regarded by most international financial institutions as a "middle income developing country,"[86] and one of the economic success stories in sub-Saharan Africa. The Gross National Product had increased from CFA 300 billion francs in 1970 to 2,000 billion in 1982.[87] Also, with an average growth rate of 6 percent between 1977 and 1982, Cameroon's economy was ranked one of Africa's most credit worthy nations, with a triple-A rating.[88] Cameroon was also described as "the paradigm for African development" and "an agricultural success story"[89] partly because of the administration's encouragement of agricultural development rather than relying more exclusively on oil production as some African countries such as Nigeria and Gabon had done following petroleum's discovery and exploitation.

Western, particularly French financial and technical support were important in the socio-economic progress Cameroon experienced during this period. In fact, the foundation of French economic involvement in Cameroon was established in a series of financial and economic agreements between France and the Ahidjo government in 1959 and renegotiated with only minor changes in 1973. These agreements allowed France to become heavily involved in providing financial and technical aid to almost every phase of the Ahidjo administration. For example, in 1960, the first year of Cameroon's independence, French aid to Ahidjo's administration totaled a Franc equivalent of US$50,000,000, representing 80 percent of the total revenue collected by the East Cameroon government.[90] Meanwhile, key positions in various government ministries and institutions, including the national university opened in 1962, were dominated by French personnel, paid by the French government. Although French business interest in Cameroon decreased from 52 percent of the total capital investment in 1973 to 23.6 percent in 1982 as a result of the government's effort to Cameroonize the economy, it was still the largest foreign factor in 1982 with 60.8 percent of the total foreign capital investment.[91]

While these investments were important in leading Cameroon closer to what DeLancey describes as "a modern capitalist industrial state," there was little progress in making Cameroon an economically independent state.[92] Part of the reason for this abnormality was due to the government policy granting special economic incentives to the advantage of foreign over indigenous businesses. A typical example involved the ALUCAM metallurgical works established in Edea in the 1950s by the French firm of Péchiney-Ugine. While the company as late as 1970 consumed about six times as much electricity as all of the rest of Cameroon, it was still paying only CFA 0.04 francs per kilowatt of electricity compared to CFA 28–30 francs for all other domestic consumers.[93]

Moreover, because of Cameroon's liberal investment policy aimed at encouraging foreign investments, much of the profits and salaries earned by French workers and businessmen (and other foreign investors) in Cameroon were repatriated to France. According to Neville Rubin, by 1967 the expatriate population in Cameroon (mostly French citizens, estimated at about one-third of one percent of the entire Cameroon population) earned about one-sixth of the total national income.[94] Therefore much of the needed capital that could be invested in Cameroon to generate further growth was lost. Furthermore, most of the industries in Cameroon such as ALUCAM, Brasseries du Cameroun and Guinness, were capital-intensive enterprises that did not create proportionate employment opportunities for Cameroonians. And most of the managerial positions in these industries were held by French citizens.[95]

Despite the relatively strong economy, the appearance of national unity and political stability in Cameroon, and Ahidjo's apparent firm control of all the instruments of power, it would be a misrepresentation to argue that Cameroon was not without serious political problems and discontent. Until his arrest in August 1970 and his eventual execution in January 1971, Ernest Ouandié, the last leader of the internal wing of the UPC, was still able to carry out isolated attacks in the Bamileke and Mongo areas, including an attack on a clinic in Loum, South West Province, in July 1970. Two nurses were killed in the incident while two others were wounded. In another incident, a cache of firearms thought to belong to the UPC was discovered on the site of a West Province company.[96]

There were other signs of opposition to the regime in the middle and late 1970s. Thousands of antigovernment leaflets were strewn around the capital city of Yaounde in 1976, and in Douala, the nation's economic capital, in January 1977. Most of the unrest was the result of economic discontent resulting from the increasing financial disparity between most Cameroonians and the small but privileged group of bureaucrats and politicians. In an unprecedented defiance of the regime, for instance, workers in Douala, the nation's economic capital, went on strike in January 1977 to protest the high rate of inflation and rising prices. The administration settled the crisis by granting the workers a wage increase of between 5 and 18 percent, but it warned against future strikes.[97]

In January 1977, the twentieth synod of the Presbyterian church in the South West and North West Provinces expressed its concern over the treatment of prisoners. Similarly, in a letter published in the *Cameroon Times* in September 1977, the Catholic bishops of Buea and Bamenda diocese condemned corruption by government officials, especially within the police force.[98] There were also reports in some foreign newspapers of an attempted military coup d'état in July 1979.[99]

However, there is no evidence that these signs of discontent were a contributory factor to Ahidjo's sudden resignation from the presidency on November 6, 1982. On the contrary, in a society where elections were simply occasions for "symbolic ratification of government policy and personages"[100] opposition hardly registered in the April 1980 presidential election, the last in which Ahidjo was the CNU's candidate. He won a fifth term from over 99 percent of those who cast their votes.

Notes

1. For a definitive analysis of the formation of the UPC and its problems with French colonial authorities see Richard Joseph, *Radical Nationalism in Cameroun: The Social Origins of the U.P.C. Rebellion* (Oxford: Oxford University Press, 1977).

2. Victor Le Vine, "Cameroun," in James Coleman and Carl Rosberg Jr., eds., *Political Parties and National Integration in Tropical Africa* (Berkeley and Los Angeles: University of California Press, 1964), pp. 134, 139; the pattern was repeated in the 1990s when Paul Biya needed to counter the SDF.

3. Ahmadou Ahidjo, *The Political Philosophy of Ahmadou Ahidjo* (Monaco: Paul Bory, 1968), p. 23.

4. Dennis Austin, *Politics in Africa* (Hanover: University Press of New England, 1984), p. 12.

5. John Chipman, *French Power in Africa* (Oxford: Basil Blackwell, 1989), p. 86.

6. Chipman, *French Power in Africa,* pp. 102–103; Guy de Lusignan, *French-Speaking Africa since Independence* (New York: Praeger, 1969), p. 23.
7. Willard Johnson, *The Cameroon Federation: Political Integration in a Fragmentary Society* (Princeton: Princeton University Press, 1970), p. 351.
8. Philippe Gaillard, *Ahmadou Ahidjo (1922–1989)* (Paris: Groupe Jeune Afrique, 1994), p. 80.
9. Victor Le Vine, *The Cameroons: From Mandate to Independence* (Berkeley and Los Angeles: University of California Press, 1964), p. 170.
10. *Africa Report,* 5, 1 (1960), p. 2.
11. Jean-François Bayart, "The Birth of the Ahidjo Regime," in Richard Joseph, ed., *Gaullist Africa: Cameroon Under Ahmadu Ahidjo* (Enugu, Nigeria: Fourth Dimension, 1978), p. 49.
12. Ahmadou Ahidjo, *Anthologies des Discours,* cited in Dieudonné Oyono, "Introduction a la Politique Africaine du Cameroun," *Le Mois en Afrique* 18 (1983), 21.
13. Richard Joseph, "The Gaullist Legacy in Franco-African Relations," in Joseph, *Gaullist Africa,* p. 23. For more information on the Franco-Cameroon agreement see David Gardinier, *Cameroon: United Nations Challenge to French Policy* (London: Oxford University Press, 1963), pp. 97–98; Ndiva Kofele-Kale, "Cameroon and Its Foreign Relations," *African Affairs* 80, 319 (1981), pp. 200–201.
14. Le Vine, *The Cameroons,* p. 167.
15. Gardinier, *Cameroon,* p. 87.
16. Gardinier, *Cameroon,* p. 88.
17. Le Vine, *The Cameroons,* p. 180.
18. Neville Rubin, *Cameroon: An African Federation* (London: Pall Mall Press, 1971), p. 97.
19. Political Bureau of the CNU, *Ahmadou Ahidjo: Ten Years of Service to the Nation* (Monaco: Paul Bory, 1968), p. 19.
20. Le Vine, *The Cameroons,* p. 180.
21. Pierre Flambeau Ngayap, *Cameroun: Qui Gouverne? de Ahidjo à Biya, l'héritage et l'enjeu* (Paris: L'Harmattan, 1983).
22. For a discussion of the debate preceding passage of the bill, see Le Vine, *The Cameroons,* p. 186–188.
23. Gaillard, *Ahmadou Ahidjo,* p. 118.
24. Richard Joseph, "Cameroon Under Ahmadou Ahidjo: The Neo-Colonial Polity," in Joseph, *Gaullist Africa,* p. 36.
25. Philippe Gaillard, *Foccart Parle: Entretiens avec Philippe Gaillard* (Paris: Fayard/Jeune Afrique, 1995), pp. 207–208.
26. Victor Le Vine, *The Cameroon Federal Republic* (Ithaca: Cornell University Press, 1971), p. 85.
27. Johnson, *The Cameroon Federation,* p. 189.
28. For excellent analyses of the Federal Constitution see Rubin, *Cameroon,* ch. 6, and Johnson, *The Cameroon Federation,* ch. 8.
29. Gardinier, *Cameroon,* p. 107.
30. Gardinier, *Cameroon,* p. 109.
31. Joseph, "Cameroon Under Ahmadou Ahidjo," p. 37.
32. Ahidjo, *The Political Philosophy of Ahmadou Ahidjo,* p. 62.

33. Gardinier, *Cameroon*, p. 124.
34. Gardinier, *Cameroon*, p. 123.
35. Gardinier, *Cameroon*, p. 125.
36. Political Bureau of the CNU, *Ahmadou Ahidjo: Ten Years of Service*, p. 22.
37. Rubin, *Cameroon: An African Federation*, p. 147.
38. Ahidjo, *The Political Philosophy of Ahmadou Ahidjo*, p. 35.
39. Frank Stark, "Federalism in Cameroon: The Shadow and the Reality," in Ndiva Kofele-Kale, ed., *An African Experiment in Nation-Building: The Cameroon Federal Republic Since Reunification* (Boulder: Westview Press, 1980), p. 117.
40. Johnson, *The Cameroon Federation*, p. 263.
41. Le Vine, *The Cameroon Federal Republic*, p. 96, called West Cameroon politics described below, 1962–1966, a "complex political ballet."
42. Colin Legum, ed., *Africa Contemporary Record 1976–77* (London: Rex Collings, 1978), p. B463.
43. Ndiva Kofele-Kale, "Class, Status, and Power in Postreunification Cameroon: The Rise of an Anglophone Bourgeoisie, 1961–1980," in Irving Markovitz, ed., *Studies in Power and Class in Africa* (Oxford: Oxford University Press, 1987), p. 161. For the record, Foncha and Muna, thirty years later the most prominent surviving parties to this rivalry, admitted its destructive impact on West Cameroon's anglophones, made their peace, and joined forces with those pushing for renewed autonomy, especially as elder statesman in a delegation sent to plead that case at the United Nations in mid-1995; see Chapter 6 below.
44. Ahidjo, *The Political Philosophy of Ahmadou Ahidjo*, p. 51.
45. Jean-François Bayart, "The Neutralisation of Anglophone Cameroon," in Joseph, *Gaullist Africa*, p. 88.
46. Stark, "Federalism in Cameroon," p. 110.
47. Claude Welch, *Dream of Unity: Pan-Africanism and Political Unification in West Africa* (Ithaca: Cornell University Press, 1966), p. 248.
48. Johnson, *The Cameroon Federation*, p. 208.
49. For a discussion of the authority of the federal inspectors, see Stark, "Federalism in Cameroon," p. 113, and Johnson, *The Cameroon Federation*, pp. 200–213.
50. Johnson, *The Cameroon Federation*, p. 210.
51. Gaillard, *Ahmadou Ahidjo*, p. 155.
52. Welch, *Dream of Unity*, p. 248.
53. Jacques Benjamin, "The Impact of Federal Institutions on West Cameroon's Economic Activity," in Kofele-Kale, *An African Experiment*, p. 201.
54. For a discussion of the economic status of West Cameroon, see the 1959 Phillipson Report on the Financial, Economic and Administrative Consequences of the Southern Cameroons on Separation of the Federation of Nigeria.
55. Stark, "Federalism in Cameroon," p. 113.
56. Kofele-Kale, "Class, Status, and Power in Postreunification Cameroon," p. 137.
57. For a discussion of these differences see Benjamin, "The Impact of Federal Institutions," pp. 191–199.
58. Mbu Etonga, "An Imperial Presidency: A Study of Presidential Power in Cameroon," in Kofele-Kale, *An African Experiment*, p. 144.
59. Legum, *Africa Contemporary Record 1972–73*, p. B499.

60. Bayart, "The Neutralisation of Anglophone Cameroon," in Joseph, p. 87.
61. Kofele-Kale, "Introduction," in Kofele-Kale, *An African Experiment*, p. xxxi.
62. Mark DeLancey, *Cameroon: Dependence and Independence* (Boulder: Westview Press, 1989), p. 57.
63. Etonga, "An Imperial Presidency," p. 150.
64. For a good discussion of how Ahidjo used the ethnic, regional and linguistic balance to create a supportive network and maintain his authority see Kofele-Kale, "Class, Status, and Power in Postreunification Cameroon," pp. 135–169.
65. Etonga, "An Imperial Presidency," p. 142.
66. DeLancey, *Cameroon*, p. 59.
67. Abel Eyinga, "Government by State of Emergency," in Joseph, *Gaullist Africa*, p. 106.
68. Ngayap, *Cameroun: qui Gouverne?*, p. 42.
69. Victor Le Vine, "Leadership and Regime Changes in Perspective," in Michael Schatzberg and William Zartman, eds. *The Political Economy of Cameroon* (New York: Praeger, 1986), p. 30.
70. The new ministers in the July cabinet included Thomas Kamga, Atanasa Eteme Onada and Andre Ngongang Wandji. Two others, William Eteki Mboumoua and Youssoufa Daouda, in charge of missions at the presidency and economic planning respectively, were "old guards" who had served Ahidjo's administration at various times.
71. For details on the "Kanga Affair" see Le Vine, *The Cameroon Federal Republic*, pp. 157–158.
72. Legum, *Africa Contemporary Record 1978–79*, p. B509.
73. Perhaps the only exception was Governor Ousman Mey of North Province, who remained in office from 1972 until he was deposed by President Biya in August 1983.
74. Nelson, Harold and Margarita Dobert, et al. *Area Handbook for the United Republic of Cameroon* (Washington, D.C.: U. S. Government Printing Office, 1974), p. 151.
75. For more details on the March 1962 law see, Philippe Lippens and Richard Joseph, "The Power and the People," in Joseph, *Gaullist Africa*, p. 114 and Appendix II(b), pp. 210–211.
76. Legum, *Africa Contemporary Record 1969–70*, p. B374.
77. Eyinga, "Government by State of Emergency," in Joseph, *Gaullist Africa*, p. 107.
78. Robert Jackson and Carl Rosberg, Jr., *Personal Rule in Black Africa* (Berkeley and Los Angeles: University of California Press, 1981), p. 155.
79. *Africa Confidential*, November 17, 1982, p. 3.
80. *West Africa*, October 3–9, 1983, p. 2273.
81. Le Vine, *The Cameroon Federal Republic*, p. 182.
82. These figures are based on data provided by the Ministry of Education in Yaounde and available in *Africa South of the Sahara* for those years.
83. *West Africa*, September 19–25, 1983, p. 2169.
84. Wilfred Ndongko, "The Political Economy of Development in Cameroon: Relations between the State, Indigenous Business, and Foreign Investors," in Schatzberg and Zartman, *The Political Economy of Cameroon*, p. 108.

85. *Africa Research Bulletin: Economic, Financial and Technical Series* 19,10 (October 15-November 14, 1982), p. 6611.//
86. Ndongko, "The Political Economy of Development in Cameroon," p. 83.
87. Gaillard, *Ahmadou Ahidjo*, p. 178.
88. *New African*, December 1982, p. 24.
89. *Africa Confidential*, November 17, 1982, p. 3.
90. République du Cameroun, Ministère des Finances, *Budget de l'Exercise 1960–61*, cited in Le Vine, *Cameroon*, p. 230.
91. *West Africa*, September 19–25, 1983, p. 2171.
92. DeLancey, *Cameroon*, p. 112.
93. Jean Suret-Canale, *Afrique Noire: De la Colonisation aux Indépendances, 1945–1960*, cited in Joseph, *Gaullist Africa*, p. 146.
94. Rubin, *Cameroon: An African Federation*, p. 176.
95. Richard Joseph, "Economy and Society," in Joseph, *Gaullist Africa*, p. 147.
96. Legum, *Africa Contemporary Record 1970–71*, p. B255.
97. Legum, *Africa Contemporary Record 1976–77*, p. B463.
98. Legum, *Africa Contemporary Record 1978–79*, p. B521.
99. Legum, *Africa Contemporary Record 1979–80*, p. B392.
100. Roger Tangri, *Politics in Sub-Saharan Africa* (London: James Currey, 1985), p. 116.

3

Biya's Early Presidency, 1982–1986

On November 4, 1982, President Ahidjo surprised the Cameroonian people by announcing his decision to relinquish his authority as head of state. Two days later, he was succeeded by Prime Minister Paul Biya, his constitutional successor. To many observers, this peaceful transfer of power buttressed Cameroon's image as one of Africa's most politically stable nations. Moreover, like the voluntary retirement in 1980 of his Senegalese counterpart and close friend, Léopold Senghor, Ahidjo's exit from power was also considered a new beginning for a continent where military coups had hitherto been the most frequent method of replacing governments in power. This perception was given further credence by the voluntary retirement of Presidents Julius Nyerere of Tanzania and Siaka Stevens of Sierra Leone in 1985.[1]

Although Biya had promised to follow in the footsteps of his predecessor when he took over the presidency,[2] he was also prepared to create a new society where there would be a greater degree of tolerance, individual liberty, and freer exchange of ideas, judging by speeches and actions following his inauguration.[3] But before the new president could securely establish himself in power and implement the policies he hoped would lead to the emergence of the new society, he was confronted with two issues that threatened to destroy his presidency. The first was a power struggle with his predecessor in the early and middle months of 1983 that endangered what seemed to have been a peaceful transition of power. The second was an attempted military coup in April 1984 that almost ended his presidency.

From Ahidjo to Biya: The Transfer of Power, 1982

While it is generally agreed that Ahidjo's exit from power was due to his failing health,[4] there are those who argue that his retirement may also have been precipitated by other factors, including a lack of support for his administration by the newly elected socialist government of French President Mitterrand[5] and rising domestic unrest in the late 1970s. The rift with the French president was not only the result of ideological differences between Mitterrand and Ahidjo, who had never concealed his admiration for Charles De Gaulle and regimes like his in France, or his disdain for the Socialist Party there.[6] It could also be traced back to the 1950s and that party's support for the UPC against Ahidjo's government. The conflict between the two leaders was also exacerbated by the refusal of Ahidjo's administration to grant an entry visa to a French socialist lawyer who had volunteered to defend Ernest Ouandié, the militant UPC leader who was captured in August 1970.[7] He was later tried in what many observers thought was a show trial aimed at strengthening the regime, found guilty, and publicly executed in 1971. The domestic unrest that might have contributed to Ahidjo's resignation included an increase in semiclandestine groups and activities and an alleged plot by noncommissioned officers in July 1979 to overthrow Ahidjo's government.[8]

Other observers of Cameroon politics including Victor Le Vine have suggested that Ahidjo had been carefully planning his retirement as early as 1975, and that the health issue may simply have precipitated his exit from power. For instance, shortly after the CNU party congress in Douala in February 1975 (christened the Congrès de la Maturité) when Ahidjo was apparently persuaded by party militants to remain as president following rumors of his impending resignation,[9] the constitution restored the office of prime minister, abolished following the creation of the unitary constitution in 1972.

Paul Biya, at that time secretary general at the presidency, was appointed to fill the post. Later, in June 1979, Article 7 of the constitution was amended, designating the prime minister rather than the president of the National Assembly as the constitutional successor in case of the president's death, resignation, or inability to carry out his functions.[10] The amendment also allowed the successor to serve out the rest of the term of his predecessor and to choose his own cabinet. This was in contrast to the previous situation in which the president of the National Assembly could only succeed the president on an interim basis until elections for a new president were conducted within a period of not more than twenty days and not exceeding fifty days after the vacancy had occurred (Article 7). Moreover, the interim president could neither amend the constitution nor alter the composition of the government (Article 7b). Since the president was also responsible for selecting members of his cabinet, including the prime minister (Article 8), it could be argued that whoever was Ahidjo's choice as prime minister was probably designated to succeed him. In other words, his retention of Biya as prime minister following the June 1979 constitutional amendment meant that Ahidjo favored him as his successor.

Perhaps in an effort to secure a peaceful transition of power and to maintain his own legacy, Ahidjo reshuffled his cabinet in January 1982, and in the process placed many of his close associates in charge of important ministries. For instance, Sadou Daoudou, who had occupied the important post of minister of armed forces since Cameroon gained its independence in 1960, became deputy secretary general at the presidency. His former ministry went to another northern Muslim and Ahidjo loyalist, Maikano Abdoulaye. Le Vine argues that by bringing Sadou Daoudou to the presidency, Ahidjo "put his own eyes and ears into the heart of the administration, a presence whose implied threat helped the transition by providing a monitor over any who might have had independent ideas about the succession."[11]

Despite the elaborate preparation, and although the constitution clearly designated the prime minister as the presidential successor, the decision to hand over power to Biya shocked even some of Ahidjo's close supporters. Many of them had seen Biya's appointment as prime minister in 1975 and his retention after the constitutional amendment in 1979 simply as a matter of political expediency, aimed at appeasing southern Christians. As a result, they were convinced that Biya would be replaced as prime minister by a northern Muslim when it was time for Ahidjo to retire. That may explain why some observers perceived Sadou Daoudou's transfer to the presidency as providing him a training ground, a prelude to his appointment as prime minister, and therefore as constitutional successor to the president.

Therefore, while Ahidjo's exit was greeted with relief by those who saw him as a dictator and a French puppet,[12] his choice of Biya, a southern Christian from the Beti ethnic group, was not popular, especially among a large section of northern Muslims within the administration and the CNU party. Apart from the regional and religious reasons for opposing Ahidjo's choice, others objected to Biya's selection because he was perceived as a weak and inexperienced leader. In fact, although he had served in various high-ranking positions since returning from his university studies in France in 1962, and as prime minister since 1975, these critics felt that he had had insufficient experience because of his noninvolvement in the nationalist struggle in the 1950s.[13] Moreover, as prime minister and head of government, he had simply carried out the wishes of the president rather than formulate or initiate policies.

Nevertheless, efforts by members of the CNU Central Committee, including Biya, to persuade Ahidjo not to resign, or at least to retain his positions as president and party chairman while Biya carried out the actual daily business of running the government, were unsuccessful. But as a compromise, Ahidjo stayed as chairman of the party. Whether Ahidjo would have surrendered both positions had he not been persuaded by the party's Central Committee not to do so, or whether Biya was genuinely sincere in trying to persuade Ahidjo to retain at least the leadership of the party, are matters of conjecture. Whatever the case, Bayart suggests that the attempt by some members of the Central Committee to persuade Ahidjo to retain the leadership of the party was not simply a demonstration of their continued loyalty to Ahidjo, but was also an attempt to limit the power of the new president, thereby depriving him of control of the party. After all, since the constitution only allowed a presidential successor to complete the term of his predecessor, these members hoped that depriving Biya of the CNU chairmanship would enable the nomination of a candidate for the presidential election scheduled for 1985 by forces not under Biya's control.[14] At the same time, it would diminish the chances that Biya (who was not yet in control of the party) could emerge as the party's candidate for the election.

Despite opposition to Biya's selection as president and the apparent split within the party's executive caused by his selection, the former president was determined to see Biya as his successor. Perhaps in an effort to support Biya's authority, Ahidjo had also orchestrated the admission of Biya into the CNU Central Committee and Political Bureau, and his election as vice-president of the party. Later, at a meeting of the Political Bureau on December 11, 1982, Biya as vice-president of the party was given the responsibility of running the affairs of the party in the absence of the president.

But was Ahidjo genuinely interested in gradually fading away from the political scene once his successor had established himself in charge, or was he simply playing for time until his health recovered and he was able to come back? Was the choice of Biya as Ahidjo's successor simply a ruse to mollify southern Christians until the opportune time when Maïgari Bello Bouba, a northern Muslim whom Biya had appointed to the post of prime minister (reportedly at the urging of the former president) and whom many observers believed to have been Ahidjo's true *dauphin* was ready to take over? Whatever the case, events in 1983 would reveal an ambiguous mixture of support, political intrigue and ultimate betrayal that contributed to strained relationships between Biya and the former head of state.

Disputed Succession and Legacy: Ahidjo and Biya, 1983–1984

One line of evidence suggested Ahidjo's determination to ensure a peaceful and successful transfer of power to his prime minister, despite opposition from some members of the CNU Central Committee. In fact, soon after his resignation, Ahidjo sent Moussa Yaya, a fellow northern Muslim and one of his closest associates, to West Province to reassure its people, particularly Bamileke businessmen, that Biya's call for greater liberalization and democratization, and his appeal for "rigor" and "moralization" in many of his inaugural speeches did not represent a change in policy. Both concepts were supposed to form the foundation of the New Deal society that Biya hoped to establish in Cameroon. However, instead of carrying out his assignment, Moussa Yaya, who had been openly critical of Biya's selection as president, used the occasion to question Ahidjo's resignation. Furthermore, Yaya extended his tour to North West Province without Ahidjo's authorization, carrying the same message. He even accused the former president of treason and raised the specter of serious domestic problems as a result of Ahidjo's resignation.

Since Ahidjo was still the chairman of the CNU, he orchestrated the January 10, 1983, dismissal from the party of Moussa Yaya and three others who had openly expressed their opposition to his resignation; this cost Moussa Yaya his post as vice-president of the National Assembly.[15] And in a further attempt to strengthen Biya's authority and demonstrate his continued support for his successor, Ahidjo toured six of the country's seven provinces from January 23–29, in an effort to rally support for Biya. Clearly, Biya owed his political fortunes to Ahidjo. And perhaps in an effort to show his appreciation and loyalty to his mentor, he maintained the cabinet he had inherited virtually intact. In fact, there were only three minor changes. Bello Bouba, minister of planning and economic affairs, became prime minister and therefore the constitutional successor to the president. Sadou Daoudou was promoted from deputy secretary general to secretary general at the presidency. Samuel Eboua, secretary general at the presidency, became minister of agriculture.

Biya also showed his continued loyalty and respect to his predecessor by constantly evoking Ahidjo's name in many public speeches during his first months in office. For instance, opening the 18th session of heads of states of the Customs and Economic Union of Central Africa in Yaounde on December 17, 1982, Biya referred to Ahidjo as "Father of the Nation." Similarly, in his speech to delegates at the Second Yaounde Medical week in Yaounde on January 24, 1983, the president referred to Ahidjo as his "distinguished predecessor." In fact, cajoling his mentor in this manner might have been politically expedient, especially at a time when Biya was still surrounded by people who remained loyal to his predecessor or owed him their political fortunes. But the frequent reference to Ahidjo nevertheless portrayed Biya's lack of independence and strong leadership, and may have led Ahidjo to think that he was still in charge and could even return to power if he chose to do so.

Indeed, a less cordial reading of their connection emerged in 1983. Beyond Ahidjo's show of mutual respect, cooperation, and support for Biya, a struggle for supremacy and control was simmering between the head of state and the chairman of the CNU. It appears that Ahidjo had fully regained his health and strength by January 1983 and was prepared to reassert his authority, either by exercising his control over Biya or by asserting the primacy of the party over the government. The latter course became apparent in an interview with Henri Bandolo, a correspondent for the government's *Cameroon Tribune,* on January 31, 1983. Responding to a question, Ahidjo intimated that the party was responsible for defining guidelines for the nation's policies, which the government was merely to implement.[16]

The apparent confusion between the role of the party and the executive in formulating national policy was due to the fact that for almost two decades Ahidjo had served as both head of state and chairman of the country's single party. In other words, there had been no clear delineation between his authority and responsibility as head of state and as head of the party. But with the separation of the two offices following Ahidjo's resignation as president, it would appear that Ahidjo's assertion of the party as the policy-making body in the country was a violation of Article 5 of the constitution. Under the latter, the president of the republic as head of state and head of government had the sole responsibility of conducting the affairs of the state.

Whatever the case, it seems that Ahidjo in his party role was still not prepared to play second fiddle to his presidential successor. In fact, a series of events in December 1982 and the early months of 1983 reveal that Ahidjo still considered his position as chairman of the party to be superior to that of the head of state. Such claims certainly contributed to the strained relationship between Ahidjo and his successor. For instance, the former president apparently without Biya's authorization took it upon himself to lead a delegation of four northern Muslim officials in the administration on an official visit to Nigeria in December 1982.[17]

Ahidjo's advocacy of the party leader's precedence over Cameroon's presidency may also have been inadvertently promoted by foreign diplomats in Cameroon, the party, and even by Biya himself. For instance, the fact that Biya as head of state was at the airport to welcome Ahidjo after the latter's visit to France, Senegal, and Côte d'Ivoire in December 1982 may have fed Ahidjo's resolve to remain in power. Moreover, during a visit to Cameroon in January 1983, the French minister of foreign affairs, Claude Cheysson, met with President Biya at the nation's capital for only an hour before flying to Ahidjo's hometown of Garoua, where he met for four hours with the former president.[18]

Another incident that seemed to undermine Biya's authority occurred on January 29, 1983, during Ahidjo's visit to the Yaounde section of the CNU party, the final stop of his six province tour. Under normal protocol, Biya should have been the last person to arrive at the reception hall. Instead, the former president was the last to do so. A similar incident occurred in March, when Ahidjo was awarded the Dag Hammarsjoëld Peace Price. Although this time the order of arrival was corrected at the last moment, the former president appeared unhappy at the fact that he preceded Biya into the hall.[19] The party also helped to assert Ahidjo's authority when the CNU Central Committee rejected Biya's suggestion in December 1982 for the introduction of multiple candidates in party elections.[20] Biya was also upset at the fact that many CNU party militants still wore uniforms imprinted with pictures of Ahidjo during the 11th anniversary of the United Republic of Cameroon on May 20, 1983.

Although these incidents may seem minor, perhaps insignificant, perceptions mattered, and clearly placed the president in a difficult situation. While Biya was prepared to work with his predecessor by seeking his advice on important political matters, he was not prepared to concede his power and authority as head of state to the chairman of the party. But to be able to assert his authority, the new president had to first establish his independence from his predecessor and gain control of the various state institutions. Unfortunately, these institutions were still controlled by people who owed their offices and political fortunes to Ahidjo and were still loyal to him. To make matters worse for Biya, most of the military command structure in Yaounde (including the Republican Guard which was responsible for the president's security) was under the control of northern officers appointed by Ahidjo.[21] Therefore the first step in trying to establish himself in charge was for Biya to replace these men with people loyal to him.

The first opportunity for Biya to establish his independence and authority occurred early in 1983, when he solicited but failed to follow Ahidjo's recommendations on a cabinet shuffle which was implemented on April 12, 1983. Not only was Ahidjo upset at this apparent slight to his authority, but he was also troubled by rumors that Biya had decided to include French security experts as members of the presidential security detail. The former president noted that in spite of his close cooperation with France on issues of security, he had never allowed French agents to occupy such sensitive positions.

Despite the growing rift between the two leaders, Ahidjo as CNU chairman actively campaigned in the May 24, 1983, legislative elections, including a radio message to party militants on May 14, urging them to vote in the forthcoming legislative elections. Shortly after the elections, however, he proposed two constitutional amendments, to make

Cameroon a *de jure* one party state and to give the CNU the authority to select the candidate for presidential elections. Ahidjo's intent must have been to further undermine Biya's authority as president, for the passage of these amendments would not only have confirmed the primacy of the party over the state, but would have diminished Biya's chances of running for re-election in 1985, since the responsibility of selecting the party's presidential candidate would have been in the hands of the CNU Central Committee. Passage of the bills would also have created the opportunity for Ahidjo to run for re-election in 1985 if he decided to contest the presidency.

Biya prevented the passage of both amendments and thus widened the rift and intensified the conflict between the two men. The conflict escalated on the eve of French President Mitterrand's visit to Cameroon, when Biya on June 18 reshuffled the cabinet, without Ahidjo's consent. In the process, several members suspected of continued allegiance to the former president were dropped from office, including some like Sadou Daoudou whom Biya had himself previously promoted. Although the cabinet change may have been an attempt by Biya to establish himself in charge and to put his own team in place before Mitterrand's visit, it was perceived by Ahidjo and his supporters as insulting and a sign of disrespect to the former president. Clearly, Ahidjo thought that he should have been consulted before such a change was made.

President Biya also used this cabinet change to address Ahidjo's claims regarding the primacy of the party over the state. Responding to a question by a journalist on the duality of authority between the president and chairman of the party, Biya indicated that the constitution clearly established the primacy of the state over political parties:

> ... as concerns the constitution which in the hierarchy of juridical instruments, is the most important, it is stated therein that it is the President of the Republic who defines policy of the nation.... The same constitution provides that political parties and groups may take part in elections. Thus this instrument, which is the fundamental law of the Nation, defines clearly enough the power of the state and the party.[22]

The war of words and actions mounted. Later the same day as the cabinet changes, and certainly upset by them, Ahidjo summoned the northern members of the new cabinet and those from Noun Division to his residence in Yaounde, where he apparently convinced most of them to sign a letter of resignation from the government, to be written by Youssoufa Daouda, the minister of public service.[23] Another Yaounde meeting called by Maikano Abdoulaye, the minister of armed forces, attended exclusively by high-ranking northern military officers, was held at the residence of Ibrahim Wadjiri, the delegate-general of the Gendarmarie, to inform them of the impending resignation of northern members of the cabinet. The officers, however, were advised not to become involved in the crisis since it was purely a political matter.[24]

The tensions at work during this episode can be gauged by Biya's refusal to attend a June 19 reception hosted by the chairman of the party for the newly elected legislators and members of the party's Central Committee, because of rumors of threats on his life by Ahidjo and his supporters.[25] Ahidjo had hoped that the northerners' resignations would not only undermine the president's authority, but would lead to the collapse of his government, just as a similar strategy by Ahidjo had contributed to the demise of André-Marie Mbida's government in February 1958.[26] If successful, Ahidjo's action might also have escalated the power struggle between the two men by infusing a regional and religious element into the conflict.

But he failed, having apparently been persuaded to call the plan off by some of his loyal supporters, including the Sultan of Foumban (informed of the plan by his son, Ibrahim Mbombo Njoya, who had been reappointed to the cabinet as minister of youth and sports) and then by French President Mitterrand during his visit to Cameroon June from 20–21.[27] Nevertheless, these events brought the conflict and struggle for supremacy between Ahidjo and his successor to a head. Apparently upset by his inability to re-establish his authority, Ahidjo went into self-imposed exile to France on July 19, 1983. His departure marked the beginning of the end of the leader whom Cameroonians had hailed as "Father of the Nation"[28] and the architect of the Cameroonian nation for over two decades. It also gave Biya a free hand to begin the process of dismantling Ahidjo's power structure, purging his loyalists, and establishing himself in charge. In fact, soon after the former president's departure to France, Biya's portrait in offices and public buildings throughout the nation replaced Ahidjo's. His name was also deleted from the CNU anthem where it recognized his contributions in building the nation.[29]

Ahidjo's departure did not, however, end the conflict between the two men. In a radio broadcast to the nation on August 22, 1983, Biya announced the discovery of a plot against "the security of the state," implicating the former president and two close aides, Major Ibrahim Oumarou and Captain Adamou Salatou. The president also used the occasion to announce administrative changes. The first was a cabinet shuffle, the third in four months; Bello Bouba and Maikano Abdoulaye were replaced as prime minister and minister of armed forces, respectively, by Luc Ayang, a northern Christian, and Gilbert Andze Tsoungui, a Christian from the president's Beti ethnic group. The second change involved civil administration. The number of provinces rose from seven to ten; the former

North Province was divided into Far North, North, and Adamawa, and the former Center-South province was divided into Center and South. In addition, six of the seven governors who were in office before the changes and thirty-six of forty-nine Senior Prefects were replaced, and forty-one of forty-nine heads of administrative departments were transferred to other positions. Third, the military authority system was restructured, placed under a single command headed by General Pierre Semengue (a Beti from the president's ethnic group), named chief of staff. Three new generals were appointed: Oumarou Djam-Yaya, a Fulbe from North Province, Nganso Sunji, a Bamileke from West Province, and James Tataw, an anglophone from South West Province, became heads of the gendarmerie, air force, and army, respectively.[30]

Although Biya's detractors tried to portray the division of the former North Province as a campaign against the North (since it was announced on the same day that the charges of a plot were made), the argument was muted by the fact that the president's province of origin (Center-South) had also been divided. According to the president and his supporters, the administrative restructuring was necessary for greater administrative efficiency and fair representation in the National Assembly.

Beyond the veil of administrative and political reforms, however, one could argue that President Biya was simply using the same policy of divide and conquer that his predecessor had so skillfully used to establish his authority, especially in the early 1960s. For instance, breaking up North Province which contained about 25 percent of the nation's population into three provinces made sense as Biya's response to the coup attempt and its aftershocks implicating leaders from the North, where Ahidjo's city of Garoua had become a regional power base. Biya may also have expected to win the support of the majority non-Fulbe ethnic groups in the region who had been dominated by the minority, albeit politically powerful Muslim-Fulbe population during Ahidjo's presidency,[31] but who could now boast of their own administrative units. The strategy seemed to have achieved both objectives. For instance, a businessman in Ngaoundere, the administrative capital of the newly-created Adamawa Province, noted that the change had made it easier for those in the new province to do business:

> ... we of the Adamawa province cannot conceal the joy we have in us, as regards the creation of the province.... Formerly, we were obliged to always go to Garoua for administrative affairs, with all the risk inherent in such journey.[32]

Meanwhile, the former president's reaction to charges of a plot against his successor was not unexpected. In an interview with Radio

France International on August 23, and in various press releases, Ahidjo accused his successor of "plotphobia" and of installing a "police state" in Cameroon. He denounced Biya for bugging the telephones of important government officials, carrying out arbitrary arrests and interrogations, and seizing the passports of ministers dismissed from his cabinet to prevent them from leaving the country. He also accused Biya of threatening the unity he had patiently and stubbornly built in Cameroon over twenty-five years and of "creating a diversion, instead of looking after the healthy social, political, and economic situation that he had inherited in Cameroon."[33] Attempts to reconcile them by President Mitterrand of France and African leaders, including Houphouët-Boigny of Côte d'Ivoire, Abdou Diouf of Senegal, Mobutu Sese Seko of Zaire, and Omar Bongo of Gabon, were unsuccessful.[34]

Finally, on August 27, while still in France, Ahidjo announced his resignation as chairman of the CNU. Less than a month later, on September 14, Biya succeeded him at an extraordinary party session held in Yaounde. This was his opportunity to reaffirm the direction he planned for the party and the nation, and his closing remarks to the delegates declared that the party must not only adapt to changing times, but must become "truly democratic" and open to free discussions and democratic choices.[35] As if to demonstrate these commitments, Biya in November, now both Cameroon's president and the CNU's chairman, announced a presidential election rescheduled for January 14, 1984, instead of 1985, the original election date. This decision was apparently meant to establish his own mandate and bury any perception that he was still Ahidjo's protégé. As expected, Biya was the only presidential candidate and was elected by 99.98 percent of those casting votes.

For those who had seen Biya simply as Ahidjo's puppet or who may have underestimated his political skills, the events of 1983 and the election of 1984 demonstrated that in spite of the president's quiet demeanor he had understood the intricacies of Cameroon politics. Not only had he frustrated, at least temporarily, any attempt by Ahidjo to resume authority in Cameroon, but he had replaced Ahidjo in both the nation's and party's leaderships.

One more episode intervened, however, before Biya's consolidation of power and road to reform were clear of Ahidjo's legacy. Despite his apparent popularity and recent successes, Biya's administration in April 1984 faced the most serious political crisis to threaten the country since the UPC uprising in the 1960s. On the morning of April 6, a faction of the military consisting mostly of members of the elite Republican Guard (also known as the Presidential Guard) and a few disloyal elements from the gendarmerie, the police, and the army, led by Colonel Saleh Ibrahim, commander of the Presidential Guard, and

Captain Awal Abassi (both from North Province) attempted to overthrow the government.

Several reasons have been advanced for the attempted coup. For instance, because most of the plotters were northern Muslims, it was suggested that the coup was an attempt to restore northern control of government and possibly to return Ahidjo to power. Others have argued that the attempted coup was also supported by the business class, who, had enjoyed many privileges under Ahidjo and who saw Biya's call for "rigor" and "moralization" as an effort to deprive them of those privileges.[36] For the latter group therefore, supporting the coup was a means of returning to the pre-Biya days of "economic freedom."

Although plans for the coup may have been developed earlier, it appears that the president's decision on April 5, 1984, to dismantle the Republican Guard (which was dominated by officers and troops from northern Cameroon) and transfer its members to other branches of the military forced the plotters to launch a premature attack.[37] In fact, it was also speculated at the time that President Biya and his chief of staff, General Semengue, were aware of the plot but allowed it to develop in order to learn the extent of Ahidjo's involvement in the conspiracy.[38]

Because the coup was hurriedly executed, poorly coordinated and also confined to Yaounde, the capital city, it was easily crushed with the help of loyal troops in Yaounde and from nearby military bases. The capital was calm by April 8, after most of the plotters had been killed, had surrendered, or had sought refuge in the largely Muslim section of the city. According to government sources, seventy people including four civilians and six loyal troops were killed, while 265 gendarmes were reported missing during the two days of fighting. Other sources, however, put the number of deaths as high as 1,000, with 1,053 arrested or taken prisoners.

Judging by the identity of the conspirators (and suspicions that Muslim merchants in Yaounde had prior knowledge of the plot)[39] the coup may have been inspired by the desire of some northerners to regain control of the presidency. Even so, Biya in an April 10 radio message refuted earlier charges by the minister of armed forces, Andze Tsoungui, and the chief of staff, General Semengue, that the plotters were 99.99 percent from the North.[40] Instead, he noted that the culprits were only a small group of ambitious officers intoxicated by power. He added that the loyal troops who had participated in putting down the rebellion were Cameroonians from every part of the country and all ethnic, regional, and religious persuasions.[41] The president's objective in refuting the northern source of the coup was twofold. First, he wanted to dispel the perception of a North-South or Muslim-Christian rivalry that some scholars have used in trying to explain political conflicts in Cameroon.[42] But perhaps more importantly, he needed to prevent any domestic political problems that might result from making such a connection.

There is a reading of this coup episode that emphasizes its dampening impact on Biya's entire presidency:

> Biya emerged a bruised man from the coup. To be sure that such an incident should never happen again, he surrounded himself with trusted friends. . . . Also Biya seemed to jettison his pet slogans of Rigour and Moralization (the cornerstone of his new deal policy), and concentrated on the art of surviving in power . . . he backpeddled on his incipient liberalization moves, muzzled the press, institutionalized presidential patronage and enhanced autocratic government. Thanks to the coup he swung full circle. . . . The smoking gun had killed the dream.[43]

But in fact, with the coup crushed, Biya had and took an excellent opportunity to reconsider the pace of the reforms he had initiated and to consolidate his authority by surrounding himself with his close and trusted friends. Perhaps because the party and its executive bodies had failed to express their support for the president during the critical period of the attempted coup, Biya was quick to make personnel changes, first within the party and then in the state apparatus. In May 1984, he replaced seven members of the CNU Political Bureau with seven new members considered his loyalists. The Central Committee of the party was also expanded from forty-eight to fifty-five members. These moves added Biya's supporters and fellow "New Deal" advocates to the CNU's inner circle. Similar changes were made in the cabinet and the upper echelon of the administration, July 7, 1984, marking the sixth cabinet reshuffle in Biya's less than two year presidency. Changes were also made in the directorship of various state and parastatal organizations, including the National Airlines Company (CAMAIR), the National Petroleum Company (SNH), the Produce Marketing Board (ONCPB), the National Electric Company (SONEL), the Urban Transport Company (SOTUC) and the Rural Development Bank (FONADER).

In making these changes, Biya was careful not only to maintain the ethnic and regional balance of North and South within the cabinet, the administration, and various parastatal corporations, but also to include experienced and trusted politicians whose services under President Ahidjo could be valuable to the success of his own administration. Both reasons explain why Joseph Charles Doumba was chosen to replace a fellow easterner Felix Sabal Lecco who was dropped from the Political Bureau, and why long-time Ahidjo loyalists such as Egbe Tabi and Solomon Tandeng Muna from the anglophone provinces were also retained as members of the Political Bureau. Doumba is the best example. Although earlier associated with the "gang of four" dismissed from the party by Ahidjo in January 1983 for objecting to his selection of Biya to replace him, Doumba's experience as a party organizer, including his position as director of the CNU party school, made him a valuable asset for Biya as he attempted to restructure the party to meet his political objectives.[44]

This work done, one of the first tasks of the expanded CNU Central Committee was to undertake a July 24–27, 1984, tour of the ten provinces, to inform the people about the new dynamism of the party and Biya's accomplishments since becoming president in November 1982. With the coup behind him and his political inner circle strengthened, Biya began the most distinctive phase of his presidency, poised now to carry out the political reforms he promised when he became president in November 1982.

The New Deal: Aspiration, Compromise, and Reality

Political reforms or political liberalization in this context are aptly defined by Michael Bratton and Nicholas van de Walle to include "any measure taken by the ruling elite to increase political competition" and also "recognize mass rights, such as freedom of political expression and freedom of political association."[45] Many argue that Biya's reform measures were shallow and tailored to prevent any serious challenges to the established regime, and we will examine that argument, and weigh the aspirations and the reality, before this chapter ends. But they were nevertheless a change from the restrictive and authoritarian rule of his predecessor.

Even before election to his own term as president in January 1984, Biya had started reforming the political process and laying the foundation for the "liberal" and "democratic" society he hoped to create. For instance, on November 18, 1983, he initiated the amendment of Article 7 of the constitution to allow multiple candidates to run in presidential elections. In other words, for the first time since the creation of the single-party system in 1966, challengers meeting certain conditions could run in presidential elections.

More than such specific constitutional or political moves signify, the early years of Biya's presidency were generally characterized by the vision of a new humane and "democratic" society that he hoped to create, once Ahidjo's shadow receded. The foundation of that society would be based on two principles, Communal Liberalism and what the president later defined as the National Charter of Freedom. The former was meant to address the socio-economic inequities by availing all Cameroonians of the fruits of the nation's economic development.[46] The latter was designed to guarantee all forms of individual and collective freedoms, particularly freedom of thought and expression, equal protection before the law, the secularity of the state, and the abolition of all forms of racial or ethnic discrimination.[47] In response to the President's call for freedom, François Sengat Kuo, the minister of information and culture, in August 1983 called on Cameroon's journalists to "write freely and report fearlessly within the law."[48] Commenting in 1985 on the emphasis on freedom evident in the National Charter of Freedom, a correspondent for the government's *Cameroon Tribune* noted that:

> For so long Cameroonians have listened to political speeches crammed with words like subversion and the like, until November 6, 1982. Since that date, the birth of the New Deal Crusade, words like freedom of expression and democracy became the centerpiece of official speeches.[49]

Independent newspapers were quick to take up the challenge and opportunity. In fact, there were over a dozen independent newspapers and magazines published in the country by 1985. Some were very ctitical of the administration, including *Le Messager*, the largest independent newspaper in Cameroon with a weekly circulation of almost 50,000 copies by the late 1980s. Foreign publications like *West Africa* and *Jeune Afrique* were allowed into the country without the excessive scrutiny every issue faced in the past to make sure they did not contain articles that the administration considered subversive. In addition, books such as Richard Joseph's *Radical Nationalism in Cameroon* and *Gaullist Africa*, Jean-François Bayart's *L'Etat au Cameroun* and Mongo Beti's highly critical *Main Basse sur le Cameroun*, banned during Ahidjo's era because of their pro-UPC sympathy, were allowed into the country and in book stores.

The new freedom also led to increased confidence among the people and an ability to express themselves without the fear of being arrested or persecuted. For instance, political discussions at the University of Yaounde were tolerated more than at any time since the creation of the institution in 1962. Professors were allowed to write articles that were critical of the administration, which previously would have led to discipline or even dismissal from the university.

Another change, at least in the first years of Biya's administration, related to efforts by the president and members of his cabinet to make themselves available to the media. Cabinet members were encouraged to appear on radio programs and respond to questions relating to their various ministries. Unlike his predecessor whom some observers considered inaccessible to his people,[50] Biya believed that this style of leadership was not only critical for understanding the nation's problems, but could help the administration in providing various services, including schools, clinics, and recreational facilities to different communities.[51] The president also hoped that the new practice would help to establish a closer relationship between the president and the people, and to gain their confidence in his leadership.

A population which had suffered for more than two decades under the oppressive and authoritarian rule of Ahidjo therefore welcomed Biya's policies. Cameroonians, including politicians, described the president in such glowing terms as "handsome," "a man who mixes," "an approachable man," and as a president "having the intent of the people at heart."[52] He was praised by Cameroonian students at home and abroad for his openness and candor. Biya's popularity was also evident by the fact that he was initiated into a number of traditional organizations throughout the country. In December 1984, for instance, he was initiated into Ngondo, the society of traditional rulers of the Duala people. In January 1985, he was bestowed the title "Fon of Fons," (making him the superior fon in the North West Province) by the fons of Bali, Kom, Bafut, Mankon and Nso.[53]

Further indication of the greater freedom and openness in Cameroon during the early years of Biya's tenure emerged at the CNU congress (christened the New Deal Congress) in Bamenda in March 1985. This was his true opportunity to establish the direction he wanted the party and the nation to pursue under his leadership. The process was not, however, easy. For three days, from March 21–24, 1985, the congress became a battlefield between two groups. Supporters of the New Deal led by young radicals such as Professor Georges Ngango and Mohamad Labarang advocated a complete break from the past, an increase in the pace of political liberalization, and even the creation of a multiparty system in Cameroon. On the other hand, there were carry-overs from the *ancien régime* such as Sengat Kuo, Doumba, and Mengueme, who not only favored caution and continuity but warned that radical changes might lead to political instability.

Unlike previous party congresses that were merely well-choreographed events to showcase party officials and endorse policies that had already been approved by the president and the party's Political Bureau and Central Committee, delegates were forthright and very critical of government policies. For example, a memorandum presented by the elites of North West Province cited the "high-handed manipulation of the constitution to the disadvantage and detriment of English culture."[54] Among the issues addressed in this text were the unilateral attempts by francophone officials to modify anglophone education without implementing similar changes in the French system of education. It also criticized the prevailing government practice of appointing English-speaking Cameroonians as assistants to French speakers even in cases where the former were better qualified.

A similar memorandum from anglophone elites of South West Province focussed on its lack of economic development since independence. The document highlighted the province's large financial contribution to the nation's economy, especially from the sale of agricultural products from the Cameroon Development Corporation (CDC) plantations, most of which were located in the province, and from the sale of oil from the coast of Limbe. The report claimed that, despite its financial contribution to the national treasury, much of the oil revenue was spent in developing other provinces. The document also alluded to the fact that huge sums of money had been spent in improving the port of Douala in Littoral Province to the detriment of Limbe in Fako Division, even though according to most experts it would not only have been cheaper to develop the latter, but the port of Limbe would have been more cost effective for exporting oil and agricultural products from the hinterland the two ports substantially shared.

In the end, Biya was forced to strike a balance between the progressive and conservative factions at the Bamenda Congress. As a compromise to those who had expected it to serve as a launching pad for multipartism in Cameroon, a symbolic gesture was made by changing the name of the party from the Cameroon National Union to the Cameroon People's Democratic Movement (CPDM). This new name and the party's new motto, "Union, Progress and Democracy," expressed the president's desire for political liberalization and the need to make the party a forum for all Cameroonians, rather than maintaining the elitist party character of the CNU.[55] The CPDM's objective was voiced by a party militant's remark that under Ahidjo the party had mainly served as a control mechanism working from the top down, but that the CPDM's purpose was to introduce democracy into the party so that the base could contribute more to its work and development.[56] While appealing for more openness within the party, however, the president also warned members that such democratic openings should not be an opportunity for those bent on sowing the demons of division and confusion within the party and the nation.[57]

As a further sign of the president's desire to accommodate both the progressive and conservative wings of the party, the Central Committee of the newly-formed CPDM was increased from the CNU's forty-one to sixty-five members, and the number of alternate members was raised from eleven to twenty. The expansion allowed progressives like Georges Ngango, Denis Ekane, and Joseph Fofe, and businessmen like James Onobiono, Pierre Tchangue, Samuel Kondo and Joseph Sack to be included, although at the same time veteran politicians such as Mengueme, Sengat Kuo and Doumba were retained. Bayart, in line with his "recherche hégémonique" analysis, has suggested that the inclusion of eminent business leaders in the Central Committee was meant to create an alliance between a strong bureaucratic regime and the business class.[58] Altogether, forty of sixty-five full members and fifteen alternates on the CPDM Central Committee were new, having held no such positions under the CNU.[59]

Despite the large number of new officers and a change in the name of the party, however, it could be argued that the veteran conservative politicians were the ultimate victors at Bamenda. That is because they were able to retain some of the most important positions in the party. For instance, Doumba and Sengat Kuo were elected organizing secretary and political secretary of the party, respectively, while young progressives such as Ngango and Ebenezer Njoh Mouelle who advocated more radical changes entered the Central Committee only as alternate members.[60] Yet this New Deal Congress, and the features introduced before and after as well as during its proceedings, marked an important juncture in Biya's presidency and political career. He phased out the last vestiges of the Ahidjo-CNU era in Cameroon by creating his own party, the CPDM. He bolstered a political image and reputation that had suffered following the attempted coup in April 1984. He secured control of the institutions of power in Cameroon, as head of state and chairman of the party, and was sufficiently confident to permit the return in February 1986 of fifty-five political dissidents who had been forced into Ghanaian exile during Ahidjo's administration. Later, in August, he granted clemency to fourteen UPC sympathizers who had been detained since December 1985 on charges of disturbing public order.[61]

The president was also in a much stronger position to introduce further political reforms, including the introduction of multiple candidates in party elections which he had first suggested soon after he took office in November 1982. He wanted at that time to introduce competitive elections so as to infuse "new blood" into the party and the administration, but the proposal was rejected by the CNU Central Committee which was still under Ahidjo's control. The Bamenda Congress, establishing his *de facto* and *de jure* supremacy, allowed Biya to introduce the new competitive system, implemented in the election of representatives to various organs of the party in 1986, then in the November 1987 municipal and April 1988 legislative elections. Under the new system of election to the legislative assembly, for example, each administrative division became an electoral district and was assigned a number of legislative seats based on its population. Whereas previously the entire country was a single electoral district and the electorate voted only for a single list of candidates, the new system allowed more than one candidate to run for the same party position or legislative seat from the same district. Candidates wishing to run for the legislature from a particular electoral district grouped themselves into two different lists (except in districts where there was a consensus list of candidates).

There were constraints. The names were submitted to the party's Political Bureau, which was responsible for screening the candidates to ensure that those on the lists were good party militants. A good militant was required to hold a current CPDM membership card, to be able to serve and lead, and to respect the laws of the land and the party.[62] The screening also provided the president and members of the Political Bureau with the opportunity to eliminate potential political rivals and to ensure that each list of candidates was a fair representation of the ethnic, religious, and economic composition of the particular constituency.[63] In fact, members of the Political Bureau could even change the composition of the lists by switching names from one list to another just to ensure that their preferred candidates or those likely to win were on the same list.[64] And although candidates on the different lists were allowed to debate each other in their respective districts, the debates were centered on how to best implement the president's New Deal policies rather than suggesting divergent policies they thought might be best for the country. In other words, the debates did not digress from policies established by the party; rather, they focussed on which list of candidates could best represent their district at the national political center.[65] As one observer noted: " . . . the elections were contests between rival local entrepreneurs who competed for votes on the basis of their claimed and demonstrated potential for servicing the local community and promoting its development."[66]

Despite shortcomings and the retention of such substantial regime discretion in the process, the new electoral system was a significant improvement on the former single-list system, where the party's Political Bureau gave little weight to the rank and file in deciding who represented the people in the legislative assembly. The impact of the new process would register in the next round of both party and national legislative elections. In the 1986 party elections, twenty-five of the forty-nine section party presidents were new members who had never held the position. At the same time, twenty-eight of the forty-nine presidents of the women's wing of the party (WCPDM) were new, as were all forty-nine presidents of the youth wing (YCPDM).[67] In the 1988 legislative elections, 85 percent of those elected to the National Assembly were new,[68] and the number of women elected rose from 15 to 22 percent of its membership.[69] Among those elected to important leadership positions were candidates previously opposed to Ahidjo or sympathetic to the UPC. They included Jean-Jacques Ekindi, elected president of the party in Wouri Division (Douala), and Professor Thomas Melone, elected to the National Assembly in Sanaga-Maritime. The fact that such men were deemed qualified to run for party and public office signified Biya's concern to reconcile differences among various factions created during the nationalist struggle in the 1950s and thereafter, a path Ahidjo took in the first years of his presidency but later reversed. It was also, as Bayart points out, a brilliant move that contributed to ending the conflict that was brewing between the progressive and the conservative wings of the party.[70]

Reaction to the new electoral process, especially from party members, was generally positive. A member of the CPDM Central Committee described the new electoral process as an appropriate means to "satisfy party activists who have always wanted to choose their leaders rather than submitting to the choice made for them." Another member described the change as "a realisation of President Biya's expressed wish, when he took power, that the party should become 'truly democratic'." Meanwhile, an editorial in the government-owned *Cameroon Tribune* was also positive on the change, describing it as "a tangible manifestation of the commitment to authentic democracy—one of the essential characteristics of national renewal."[71] Some Western observers hailed the 1986 party elections as "a step toward greater democracy in Cameroon."[72] And even Pius Njawe, editor of the independent weekly *Le Messager* and one of his ardent critics, would later acknowledge that Biya had mobilized political and economic liberalization policies in Cameroon before Mikhail Gorbachev came to power in the Soviet Union and launched "perestroika" there in the later 1980s.[73] In fact, as a poem in the January 1992 edition of *Cameroon Life* magazine indicates, some Cameroonians not only saw Biya as a reformer, but also as a Messiah who had come to save them from the tyranny of the previous administration:

> My fellow brethren
> The Messiah is born
> And with him
> Our new nation too
> Overflowing
> With heavenly ideals.

Not all Cameroon's mid-1980s political paths, however, pointed to reform or reflected pragmatic compromise between progressives and conservatives. There was evidence of a power calculus as dominating as Ahidjo's at the heart of governance, which must enter any reckoning of Biya's New Deal.

One line of evidence noted earlier, the November 18, 1983, constitutional amendment permitting multiple presidential candidacies under "certain conditions," was characteristic. In fact, the conditions were so difficult that they effectively eliminated the chances of anybody qualifying. A potential challenger was expected to present a petition with 500 signatures (fifty from each of the ten provinces in the country) from important elected or government officials such as legislators, governors, important traditional chiefs, divisional officers, members of the Central Committee of the CPDM, and municipal councillors. In addition, any potential candidate was required to have had at least five years of continuous residency within the national territory on the date of filing his/her petition for election.

One of the difficulties in fulfilling the first condition was the fact that under the regime established by Ahidjo and continued with a new name, personnel, and procedures by Biya, almost every element of the government apparatus was either controlled by the party or the state. With the same person dominating both governance channels, all nominations to important party and administrative positions—cabinet members, all high-ranking administrators—were made by the president. Therefore it would have been a sign of disloyalty to the president and a form of political suicide for any official who was qualified to sign the petition of a potential challenger to do so. Additionally, the five year residency requirement for a potential candidate was perceived by many observers of contemporary Cameroon politics as an attempt to exclude exile groups such as the UPC or the Cameroon Democratic Party, which had strong bases of support in France and Britain, respectively from presenting candidates in presidential elections. Nevertheless, Fouman Akame, the minister of territorial administration, described the decision to change the system as symbolic of the political maturity of Cameroonians.[74] But nothing in all this challenged or impaired Biya's path to virtually 100 percent support in the 1988 presidential ballot.

Shortly after the January 1984 presidential election, the president introduced a series of further amendments modifying various articles of the constitution, signed into law on February 4 after debate and approval by the National Assembly. Among the changes were the country's name, from the United Republic of Cameroon to the Republic of Cameroon (Article 1), the abolition of the office of prime minister (Article 5), and the latest modifications of Article 7 relating to presidential succession, covered earlier in this chapter but now again relevant because Biya closed the door to a successor with some authority which Ahidjo had opened for him. On the first point, Biya argued that the name change was not only a demonstration of the political maturity of the Cameroonian people after twenty-five years of independence, but a sign that the people had finally overcome divisions caused by seventy years of European colonization.[75] But what he considered maturity, many anglophones considered the boldest step yet taken toward their assimilation and disappearance as a distinct founding community in 1960–1961; they protested at the 1985 Bamenda Congress (as we saw) and intensified the critique in the 1990s as we will discuss later. On the second and third points, with the prime minister's office abolished and the return to the president of the National Assembly as first in the line of presidential succession, the presidency's powers were again augmented. For unlike previous provisions for the prime minister to serve the remainder of the president's term in case of the latter's inability to perform his duties, the amendment allowed the president of the National Assembly to assume the office only on an interim basis. Moreover, the election to fill the presidency was to be held within a period not exceeding forty days. Additionally, the president of the National Assembly could not amend the constitution, change the composition of the government during the interim period, or be a candidate for election to fill the vacant office of president.[76] In these ways, and in the use of rhetoric which blurred reality, Biya's New Deal political changes, intended to portray him as a reformer, operated in such a way that they did not jeopardize his control of the institutions of power and authority.

We have surveyed a complex transfer of the presidency from Ahidjo and an intricate mixture of intentions, words, and deeds during Biya's early years in office. While it is generally agreed that he instituted significant political changes when he took over the presidency in November 1982, it can equally be argued that Biya's changes were primarily designed to ensure his continuation in power. Freedom of expression was tolerated only to the extent that it did not challenge the existing orthodoxy. And while multiple candidates could run for party and public elective offices, the CPDM Political Bureau still had a primordial role in selecting the candidates. Ultimately, those who were elected to the legislature and the party hierarchy were not the authentic representatives of the people but those favored by the party's core leaders at its center. By two measures we have applied previously, the structures and workings of Ngayap's "classe dirigeante" and Bayart's "recherche hégémonique," more continuity than change, more growth than reduction, marked the structure as distinct from the personnel in Cameroon's governance. One fundamental Ahidjo principle, the alliance of northern and southern elites at the regime's heart, remained in place for Biya through the mid-1980s, though imperilled by the 1984 coup attempt and the notables from the North killed or jailed at the time. Many of the latter were released only in the early 1990s, and their regional resentment, as with anglophones, would become an issue that we will analyze later, in Chapter 6.

In truth, looking beyond the regime networks and leaders, most Cameroonians who had hoped for radical political changes from Biya's presidency became increasingly disenchanted with his failure to fully implement the freedom and democracy he had promised. Soon, this discontent would also encompass Cameroon's worsening economic problems, resulting from the decline in world market prices for Cameroon's exports and from the administration's corruption and mismanagement.

At the same time, the end of the Cold War and the introduction of political reforms in the former Communist nations of eastern Europe would provide the opportunity for Cameroonians from all sectors (like most of their African counterparts) to demand greater political reform. Chapter 4 deals with an increasingly troubled Cameroon through 1990, when those bulwark features of political class solidarity within state and party faced their first comprehensive challenge since the 1960s.

Notes

1. Harry Ododa, "Voluntary Retirement by Presidents in Africa: Lessons from Sierra Leone, Tanzania, Cameroun and Senegal," *Journal of African Studies* 15, 3/4 (1988), pp. 94–99.

2. See Biya's speech to members of the National Assembly when he took the oath of office on November 6, 1982.

3. Paul Biya, *Communal Liberalism* (London: Macmillan, 1987), pp. 36–37.

4. Jean-François Bayart, "Cameroon," in Donal Cruise O'Brien, John Dunn and Richard Rathbone, eds., *Contemporary West African States* (Cambridge: Cambridge University Press, 1989), p. 31.

5. *West Africa*, January 2–8, 1984, p. 13.

6. For evidence of Ahidjo's perception of the French Socialist Party, see Philippe Gaillard, *Ahmadou Ahidjo (1922–1989)* (Paris: Groupe Jeune Afrique, 1994), pp. 83, 194.

7. Victor Julius Ngoh, *Cameroon: A Hundred Years of History, 1884–1985* (Limbe, Cameroon: Navi-Group Publications, 1988), p. 282.

8. *Africa Confidential*, November 17, 1982, p. 1.

9. It is not clear whether Ahidjo had actually planned to resign at the Douala congress or whether the rumored resignation was simply a ruse by Ahidjo himself to find out whether he still enjoyed support in country.

10. For more details see Victor Le Vine, "Cameroon: The Politics of Presidential Succession," *Africa Report*, 28, 3 (1983), pp. 22–26.

11. Le Vine, "The Politics of Presidential Succession," p. 24.

12. Ngoh, *Cameroon: A Hundred Years of History*, p. 305.

13. Soon after returning to Cameroon in 1962 from his university studies in France, Biya was appointed chargé de mission at the presidency. Two years later, in January 1964, he was appointed director of cabinet to the minister of national education, youth and culture, then promoted to the rank of secretary general in the same ministry in July 1965. He returned to the presidency in December 1967 as director of civil cabinet and, keeping that post, he also became secretary general at the presidency in January 1968. In August 1968, his office of secretary general at the presidency was elevated to a cabinet position, and in June 1970 he was again raised to a senior cabinet position as minister of state. Following the creation of the United Republic in May 1972 and the restoration of the office of prime minister in June 1975, Biya was appointed to fill that office. He retained the office after the constitutional amendment of 1979.

14. Bayart, "Cameroon," p. 36.

15. The other three members who were dismissed from the party were El Hadj Ibrahim Ninga Songo, Bienvenue Atemengue Nkoulou and Prosper Mbassi.

16. *Cameroon Tribune*, February 2, 1983, p. 9.

17. Henri Bandolo, *La Flamme et la Fumée* (Yaounde: Editions Sopecam, 1985), p. 23. The delegation consisted of Maïgari Bello Bouba, the new prime minister; Oumarou Aminou, minister delegate at the ministry of foreign affairs; Mohaman Lamine, Cameroon's ambassador to Nigeria, and Ousman Mey, the powerful governor of North Province.

18. *Africa Confidential*, July 22, 1983, p. 7.

19. Bandolo, *La Flamme et la Fumée*, p. 90.

20. For more examples of Ahidjo's effort to assert his authority over that of the head of state see Bandolo, *La Flamme et la Fumée*, especially pp. 37–190.

21. Bandolo, *La Flamme et la Fumée*, p. 67.

22. Cited in Cameroon National Union, *The New Deal: Two Years After* (Yaounde: n.p., 1985), p. 15.

23. Whether all the members present at the meeting signed the letter of resignation or not is a matter of contention. But in an interview in July 1983 with Henri Bandolo, the prime minister Bello Bouba who was present at the meeting in Ahidjo's residence confirmed that all the members signed the letter. For details on the interview see Bandolo, *La Flamme et la Fumée*, p. 141.

24. *Le Messager*, April 6, 1995, p. 3.

25. Bandolo, *La Flamme et la Fumée*, p. 175.

26. Bayart, "Cameroon," p. 33.

27. Gaillard, *Ahmadou Ahidjo*, p. 226.

28. For a discussion of the concept of "Father of the Nation" see Michael Schatzberg, "The Metaphor of Father and Family," in Michael Schatzberg and William Zartman, eds., *The Political Economy of Cameroon* (New York: Praeger, 1986), pp. 1–19.

29. *Africa*, September 1983, p. 37.

30. *Africa Confidential*, October 5, 1983, p. 7.

31. Mario Azevedo, "The Post-Ahidjo Era in Cameroon," *Current History* 86, 520 (1987), 217.

32. Ecole Supérieure des Sciences et Techniques et de l'Information (ESSTI), *Paul Biya, 5 Ans Apres ... Le Camerounais Jugent leur President* (Yaounde: n.p., 1987), p. 24.

33. *West Africa*, August 29–September 4, 1983, p. 1988.

34. *Africa Research Bulletin: Political, Social and Cultural Series* 21, 3 (March 1–31, 1984), p. 7183.

35. *Africa Research Bulletin: Political, Social and Cultural Series* 20, 9 (September 1–30, 1983), p. 6976.

36. Colin Legum, ed., *Africa Contemporary Record 1984–85* (London: Rex Collings, 1985), p. B105.

37. *Africa Confidential*, April 25, 1984, p. 5.

38. *Africa Confidential*, July 4, 1984, p. 5.

39. Bayart, "Cameroon," p. 33.

40. *West Africa*, May 14–20, 1984, p. 1014.

41. *Africa Research Bulletin: Political, Social and Cultural Series* 21, 4 (April 1–30, 1984), p. 7210.
42. Ndiva Kofele-Kale, "Ethnicity, Regionalism, and Political Power: A Post-Mortem of Ahidjo's Cameroon," in Schatzberg and Zartman, eds., *The Political Economy of Cameroon*, pp. 54–82.
43. Cited in *Cameroon Life*, November/December 1991.
44. A decade later, Doumba remained a key Biya supporter as party General Secretary.
45. Michael Bratton and Nicolas van de Walle, "Popular Protest and Political Reform in Africa," *Comparative Politics* 24, 4 (1992), p. 422.
46. Biya, *Communal Liberalism*, pp. 13–14.
47. Biya, *Communal Liberalism*, pp. 42–43.
48. *West Africa*, September 5–11, 1983, p. 2049.
49. *Cameroon Tribune*, March 23, 1985.
50. Le Vine, "The Politics of Presidential Succession," p. 26.
51. Biya, *Communal Liberalism*, p. 127.
52. *West Africa*, November 9, 1987, p. 2213.
53. Azevedo, "The Post-Ahidjo Era in Cameroon," p. 220.
54. *West Africa*, May 6–12, 1985, p. 877, for details of this and the next paragraph.
55. Frederick Scott, "Biya's New Deal," *Africa Report*, 30,4 (1985), p. 59.
56. *West Africa*, April 1–7, 1985, p. 605.
57. *West Africa*, April 1–7, 1985, p. 605.
58. Jean-François Bayart, *L'Etat en Afrique: La Politique du Ventre* (Paris: Fayard, 1989), translated by Mary Harper, Christopher and Elizabeth Harrison as *The State in Africa: The Politics of the Belly* (London: Longman, 1993), p. 94.
59. Bayart, "Cameroon," p. 42.
60. Bayart, "Cameroon," p. 43.
61. *West Africa*, January 5–11, 1987, p. 13.
62. *West Africa*, April 28–May 4, 1986, p. 873.
63. *West Africa*, May 9–15, 1988, p. 829.
64. Albert Mukong, *My Stewardship in the Cameroon Struggle* (Enugu, Nigeria: Chuka Publishing, 1992), p. 6. Interviews in and documents from North West Province in 1995 disclosed long, bitter memories about the tailored selection process in 1987–1988. Consequences included the move toward independent parties by those the CPDM frustrated.
65. *West Africa*, May 9, 1988, p. 829.
66. Tatah Mentang, "New Deal Electoral Politics in Cameroon: A Political Analysis," *ESSTI* 3 (1986), 29.
67. The main reason why all the presidents of the youth wing of the party were new was because the age of those classified as youths was lowered from thirty-five to twenty-five years. Therefore the change in age limitation disqualified many candidates who had previously held the office or could be elected president.
68. One reason for the high percentage of new members in the legislative election was the fact that many incumbents had either decided not to run for re-election or had not received the party's approval to run.
69. *West Africa*, May 9–15, 1988, p. 830.

70. Bayart, *The State in Africa*, p. 166.

71. Cited in *Africa Research Bulletin: Political Series* 23, 2 (February 1–28, 1986), pp. 7933–7934.

72. *African Recorder*, July 30-August 12, 1986, p. 7073.

73. *World Press Review*, January 1992, p. 51.

74. *Africa Research Bulletin: Political, Social and Cultural Series* 20, 11 (November 1–30, 1983), p. 7042.

75. Biya, *Communal Liberalism*, p. 6.

76. Since Solomon Muna, an anglophone from North West Province was the president of the National Assembly, the bar to his more permanent succession to the office was perceived by many English-speaking Cameroonians as an effort to prevent an anglophone from becoming president.

4

The State Under Pressure, 1986–1990

When Paul Biya assumed Cameroon's presidency in 1982, some considered him a savior who would "take the masses away from the darkness of two decades of dictatorship to the brightness of political liberalism and economic prosperity" and compared him to other young progressive presidents such as Abdou Diouf in Senegal and Thomas Sankara in Burkina Faso, whose rise to power had led to significant political changes in their countries.[1] Yet, despite the reforms we have seen to 1986, including a greater degree of individual freedom and the introduction of multiple candidates in party elections, Cameroonians became increasingly disenchanted at their slow pace and at the reluctance or inability of the president to implement many of the broader social and economic changes he had promised:

> The program of action of the CPDM which was initially stated at the Bamenda Congress and later developed fully in the book *Communal Liberalism* won national and international acclaim upon its publication. It raised the hope that Cameroon would soon be, if not the first, then at least among the first African countries to join the club of liberal democratic states. Once this happened, Cameroonians, it was expected, would live in freedom, dignity, justice; they would live with a sense of rigour in the management of resources, moral rectitude and rediscovered communal solidarity, all of this in a permanent endeavour to develop our country.
>
> Since then the party's leadership has done little to implement such strategic watchwords as internal dialogue, the free exchange of ideas, tolerance of opinion, promotion of our national culture, the local economy and social justice.[2]

In other words, it became apparent to many Cameroonians by the late 1980s that Biya's early reforms were more concerned with consolidating his power than advancing the interest and welfare of the people. Once the former was achieved, he was reluctant to implement additional political reforms, and this generated opposition. The combined impact of political criticism, economic deterioration beginning in 1986, and the example of political reforms in Eastern Europe following the end of the Cold War led to demands in Cameroon, as in most African states, for larger scale political change.

Failed Promises and the Illusion of Reforms

In the preceding chapter we examined some of Biya's political reforms and how they led to significant changes, especially in the composition of the legislative assembly. He was highly praised for early reformist inclinations prior to the wind of change that swept the continent in the late 1980s. But a close scrutiny reveals the limits, even the hypocrisy, of many of the reforms designed in Biya's early years to prevent Ahidjo's control behind the scenes or his actual return to power, then to sustain his authority as president and his control of the various state institutions.

Thus, Biya did not dismantle many of the repressive structures and institutions such as the laws of March 12, 1962, and June 19, 1967, restricting freedom of expression and association, respectively, inherited from Ahidjo. Nor did he abolish the Direction Général d'Etudes et de la Documentation (DIRDOC), formerly the CND, with its intricate and omnipresent network of spies, and the BMM, responsible for the interrogation and torture of antigovernment suspects. These laws and institutions were used in suppressing dissenting views and in intimidating critics of the administration.

Despite the momentary relaxation of censorship following Biya's accession to power, for instance, the minister of territorial administration on November 8, 1983, issued Law No. 3292/L/MINAT/DAP/LP retaining prepublication censorship of newspapers and the confiscation by police of newspapers that did not conform to the law. According to administration sources, passage of the law was motivated by the callousness and exaggerated manner the independent press used when reporting the rift between the president and his predecessor.[3] Consequently, the independent press became a major casualty of this policy.

In fact, although the president and members of his cabinet had on many occasions encouraged Cameroonians to express themselves freely and become more involved in the affairs of the state, it soon became evident that such freedoms were tolerated only to the extent that government officials considered them no threat to the presidency. Journalists could be arrested and newspapers banned for publishing whatever the authorities considered inappropriate. In 1985, Paul Nkemayong, editor of the *Cameroon Times*, was briefly detained for refusing to release confidential documents on a story detailing corruption and fraud at the Société Nationale de Raffinage (SONARA).[4] Similarly, a group of journalists was arrested in 1986 for using "bad language" and for attacking individuals and institutions in a manner not compatible with their status as journalists.[5] In 1988, two journalists were sentenced to four months in prison for publishing an article critical of Giscard d'Estaing, the former French president who was on a private visit to Cameroon.

Even journalists of the government's *Cameroon Tribune* and the other official news media were not immune to government control and censorship. In fact, because these journalists were civil servants, they were expected to be less critical of the administration and to represent the regime's views. Those who failed to adhere to these principles were punished in a variety of ways, including arrests, detentions, and transfers to less desirable positions or regions of the country. In June 1986, for example, several anglophone journalists were arrested for reportedly making derogatory remarks about government officials. Similarly, on January 22, 1987, the director of the Cameroon Press and Publishing Company (SOPECAM) and the editor of *Cameroon Tribune* were arrested for publishing a presidential order without official authorization. Their action was in violation of an established order prohibiting such publication so long as the decision was still considered confidential.[6] Popular radio programs such as *Cameroon Report* and *Minute by Minute* which often examined critical issues facing the nation were taken off the air because they were considered irresponsible in their criticism of the administration.[7] Commenting on the freedom of the press under Biya's administration, one observer noted that it "starts where condemnation of the Ahidjo regime is concerned and ends where criticism of the Biya era begins."[8]

Meanwhile, the 1962 antisubversion decree and the 1967 law restricting freedom of association were also used in silencing other administration critics. In January 1985, for example, Dr. Joseph Sende, a medical practitioner in Yaounde, was arrested for filing a suit requesting the administration to repeal the ban on the UPC in Cameroon. Although the suit was ultimately dismissed by the Supreme Court, Dr. Sende was arrested again on February 7, and detained without trial for several months. Also, on December 1985, in a case noted earlier, fourteen supporters of the UPC were arrested and detained for six months on charges of distributing leaflets on behalf of the party. On releasing them in August 1986, government authorities brushed off the irregularities of the legal process; instead, they publicized and praised the release as symbolic of President Biya's liberal policies.

Like Dr. Sende, Fongum Gorji Dinka, a prominent North West Province anglophone lawyer, was arrested May 28, 1985, for publishing and distributing a pamphlet titled *The New Social Order* in which he questioned Biya's constitutionality and authority in unilaterally changing the name of the country from the United Republic of Cameroon to the Republic of Cameroon. Although he was charged under Article 152 and 153 of the penal code with contempt of the president and other dignitaries, he was never brought to trial, and was eventually released in January 1986.

A pattern was clear. Although Cameroon's legal code prohibited the arrest and detention for more than forty-eight hours of anybody suspected of committing nonpolitical crimes without a court order, authorities either simply ignored the law or framed the alleged offenses to fit a political crime. For instance, in May 1988, the musician Koko Ateba was arrested and charged with singing a song deemed insulting to the presidential couple at the inaugural ball. Albert Mukong, a prominent anglophone whose politics led to years in jail after 1970, was then arrested in June 1988 following a British Broadcasting Corporation interview that criticized frequent constitutional changes in Cameroon and attributed the country's economic problems to embezzlement of state funds by government officials. He was charged with using subversive language detrimental to the government and the head of state. Like the others, Mukong was detained without trial and eventually released in May 1989; we will see his important role again after 1990.[9]

Besides restricting individual freedoms, both the 1962 and 1967 laws were also important in constraining groups and organizations whose ideals did not conform with the regime's. That appears to have been the case in December 1984, when several members of the Jehovah's Witnesses sect were arrested at their annual conference and held without trial until May 1985. Although their arrest was contrary to the spirit and the letter of the president's call for freedom of assembly and association as articulated in his National Charter of Freedoms, the episode was nevertheless justified under the law of June 12, 1967, prohibiting any association that had exclusive membership or whose objectives were contrary to national unity.[10] Also, despite President Biya's professed opposition to the apartheid regime in South Africa, his administration refused to authorize an "anti-apartheid" conference by the Association of African Jurists that had been scheduled in Yaounde from January 15–18, 1985. The real reasons for banning the conference are a matter of conjecture. But according to official government sources, the conference was banned because its organizers proceded without proper authorization as required by law.[11]

These episodes where regime politics imposed constraints on individual liberties of expression and action through the mid- to late 1980s accumulated alongside the larger institutional practices which Chapter 3 disclosed. These can be summarized in the judgment that Biya sought the reformer's mantle in appearance at the secondary political level of legislative and rank and file CPDM affairs, but permitted nothing to jeopardize his control of the executive structures of government and the party's inner circle, where politics and statecraft remained focussed. In fundamental respects, what truly mattered was in the president's and the CPDM Central Committee's and Political Bureau's domain of decisions, which lesser, rubber stamp bodies approved and carried out. Regime personnel appointments reflected these conditions, as Ahidjo-CNU veterans, earlier dismissed, were recalled and given important positions in the party or administration after they had apparently, in Victor Le Vine's words, re-established their credentials with Biya.[12] The best of many examples was Jean Fochive, the powerful head of Cameroon's intelligence services under Ahidjo, appointed ambassador to China in 1983, then brought back in 1989 to head the new intelligence bureau, the Centre National des Etudes et des Recherches (CENER). It was not just a question of the rank, power and perquisites which ministerial level appointments conveyed; all the party, administration, and parastatal posts were lucrative in the way the literature on African prebends indicates. "La classe dirigeante" remained in power and in business five years after Ngayap's analysis appeared, although a more circumscribed network developed, creating problems for the smaller inner circle.

Although the president's dependence on these men and their inclusion in his administration was politically expedient and made use of their administrative and political skills to enhance his administration, many Cameroonians perceived a betrayal of the spirit and letter of the New Deal. Kofele-Kale may have voiced the opinion and disappointment of most Cameroonians when he noted that instead of bringing into his administration new and fresh faces and a shared vision of a free and democratic Cameroon, Biya relied heavily on Ahidjo loyalists who had "turned the first two decades of Cameroon's postreunification history into a painful nightmare."[13]

Another betrayal of the New Deal, with strong repercussions on the balance of regional and ethnic forces at the regime's center, and therefore its stability and endurance, was Biya's failure to eliminate "tribalism," favoritism, and all forms of divisions that, according to the president himself, threatened national unity. In many speeches during his maiden tour of the provinces in 1983, Biya had urged Cameroonians to put the interest of the nation before their personal or group interests. Citing his predecessor at Garoua on May 4, 1983, for example, the president reminded his audience that:

> ... no one tribe can claim that it has been called upon to dominate the others, nor can any one tribe claim to be vested with any legitimacy whatever

to govern the others. . . . It would be a dangerous illusion to contend that anything profitable and durable can be done for the national community by depending solely on one ethnic group or on one particular region.[14]

Similarly, in his closing remarks at the second extraordinary congress of the CNU September 14, 1983, Biya declared the primacy of the Cameroon nation over all particularist tendencies:

> National unity, for its part, is increasingly being consolidated. This unity, which is fraught with diversity and complementarity, solidarity and faith in a common destiny, transcends all forms of particularisms, especially geographical, historical, linguistic, tribal and religious, making Cameroon a modern and powerful state, where there is stability in justice, and equality of all, in respect of the duties and benefits of public services.
>
> This means that Cameroonians are first of all Cameroonians, before being Bamiléké, Ewondos, Foulbés, Bassas, Boulous, Doualas, Bakweris, Bayas, Massas or Makas. This means that Cameroonians are first of all Cameroonians, before being English-speaking or French-speaking, Christians, Muslims or animists.[15]

The speech in the hometown of the former president may have been very important in light of the fact that several northerners had opposed Ahidjo's decision to hand over power to a southern Christian, and the speech at the party congress with its emphasis on fairness and equality was particularly welcomed by anglophone Cameroonians and other groups in the country who felt they had not been fairly treated under the previous administration.

But whereas the president may have been genuinely interested in promoting a society where national concerns transcended all ethnic, religious, and regional interests, some members of his ethnic group saw the transfer of power simply as an opportunity to promote what Bayart, using a vocabulary already familiar in Cameroon, called "ethnofascism."[16] Biya's presidency was seen by some members of his Beti group as a means of benefitting from the state's patronage and directing its affairs. In fact, Emah Basile, mayor of Yaounde and Biya's close associate, even argued that it was the Beti's turn to enjoy the fruits and power of the state which northerners had monopolized during Ahidjo's administration.[17] Such advocacy mirrored in Cameroon the common African belief and practice translating the president's position into rights for his ethnic group or regional affiliates to monopolize the power and resources of the state.

Biya's response to such pressures from various Beti factions made matters worse. He was unwilling or unable to resist them, especially after the failed 1984 military coup.[18] By August 1991, according to one source, thirty-seven of the forty-nine préfets or senior divisional officers (administrative heads of divisions), three-quarters of the directors and general managers of parastatal corporations in the country, and twenty-two of the thirty-eight senior administrative personnel in the newly restored (June 1991) office of the prime minister were from the president's Beti ethnic group.[19]

Therefore, instead of constructing a society where ethnic and regional interest was irrelevant, as he had promised under the New Deal, Biya seemed unable to maintain the ethnic and regional balance that had been so critical under the previous administration in maintaining political stability in Cameroon. This is how one commentator describes the situation Biya created:

> ... a sentiment of disappointment gradually replaced that of hope as it dawned on Cameroonians that qualification, competence and merit was the preserve of the President's tribesmen. Slowly, but surely, they started taking over all strategic appointments once held by people of different tribal horizons. In an expansive and greatly populated division like Noun, disappointment soon made way for bitterness as the division was suddenly taken over by the Betis- S.D.O., five D.Os, three [Chief] police officers, M.O., Chief Magistrates, Prison Superintendents etc. The story was the same in every other province of Cameroon as a top civil servant once inadvertently confirmed on T.V. by saying that thirty seven of the forty nine divisions were headed by Betis.[20]

Even William Eteki Mboumoua, who held a number of high-ranking positions under the previous regime, was secretary general of the Organization of African Unity from 1974 to 1980, then Biya's foreign minister from July 1984 until January 1987, accused the president of establishing a corrupt and "tribal" state instead of pursuing the objectives of the New Deal.[21]

One group particularly disappointed at the turn of events was the anglophone population. Many English-speaking Cameroonians had seen the emergence of President Biya and his New Deal agenda as signalling an end to the continued marginalization of the two anglophone provinces (North West and South West) and an end to the "second class" status anglophones had endured since reunification.[22] Unfortunately, the optimism soon turned into despair, as nothing appeared to change under Biya, including efforts to undermine the English language, a point made evident in December 1983 when the minister of national education issued an order designating French as a compulsory subject for English-speaking students taking the General Certificate of Education (GCE) examination. By contrast, English was not a required subject for French-speaking students taking the baccalaureate exam. Although the order was eventually rescinded by the president following eleven days of demonstrations and a boycott of lectures by students in the two anglophone provinces and by English-speaking students at the University of Yaounde, the minister's initial decision was seen by many anglophones as yet another attempt by the administration to undermine anglophone culture and the English language in Cameroon. Some members of the administration insisted that there was no attempt by francophones to impose their "Frenchness" on anglophones,[23] but anglophones continued to fill subordinate positions in the administration and in many of the parastatal corporations. As one analyst put it:

> To be an Anglophone in Cameroon bars you the way to be a Minister of Defence, Minister of Finance, Minister of Territorial Administration and above all the President of the country- no matter how brilliant and gifted you are. The contrary is true with Francophones. You can ascend to the highest office of the country by virtue of the fact that you are a Francophone no matter what a nonentity you may be.[24]

These factors inevitably reminded many anglophones of Law No. 84/001 of April 4, 1984, changing the country's name from the United Republic of Cameroon to the Republic of Cameroon, as yet another attempt at "assimilating" anglophone Cameroonians. Not only did many English-speakers consider the change of name unconstitutional. It seemed to them that by taking the name of the former French territory when it gained its independence in January 1960, the president had completely ignored the fact that the Republic was born out of the union of what anglophones considered to be two separate entities, and that he was completing the annexation of the former Southern Cameroons.[25] This perception was corroborated by anglophones' experience of francophones sometimes referring to them as "les assimilées,"[26] and even some government officials' (including Yaounde political boss Emah Basile's) use of the term "les enemies dans la maison." Perceived marginalization and treatment as "second class" citizens may explain why anglophones became the vanguard demanding fundamental political change in Cameroon in the 1990s; we will see, for example, their rallies under the two star flag.

Other groups, including the Bamileke of West Province and many northern Muslims, were equally disenchanted with the Biya administration. The Bamileke who form the dominant indigenous business group in the country accused the administration of a deliberate attempt to promote opportunities among Beti businessmen at their expense. For instance, entrepreneurs from the president's ethnic group, described by

Nantang Jua as "va-nu-pieds subitement devenus millionaires/bare foots suddenly turned into millionaires," were granted import-export licenses in an effort to assure them a considerable share of the nation's lucrative import market.[27] Similarly, many northern Muslims (who had played an important role in the nation's political life and held important positions during Ahidjo's regime) felt that they were being gradually excluded from the political process and from important administrative posts, especially since the failed April 1984 military coup d'etat masterminded by northern officers, with its residue of many civilian northern notables prosecuted and held in prison throughout the 1980s.

While disillusionment with Biya's inability or lack of fortitude to implement changes promised when he became president may account for increased demands for political reforms beginning in the late 1980s, the situation was simultaneously exacerbated by Cameroon's declining economy and the shifts in global politics.

Cameroon's Declining Economy

President Biya inherited a relatively strong and healthy economy in November 1982. With an annual growth rate of 7 percent, an external debt of about US$2.3 billion in 1982 (one of the lowest in sub-Saharan Africa), an excellent development track record since independence, and a history of political stability, Cameroon was seen as one of the safest countries for foreign investors not only in Africa, but also in the Third World.[28] A major reason for the strong economy, especially from the mid-1970s, was the increased revenue generated from the sale of Cameroon's oil. It contributed less than 5 percent of the nation's revenue base in 1980 but 40 percent in 1984.[29]

Together with the increase in oil revenue was the judicious use of additional earnings. Unlike many other African states, such as Nigeria and Gabon, for example, where the production of oil had led to the an overreliance on oil production to the detriment of other sectors of the economy, President Ahidjo emphasized agricultural development. The government's slogan that "before oil there was agriculture and after oil there will still be agriculture" remained the guiding principle in the nation's economic development. The policy encouraging agriculture continued into Biya's presidency, for funds devoted to agricultural and rural development were increased from 23.7 percent in the Fifth Development Plan (1981–1986) to 26.1 percent in its successor (1986–1991). Moreover, so as to prevent the over-reliance on or exhaustion of oil revenue, Ahidjo saved much of that revenue in separate, extrabudgetary accounts in foreign banks outside the franc zone and bequeathed about CFA300 billion francs from this source to his successor when he resigned in November 1982.[30]

Biya used much of the money from the extrabudgetary account and from the sale of other exports to finance New Deal programs, including rail and road construction, and the building of more than 12,000 low-cost housing units during the first five years of his administration.[31] The additional revenue was also important in sustaining a bloated government bureaucracy, which doubled from about 80,000 in 1982 to 160,000 in 1987.[32] In 1985, about CFA180 billion francs was withdrawn from the extrabudgetary account to supplement the national budget.[33]

Unfortunately, by 1986, Cameroon like many African oil producers started experiencing a decline in its foreign earnings because of a fall in the world market price. A barrel of crude oil earned US$29 in 1984, but only US$10 in 1986. This meant a decline in the country's foreign oil earnings from US$694 million in 1984–85 to US$243 million in 1986–87.[34] Complicating the problem was a decline in the value of the dollar for which oil was sold, from CFA500 francs to the US dollar in 1985 to CFA300 francs in 1987. The net result was Cameroon's drastically lower revenue from oil in 1987 than in 1985.[35]

Other export items were equally affected by the fall in prices. Cocoa, a major export crop, fell from CFA940 francs per kilogram in 1985 to CFA700 francs in 1986, cotton from CFA780 francs per kilogram in 1984 to CFA331 francs in 1986.[36] The problem was exacerbated by drought in the early 1980s and a consequent fall in the volume of agricultural production. All in all, there was a drop in the total value of Cameroon's exports from CFA1,000 million francs in 1984–85 to CFA575 million in 1986–87 and a rise in the external debt from US$2.8 billion in 1984–85 fiscal year to US$3.5 billion in 1987–88.[37]

Compounding Cameroon's economic problem was the flight of capital, facilitated by the ease with which foreign corporations and expatriate workers could take profits and other forms of income overseas. This liberal approach was part of Ahidjo's overall strategy to encourage foreign investment.[38] Consequently, instead of investing their profits in Cameroon where they could be used to generate further economic development and perhaps higher employment, most expatriates and foreign-owned firms preferred to invest their money at home, where there was greater political stability and greater return for their investments.

The problem of capital flight was further intensified by corruption among civil servants and misappropriation of government funds by high-ranking government officials. A review of the civil service registry code-named "Operation Antelope" in 1986 revealed the names of about 20,000 ghost workers (civil servants who had either died, retired, or were studying overseas) whose combined monthly salaries were costing the administration millions of CFA francs.[39] And Nantang Jua has documented both the enormous amount of money used to sustain the bureaucracy and the middle and high-ranking bureaucrats' connivance with businesses to siphon hundreds of millions of CFA francs from the government. A common practice involved over billing the cost of a particular item or items, then sharing the extra money between the bureaucrat and the businessman.[40] Most of the funds embezzled by senior government officials were either placed in European and Caribbean banks or used in purchasing homes and apartments overseas.[41]

Despite attempts by the administration to restrict the transfer of money abroad by Cameroonians, such transfers of funds (mostly in suitcases and by other illegal means) may have increased by about 1,000 percent between 1974 and 1987. It is estimated that during Ahidjo's twenty-two years in office Cameroon lost a total of CFA965 billion francs in bad investments, embezzlement, and other forms of fraud, whereas in just four years (1982–1986) of Biya's New Deal administration, the country had lost nearly CFA650 billion CFA francs.[42] Some reports attributed much of the blame for the capital flight to members of the president's ethnic group:

> It must be noted that Mr. Biya's desire to pander to the needs of his clansmen and women opened the floodgates of fraud which manifested itself in capital flight, embezzlement, the collapse of once viable state parastatals such as Cameroon Bank, Societé Camerounaise de Banque, SCB, National Produce Marketing Board (NPMB), BIAO etc.
>
> The horrible truth that soon became apparent was the fact that the President of the Republic could no longer control his own people. He could not even control his wife who depleted one of the banks to build villas, a palace, plantations and other businesses for the Presidential couple and their numerous relations.[43]

Despite efforts by the president in early 1986 to allay the public's fear of an impending economic crisis by announcing that "le Cameroun se porte bien/Cameroon is healthy," and assertions by a World Bank delegation that Cameroon had no problems repaying its external debt,[44] all was not well with the nation's economy. In fact, the president made an about face from his early statement when, in his traditional New Year's message at the end of 1986, he called on Cameroonians to be prepared to make sacrifices as a necessary step toward solving the nation's economic problems.

Despite these problems, Cameroon was still receiving praise for its exemplary economy. For instance, on a visit to Cameroon in 1987, Jacques Chirac, then France's prime minister, described Cameroon's difficulties as a temporary economic crisis. Similarly, in a speech to representatives of the World Bank and the IMF, President Ronald Reagan cited Cameroon as an example of countries in Africa having undertaken significant economic reforms.[45]

Meanwhile, in an effort to close the gap between such illusion and reality and to address the country's economic problems, the administration introduced a series of reforms in 1986, including the introduction of a new computerized system within the customs service designed to curb corruption and improve the collection of tariffs. Instead of achieving its intended objectives, however, the new system led to a decrease in import revenue and a shortage of imports. Under the old system where there was no accurate check for accountability, businessmen could make special deals with individual custom officers (which often involved underpaying or in some cases not paying tariffs at all) in order to claim their goods. Much of the revenue collected in such deals went into the pockets of customs officers; it is estimated that between 1985 and 1987 the administration lost about CFA500 million francs in revenue because of customs fraud.[46] But instead of solving the problem, the reporting and collecting system's change curtailed the formerly brisk business and compounded the loss of revenue to illicit practices.

In yet another attempt to address the nation's growing economic problem, the investment budget was reduced by 26.5 percent, from CFA340,000 million francs for fiscal 1986–87 to 250,000 million for 1987–88. The decrease further intensified the nation's burdens because it meant greater unemployment and an end to many of Biya's New Deal development projects. The economic situation had repercussions in financial spheres. There was an exodus of many foreign businesses and financial institutions such as Chase Manhattan and Boston Banks, located in the country since the boom days of the early 1980s. By 1989, domestic institutions began to fail. Cameroon Bank and the Banque Camerounaise de Développement, owned entirely by the Cameroon government, were forced to close their doors because of financial mismanagement, and two more banks with government ownership of 70 and 60 percent of the shares folded, the Société Camerounaise de Banque (SCB) and Paribas-Cameroun.[47] The collapse of the SCB was the result both of mismanagement and of the presidential couple's use of the bank as their channel to procure money for themselves and their close friends and relatives without secured guarantees. According to the French magazine *Jeune Afrique Economie,* in just five months, March–July 1988, CFA1,750,000,000 francs from the bank's fund was used for the construction of a presidential palace in his hometown of Mvomeka'a.[48] The combined impact of structural defects and "privateer" opportunism compounded Cameroon's financial crisis, so much so that by the late 1980s many potential investors, especially Bamileke businessmen who had seen their domination of the economy gradually eroded in favor of businessmen from the president's Beti ethnic group, decided to utilize informal saving clubs (*tontines*) rather than put their money in banks.

In 1987, in a desperate attempt to save the economy, the administration introduced a series of austerity measures. For the first time since independence and reunification, the national budget for the ensuing year was reduced, by 18 percent, from CFA800 billion francs for the 1986–87 fiscal year to 650 billion for fiscal 1987–88.[49] In addition, civil servants who had either reached the mandatory retirement age or had served a total of 35 years in government were forced to retire. The administration also eliminated or cut back the use of telephones, service cars, water, electricity, housing, and other free services or allowances that many civil servants had previously enjoyed. Like the other measures, what most Cameroonians knew as "the war on Pajeros" (referring to the luxurious Japanese vehicle that most senior civil servants rode in) was intended to address the nation's financial shortfall. It was estimated that the cuts would result in savings of CFA30 billion francs by the end of the 1987–88 fiscal year and CFA56.9 billion by 1988–89.

In addition to these austerity measures, strategies to generate revenue included increases on import tariffs for items such as motor vehicles, meat, perfumes, cigarettes, and tobacco. Also in 1987, the government imposed a 300 percent tax increase on local products such as beer, soft drinks, and petrol. The projected new revenue on petrol alone was CFA12.5 billion francs in fiscal 1987–88 and CFA30 billion francs the following year, with beer and soft drinks expected to yield CFA8 billion and 20 billion francs during the same periods.[50] In 1988, to further rescue the situation, the administration was forced to seek additional financial support from the World Bank and the IMF, after initially refusing to do so. In return, Cameroon was required to implement an IMF Structural Adjustment Program (SAP), which among other things included more budget cuts, restructuring of the banking system, and privatization of many of the parastatal corporations, all aimed at reintroducing fiscal sanity in the economy.

While these measures may have been necessary to prevent the country from sliding into further economic crisis, some of them were very unpopular. For instance, the increase in the price of some imports and domestic products through tariffs and taxes resulted in their abandonment by most ordinary citizens. By contrast, the political elites and members of the upper echelon of government not only had the wherewithal to keep buying these items, but still retained most of their privileges.[51]

Confronted with the nation's increased unemployment and deteriorating economic situation, President Biya and his New Deal encountered popular backlash by the late 1980s and early 1990s:

The Biya government, which has the responsibility of providing health care, decent education, employment, pipe-borne water, electricity and roads has failed in its obligations. That . . . is the cause of today's and will be the cause of tomorrow's violence.

Its genesis is despair. Instead of investing in areas of social development, the Biya government put out money into state-owned corporations, white elephants and failed businesses. Its members stole what part of it was left to build castles in Europe, open bank accounts and erect sumptuous palaces at home. State patronage of the economy was pushed so much to the detriment of private initiative that with the "government economy" now out of shape, everything is out of shape.[52]

What one Biya critic as early as 1986 had voiced in reference to the economic stability Cameroon enjoyed under former president Ahidjo, stating he would be given "a hero's welcome" if he returned,[53] was now an increasingly national sentiment.[54]

Global Politics and Domestic Demands for Reform

While it is likely that the severe and painful economic conditions Cameroon and many African states experienced since the mid-1980s would have independently led to demands for change,[55] the process was accelerated by global political developments toward liberalization and regime transition in the late 1980s. After almost three decades of single party authoritarian rule in most African states, the end of the Cold War and the shift to more democratic governments in the former Communist nations of Eastern Europe served as catalysts for Cameroonians and other Africans to demand reforms, including the legalization of multi-party politics.

Also important in the campaign for political change in many African states, including Cameroon, were pressures from France (Cameroon's major trading partner) and the other Western capitalist states and financial institutions. It was politically expedient during the Cold War for these states and institutions to support authoritarian regimes in Africa, financially and otherwise, as a means of preventing the spread of Communism and Soviet influence on the continent. But with the end of the Cold War and the collapse of the Soviet empire, such support was no longer necessary. As one commentator noted:

Following the Cold War, dictators who had previously relied on the unquestioning support of the United States, the Soviet Union or France suddenly found themselves as clients in search of patrons. As the value of these dictators as pawns in the global chess game diminished, the former patrons were unwilling to underwrite authoritarian, warlike and abusive governments. The withdrawal of international support forced these dictators to confront internal pressure for change.[56]

The "new" attitude towards African nations was evident in a policy statement by Douglas Hurd, the British foreign and commonwealth secretary, at a conference on the prospects for Africa in the 1990s held at the House of Commons June 9, 1990, in which he indicated that future Western aid would go only to nations that "tended towards pluralism, public accountability, respect for rule of law and human rights."[57] Similarly, at the Franco-African summit at La Baule, France in 1990, President Mitterrand linked future economic relations with African partners to political reform and respect for human rights. One year later, at the fourth summit of francophone heads of states and governments held at the Chaillot Palace in Paris, democracy was again an important topic of discussion. The conference's final resolution, the Chaillot Declaration, declared support for human rights and the democratic process as an essential prerequisite for economic prosperity.[58] The World Bank and the IMF also indicated that future aid to Africa would not only be linked to economic restructuring, but also to political liberalization.[59]

Perhaps not surprisingly, President Biya, like most of his African counterparts, was opposed to any changes that threatened his authority and control of power. Although Biya had promised to create a "democratic" society when he took over the presidency, he believed that such a democracy should work within the parameters of a single party structure, adopting a familiar argument he had made before, that its demise would threaten national unity and lead to political chaos similar to that plaguing other African countries since independence.[60] He rejected calls for multiple parties, which he described April 9, 1990, as "a distasteful passing fetish" and "manoeuvres of division, intoxication and destabilisation."[61] He remained until the end of 1990 consistent with views published in 1987 that one party governance was "the best laboratory for a truly pluralistic democracy in Cameroon, the necessary prelude to multipartism."[62]

However, Biya's opposition to the establishment of a multiparty system failed to extinguish further demands for political change in Cameroon. In fact, while the president rejected demands for multiparty politics, a group of Cameroonians led by Yondo Mandengue Black, a lawyer and former president of the Cameroon Bar Association (CBA), had been meeting surreptitiously to explore possible ways of creating a nonpartisan group to promote the creation of a multiparty system.[63] The result was the birth of a group called the National Coordination for Democracy and a Multi-party System (NCDM).[64] But before the group could achieve its objectives, its members were arrested in February 1990 by agents of the security police (CENER).[65] They were charged with holding clandestine meetings, inciting revolt, abusing the president, and producing and distributing pamphlets that were hostile to the regime, all in broad violation of Article 2 of Law No. 62/OF/18 of March 1962.

The arrest and eventual trial of Yondo Black and his associates, the "Douala ten" as they became known in the independent press, produced Cameroon's greatest "show trial" since Bishop Albert Ndongmo of Nkongsamba and Ernest Ouandié (the last true leader of the UPC resistance movement) were tried in 1970. It brought severe criticism of the administration from members of the CBA, various international human rights organizations, and the international community. The Bar Association called for the accused to be released and condemned the search of Black's office as a violation of Law No. 87/018 of July 15, 1987, which prohibited the search of a lawyer's office unless the lawyer was the subject of a formal judicial investigation.[66] The lawyers also threatened to boycott all other court matters during the trial.

By contrast, CPDM militants and government officials denounced the accused. In a March 10, 1990, statement, the administration refuted reports by some foreign news media that the detainees had been arrested for trying to form a political party, which was legal under Article 3 of the constitution. Rather, the statement noted that they had been arrested for "holding clandestine meetings and for printing and distributing leaflets hostile to the government, [which] showed contempt for the President of the Republic and for inciting revolt."[67]

In a massive show of support for the administration reminiscent of the Ahidjo era, CPDM cells throughout the country organized progovernment demonstrations in support of the trial. In each of the demonstrations, the defendants, their lawyers, and others sympathetic to their cause were portrayed as "trouble-makers," "adventurers," and "selfish demagogues."[68] One of the largest demonstrations by the Yaounde section of the party heard speaker after speaker denounce the "Douala ten" as agents of foreign powers, and call on the president to continue with the democratization process he had started in November 1982.[69]

After almost six weeks in detention, the accused were finally brought to trial before a special military court in Yaounde on March 30, presided over by Lieutenant-Colonel Adamou Nchankou and assisted by two assessors, Lieutenant-Colonels Antoine Onana and James Chi Ngafor. The proceedings were suspended after complaints by defense lawyers about insufficient time to prepare their cases. When the trial reopened three days later, defense lawyers were given little opportunity to defend their clients. In the end, despite insufficient evidence by the prosecutors, three defendants were found guilty and given various prison sentences, two others were given suspended prison sentences, while the rest were acquitted.[70] Amnesty International described the trial as a mockery and "an abuse of the judicial process in order to provide some legal basis for the government's determination to punish supporters of the multi-party system."[71]

Rather than intimidating other potential political opponents, the arrest and trial of the "Douala ten" served as a further catalyst for the democratic movement in Cameroon.[72] Most crucially, in response to a March 10 statement by the administration that it had never forbidden the formation of any other party in the country, John Fru Ndi, a book store owner in Bamenda, North West Province, and his supporters seized the opportunity to file a petition with local government authorities requesting official recognition for a new party, the Social Democratic Front (SDF), claiming rights under Article 5 of Law No. 67/LF/19 of June 12, 1967.[73] They also indicated that they would go ahead with the decision to launch the party and consider it legally constituted under Article 3 of the constitution if officials did not respond to their request within two months.

However, rather than granting the request, the minister of territorial administration and other government officials attempted to dissuade Fru Ndi from implementing his plans. But after every effort to stop the launching had failed and no permit had been issued, and after the party's delays in an effort to avoid violence, Fru Ndi went ahead May 26, 1990, despite the presence of 2,000 troops dispatched to Bamenda in an open show of deterrent force. A crowd estimated by its supporters at between 30,000 to 40,000[74] gathered at Ntarinkon Park near Fru Ndi's compound, two kilometers from Commercial Avenue where the action was originally scheduled before troops sealed off the site and forced the diversion.

Apparently frustrated at having been outfoxed, soldiers fired tear gas on crowds attempting to enter Commercial Avenue following the successful launching of the SDF at Ntarinkon Park. When the crowd failed to disperse and threw rocks, troops opened fire and bullets killed six young adults. A pro-SDF demonstration by students at the University of Yaounde on the same day, mostly anglophones but joined by francophones, resulted in several arrests. According to some reports, three of those arrested were tortured to death, while a female student was raped by the soldiers.[75]

The killing of "the Bamenda six" drew both domestic and international condemnation and prompted Ahidjo's partner in reunification, still the anglophone elder statesman as vice-president of the CPDM, John Ngu Foncha, to resign that post. In his letter of resignation to Biya, Foncha noted that he had become an "irrelevant nuisance," ignored and ridiculed by the administration. He also noted that the events of May 26 in Bamenda had reinforced that irrelevancy, despite his historic role in bringing about the reunification of Cameroon:

... In order to completely disgrace me and show my irrelevance, troops were deployed in Bamenda in large numbers to harass everyone in Bamenda of Bamenda origin, myself not excluded, and prevent the launching of a new political party. This was done in spite of my protestations and advice in order to completely ridicule me in my own constituency and destroy all I have stood for all my life *ie* democracy, respect for human rights, avoidance of violence and respect of the constitution, in order to allow peace and progress which God has bestowed in Cameroon on a platter of gold.[76]

Despite these condemnations of its use of force and the six deaths, the administration justified the actions taken by the soldiers. For instance, the minister of territorial administration, Ibrahim Mbombo Njoya, described the launching of the SDF as illegal, in violation of Article 7 of the June 1967 law prohibiting any association or party from functioning before it had been officially approved. He also accused the SDF of being "clannish" because a majority of SDF activists were from the same division, a violation of Article 4 of the same law prohibiting the formation of associations having an exclusively clannish or tribal character, or those eliciting causes or objectives contrary to the laws or good customs, or endangering the government or the national integrity. Finally, the minister charged outside forces with financing the SDF and other political movements demanding multipartism in Cameroon.[77] Meanwhile, in an effort to extricate the administration from responsibility for the Bamenda deaths, an editorial in the government's *Cameroon Tribune* stated that the victims had not been shot by soldiers but were trampled upon by the crowd as security forces attempted to establish normalcy on the streets of Bamenda following the unauthorized launching of an illegal party.[78] CPDM meetings and rallies throughout the country also condemned the "unauthorized" launching of the SDF. In a May 29 meeting, for example, the executive of the Fako, South West Province, section of the party appealed to all CPDM militants there not to "open their doors to the unimaginary [sic], hypocritic tendencies of foreign democracies, and their lackeys and stooges." It also condemned attempts to introduce multipartism in Cameroon because such a move would only lead to "tribalism, favouritism, sectionalism and demagoguery." In a similar message of support, the CPDM executive of Donga and Mantung, North West Province, section not only expressed its support for the president and the policies he had pursued since November 1982, but described those responsible for the events of May 26 as "misguided citizens" whose pursuit of multiparty politics in Cameroon was simply fuelled by personal ambitions.[79]

Nevertheless, the events of May 26 created greater political awareness among Cameroonians and increased demands for the legalization of a multiparty system in Cameroon. A pastoral letter signed by nineteen

Catholic bishops and published on June 3 noted that the church could no longer remain indifferent in light of the ordeal that the nation's economic crisis had brought on innocent individuals and families.[80] Although that letter made no reference to multipartism, Christian Cardinal Tumi, the archbishop of Garoua and Cameroon's primate, used a June 11 press conference to condemn the Bamenda killings and call for the creation of a multiparty system. He also appealed to the administration for an end to censorship and a change in the policy which had led to the nation's economic crisis.[81]

Perhaps in response to the events of May 26 and demands for multiparty democracy in Cameroon, Maïgari Bello Bouba, Biya's first prime minister, in Nigerian exile following his alleged involvement in coup attempts against Biya and departure from government in 1983, announced the formation of a new party, the National Union for Democracy and Progress (NUDP). Then on June 25, 1990, the Cameroon Democratic Front (CDF) was formed in Paris by Jean-Michel Tekam, who had been sentenced in absentia to five years in prison for his involvement in the Yondo Black affair.

Public demands for the creation of a multiparty system in Cameroon continued throughout 1990, especially following the legalization of multiparty politics in neighboring countries. In fact, demands for change in Cameroon seem to have renewed the rift we saw earlier at the 1985 Bamenda Congress between progressive and conservative wings of the CPDM.[82] Although it is unclear whether progressives supported the establishment of multiparty politics in Cameroon at the time, they did favor greater political reform than Biya's thus far. By contrast, conservatives led by many former members of Ahidjo's administration and officials from Biya's Beti ethnic group were against any political changes that could upset the current balance of power or threaten their authority. In fact, while most of the progressive members of the party perceived the nation's economic problems as a manifestation of the country's political problems, the conservatives saw the crisis as exclusively economic, and dismissed suggestions that the problems could be resolved by instituting greater political reforms.[83]

These public movements and the wide party division clearly showed that Ngayap's "classe dirigeante" for the first time in many years faced a challenge neither rising on narrow bureaucratic or technocratic grounds nor easily amenable to internal management. There were strains in the comprehensive Ahidjo alliance which had assembled regions and ethnicities, circulated opportunities in the political domain, and created the key links between North and South, francophone and anglophone Cameroon.

The persistent pressure for change finally appeared to have an impact on the president. Although Biya continued his opposition to a multiparty system in Cameroon, he was willing to institute some policies which he hoped might mollify demands for more radical change. His opening remarks to participants at the First National Congress of the CPDM, June 28–30, 1990, in Yaounde (labelled the "Congress of Freedom and Democracy"; Bamenda's 1985 New Deal Congress was summoned while the CNU still operated), marked the first such occasion. While he dismissed influences for liberalization and reminded delegates that Cameroon's democratization efforts preceded eastern Europe's, Biya indicated that he intended to repeal certain aspects of the March 12, 1962, anti-subversion law and the June 1967 law restricting freedom of association. Other proposed reforms included guarantees for freedom of the press, abolition of exit visas for Cameroonians wishing to travel abroad, and the formation of a government human rights committee to monitor domestic abuses. Also, the CPDM Central Committee was increased from sixty to 120 members, with twenty alternates.[84]

Among new members elected to the Central Committee at this congress were four women and a number of prominent chiefs from the two anglophone provinces. While an increase in the membership of the Central Committee may have been intended to erase the image of the Central Committee as an "elitist club" consisting of people who had lost touch with many ordinary Cameroonians, the inclusion of many prominent anglophone chiefs was perceived by some analysts as an effort by the regime to neutralize support for the SDF in the two provinces. Perhaps the most significant outcome of the Yaounde congress and the first indication that the president might be willing to authorize the formation of other parties in Cameroon came when he informed party members that the CPDM must be prepared to face competition in the future.[85]

But despite the proposed reforms, the expansion of the Central Committee, and the release from jail of members of the "Douala ten" on August 12, Biya failed to stem the tide of discontent. There were antigovernment demonstrations in most major cities and towns protesting the failure to grant full amnesty to political prisoners implicated in the April 1984 attempted coup. In a letter to the president, a group calling itself "the intellectuals" of Cameroon called on Biya to free all political prisoners, close all detention centers, legalize new political organizations, grant amnesty to political exiles, and convene a national debate on the political future of the country.[86] Most of the demonstrations and protests were brutally suppressed by the police and the military. But as they became more frequent, and as external pressure from France and other Western nations and financial institutions also intensified, Biya was forced to make further concessions, not for the first or last time appearing to yield grudgingly to initiatives from others, rather than asserting his own leadership.

Finally, on December 5, the National Assembly approved a series of bills introduced by the president, who signed and proclaimed them "Liberty Laws" on December 19.[87] Close to 100 laws came into force that day, altering Cameroon's political and economic life in both substantive and technical ways, following hundreds of decrees and ordinances issued earlier in the year by the presidency. Two areas of the December 19 enactments were crucial, touching the press and parties.

Regarding newspapers, Law No. 90/046 repealed Ordinance No. 62/OF/18 of March 1962 with its power to "repress subversive activities," thereafter used to restrict freedom of expression, especially the press. Even so, Section 14 of the companion 1990 Law No. 90/052 labelled "Freedom of Mass Communication" required newspapers to submit two signed copies or brush-proofs to administrative authorities where the paper was published, two hours before dailies and four hours before weeklies were placed on sale. Material considered subversive or politically inappropriate was removed or rendered illegible. With so little time between scrutiny and sale, it was common to find blank and darkened areas, indicating censorship. Section 17 also gave the minister of territorial administration the authority to seize or ban any publication which defied the review process. Publishers could appeal such action to the courts, but rulings took at least a month, which rendered the material obsolete.

December 19's *most* fundamental landmark, Law No. 90/056, finally reversed Biya's stand against multiparty politics and created procedures to enable new parties. But here too, as in the press laws, there were limits, permitting no participation or support activities from outside Cameroon, and prohibiting parties with a single ethnic, linguistic, religious or provincial base. These provisions appealed to unity on their face and resembled political party law elsewhere in Africa, but in fact challenged the party formation already under way in Cameroon. One provision did remove the impasse created when the regime failed to recognize the SDF, and the bloody consequences; new parties were authorized to start functioning three months after filing petitions with the minister of territorial administration, whether or not they were approved.

This legislation closed the most turbulent year in Cameroon since the coup attempt and its aftermath in 1984, and before that the years of regime-UPC armed struggle. Biya's retrospective view of 1990 and his preferred interpretation of its events can be surmised from his remark when visiting France in April 1991. Referring to President Mitterrand's call at the La Baule meeting in 1990 for francophone African leaders to pursue the course of democracy in their respective countries, and the "Liberty Law" developments in Cameroon which the regime now called "Advanced Democracy," Biya described himself as one of President Mitterrand's "best pupils" in democratization. We will see that not all

Cameroonians agreed with their president on this, the first of his many notable remarks as Cameroon's crisis intensified in 1991 and continued in the following years.

Notes

1. *Cameroon Life*, November/December, 1991.
2. *Cameroon Post*, August 12–19, 1991.
3. Colin Legum, ed., *Africa Contemporary Record 1983–84* (London: Rex Collings, 1984) p. B333.
4. *West Africa*, April 22–28, 1985, p. 778.
5. *West Africa*, September 22–28, 1986, p. 1972.
6. *Africa Research Bulletin: Political Series* 24, 2 (February 1–28, 1987), p. 8400.
7. Joseph Takougang, "The Post-Ahidjo Era in Cameroon: Continuity and Change," *Journal of Third World Studies* 10, 2 (1993), p. 292.
8. *West Africa*, April 22–28, 1985, p. 778.
9. U.S. Department of State, *Country Reports on Human Rights Practices for 1988* (Washington, D.C.: U.S. Government Printing Office, 1989), p. 48 for the Ateba and Mukong cases.
10. Jean-François Bayart, "One-Party Government and Political Development in Cameroon," in Ndiva Kofele-Kale, ed., *An African Experiment in Nation-Building: Tha Cameroon Federal Republic Since Reunification* (Boulder: Westview Press, 1980), p. 163.
11. *Africa Research Bulletin: Political Series* 22, 1 (January 1–31, 1985), p. 7493.
12. Victor T. Le Vine, "Leadership and Regime Changes in Perspective," in Michael Schatzberg and I. William Zartman, eds., *The Political Economy of Cameroon* (New York: Praeger, 1986), p. 42.
13. *West Africa*, December 12–18, 1983, p. 2872.
14. See Biya's speech in Garoua on May 4, 1983, in CNU Political Bureau, *The New Deal Message* (Yaounde: Sopecam, 1983), p. 446.
15. President Biya's closing speech to the second extraordinary congress of the CNU, September 14, 1983 in CNU, *The New Deal Message*, p. 525.
16. Jean-François Bayart, "Cameroon," in Donal Cruise O'Brien, John Dunn and Richard Rathbone, edc., *Contemporary West African States* (Cambridge: Cambridge University Press, 1989), p. 34. From about 1980, surfacing first at the university, then spreading into party circles and the press, debates increasingly keyed to such language registered Cameroon's rising tensions, with ethnicity at their core, especially in the polemical writing of two philosophers, the Beti Hubert Mono Ndjana and the Bamileke Sindjoun Pokam. These exchanges continue in the mid-1990s.
17. *Africa Confidential*, June 15, 1990.
18. There were reportedly four different Beti pressure groups vying for influence. One was led by General Benoît Asso'o Emane, commander of the military garrison in Yaounde, a second was led by Henri Damase Omgba, nicknamed "the vice president" because of his tremendous influence on the president, a third was

led by Gilbert Andze Tsoungui and Titus Edzoa, minister of armed forces and special advisor to the president, respectively. A final group was led by the president's wife, Jeanne-Irène Biya.

19. *Le Messager*, August 3, 1991.
20. *Le Messager*, March 16, 1992.
21. *Cameroon Life*, September, 1991.
22. For discussion of the anglophone problems see Stark, "Federalism in Cameroon: The Shadow and the Reality," in Kofele-Kale, *An African Experiment*, pp. 101–132; Bayart, "The Neutralisation of Anglophone Cameroon," in Richard Joseph, ed., *Gaullist Africa: Cameroon Under Ahmadu Ahidjo* (Enugu, Nigeria: Fourth Dimension, 1978), pp. 82–90; Jacques Benjamin, "Le Fédéralisme Camerounais: L'Influence des Structures Fédérales sur L'Activité Economique Ouest-Camerounaise," *Canadian Journal of African Studies* 5, 3 (1971), pp. 281–306.
23. *West Africa*, April 29–May 5, 1985, p. 829.
24. *Le Messager*, July 13, 1992.
25. For a discussion of this point of view see Albert Mukong, *The Case for the Southern Cameroons* (USA: CAMFECO, 1990), pp. 98–99.
26. *West Africa*, May 6–12, 1985, p. 876.
27. Nantang Jua, "State, Oil and Accumulation," in Peter Geschiere and Piet Konings, eds., *Pathways to Accumulation in Cameroon* (Paris:Karthala; Leiden: Afrika-Studiecentrum, 1993), p. 155.
28. *West Africa*, September 12, 1983, p. 2017.
29. Nancy C. Benjamin and Shantayanan Devarajan, "Oil Revenues and the Cameroonian Economy," in Schatzberg and Zartman, *The Political Economy of Cameroon*, p. 165.
30. Philippe Gaillard, *Ahmadou Ahidjo (1922–1989)* (Paris: Groupe Jeune Afrique, 1994), p. 183.
31. *West Africa*, March 2–8, 1987, p. 415.
32. *Jeune Afrique* cited in Nantang Jua, "Cameroon: Jump-starting an Economic Crisis," *Africa Insight* 21, 3 (1991), p. 164.
33. Mark DeLancey, *Cameroon: Dependence and Independence* (Boulder: Westview Press, 1989), p. 141.
34. *West Africa*, June 27–July 3, 1988, p. 1158.
35. The extrabudgetary treatment of oil revenue became a political issue for Biya, as his other defects turned attention to the secrecy in this discretionary reserve fund, with suspicions about its use.
36. *West Africa*, May 18–24, 1987, p. 954.
37. *West Africa*, June 27–July 3, 1988, p. 1158.
38. For example see Richard Joseph, "Economy and Society," in Joseph, *Gaullist Africa*, p. 144. Also see Neville Rubin, *Cameroon: An African Federation* (London: Pall Mall Press, 1971), p. 176.
39. *New York Times*, September 7, 1987.
40. For a detailed discussion of this system of fraud and other government excesses, see Jua, "Cameroon: Jump-starting an Economic Crisis," pp. 164–167.
41. *Africa Confidential*, March 9, 1990.
42. Cited in *La Gazette*, July 31, 1990, p. 15.

43. *Le Messager*, June 9, 1992.
44. *West Africa*, January 5–11, 1987, p. 14.
45. *West Africa*, March 7–13, 1988, p. 411.
46. *West Africa*, August 7–13, 1989, p. 1288.
47. Jua, "State, Oil and Accumulation," in Geschiere and Konings, *Pathways to Accumulation*, p. 153. Further vulnerability surfaced when the Bank of Credit and Commerce International collapsed in 1991, for the government had a 3 percent equity stake in its Cameroon operation; *The Wall Street Journal*, August 6, 1991.
48. *Jeune Afrique Economie*, May 1992, p. 126; we will meet one of the story's sources, Célestin Monga, again below.
49. *West Africa*, November 9–15, 1987, p. 2212. There was another steep decline in 1990–91, although French assistance thereafter kept total budget expenditure levels constant in the 500 billion range through the mid-1990s.
50. *West Africa*, March 7–13, 1988, p. 410.
51. *Cameroon Life*, January 1992.
52. *Le Messager*, August 3, 1991.
53. *West Africa*, October 19–25, 1986, p. 2062.
54. To anticipate later coverage, Ahidjo praise effigies appeared far from his northern base of power, indeed in presumptively hostile North West Province, when anti-government demonstrations broke out in 1991.
55. *West Africa*, May 17–23, 1993, p. 810. See also Jean-François Bayart, *The State in Africa: The Politics of the Belly* (London: Longman, 1993), p. xi.
56. *Africa Watch Overview*, 1992, p. 33.
57. *West Africa*, June 25–July 1, 1990, p. 1077.
58. *Africa Research Bulletin: Economic Series* 28, 11 (November 16–December 15, 1991), p. 10610.
59. *Time Magazine*, May 21, 1990, p. 34.
60. Biya, *Communal Liberalism*, pp. 44–45.
61. *West Africa*, July 9–15, 1990, p. 2065, for this text and the account of his next critique of multiparty challenges.
62. Biya, *Communal Liberalism*, p. 46.
63. See Amnesty International Report on the "Yondo Black Affair," July 23, 1990, p. 3. In a lengthy article marking the fifth anniversary of the arrest, Francis Kwa-Moutome confirmed that one of the objectives of the group was to form a political party, but because of the ideological differences between the members, they decided instead to form a non-political group calling for democracy; *La Nouvelle Expression*, February 21–24, 1995.
64. Albert Mukong, *My Stewardship in the Cameroon Struggle* (Enugu, Nigeria: Chuka Publishing, 1992), p. 128.
65. Besides Black, others arrested were Anicet Ekane, Charles René Djon Djon, Rudolphe Bwanga, Albert Mukong, Henriette Ekwe, Gabriel Hamani, Francis Kwa-Moutome, Vincent Feko and Julienne Bejem. Another member of the group, Professor Jean-Michel Tekam, had left for France just prior to the arrests.
66. See Amnesty International Report on the "Yondo Black Affair," p. 5.
67. *Cameroon Tribune*, March 16, 1990; see Appendix 1 for details of the letter.
68. Amnesty International Report on the "Yondo Black Affair," p. 6.

The State Under Pressure, 1986–1990

69. See *Cameroon Tribune* of April 1–2, 1990, for some of the speeches.
70. Yondo Black and Anicet Ekane were given three and four years prison sentences, respectively, while Jean-Michel Tekam was sentenced in absentia to five years in prison. Charles René Djon Djon and Rudolph Bwanga were given suspended prison sentences.
71. See Amnesty International Report on the "Yondo Black Affair," p. 9.
72. *La Nouvelle Expression*, February 21–24, 1995.
73. Mukong, *My Stewardship*, p. 135. This is one source among many for the parallel, convergent, and divergent paths of the Yondo Black and SDF stories, which need close study in order to reach a definitive account of emerging multiparty politics in Cameroon.
74. *Africa Confidential*, June 15, 1990. On the other hand, the administration estimated the crowd at about 20,000.
75. *West Africa*, September 3–9, 1990, p. 2398. Similar demonstrations by students at the University of Yaounde became a common occurrence throughout the 1991 academic year.
76. Foncha's letter of resignation was widely published in Cameroon and summarized in *West Africa*, June 12–18, 1990, p. 1090.
77. *Cameroon Tribune*, June 1, 1990.
78. *Cameroon Tribune*, May 29, 1990.
79. *Cameroon Tribune*, June 1, 1990, for these passages.
80. *Marchés Tropicaux*, June 8, 1990, p. 1609.
81. *Africa Confidential*, June 15, 1990. Perhaps not surprisingly, Archbishop Jean Zoa of Yaounde and a member of the President's Beti ethnic group, indicated in an interview with the *Cameroon Tribune* that Cardinal Tumi was expressing his personal opinion and not that of the Catholic Church in Cameroon.
82. CPDM progressives now included Jean-Jacques Ekindi, president of its Wouri section, Thomas Melone, member of the Legislative Assembly, Sadou Hayatou, secretary-general at the presidency; Joseph Fofe, minister of youth and sports and (in a surprising role we will see again) François Sengat Kuo of the CPDM Political Bureau. The conservatives, mostly from Biya's ethnic group, included Gilbert Andze Tsoungui, minister of territorial administration, Titus Edzoa, special advisor to the president, Lieutenant-Colonel Benae, military chief of staff at the presidency; General Benoît Asso'o and Colonel Ebogo.
83. *Cameroon Analysis*, March 20, 1991, p. 7.
84. The 120 members included all forty-nine of the party's section presidents, six permanent members (membre de droit), and fifteen members appointed by the president, with the rest to be elected at the congress.
85. *Africa Research Bulletin: Political Series* 27, 7 (July 1–31, 1990), p. 9760.
86. *West Africa*, July 30–August 5, 1990, p. 2210.
87. *Republic of Cameroon. Rights and Freedoms: Collection of Recent Texts* (Yaounde: SOPECAM, 1991) provides the texts of key legislation at the end of 1990, from which the following section draws.

5

Crisis Years, 1991–1992

We have established how decisive the year 1990 was for Cameroon. After more than a quarter century marked by the governance which Ngayap's "classe dirigeante" and Bayart's "recherche hégémonique" best identify for goals, personnel and institutions, events in 1990 and their consequences revealed quite different forces at work. Two Cameroons were juxtaposed, with patterns of fundamental continuity before 1990 facing the pressure for change ushered in that year. The CPDM's 10th anniversary since its formation in Bamenda in 1985, observed late in March 1995, provoked commentary across the decade's spectrum of experience and opinion. Two competing March 1995 versions of what had happened under CPDM rule help introduce all the text which follows, as we begin here to assemble details about state and civil society which reveal Cameroon as a pivotal case study in Africa's democratization struggle of the 1990s.

A CRTV radio interview March 23, 1995, stated masterfully, in five minutes, the case supporting the CPDM. The setting was worth specifying, a "hot line" call-in radio show in English on the national network, livelier in its conduct, freer in its content than any pre-1990 offering from government's telecommunications monopoly. A Yaounde CPDM official argued that after twenty-five years of authoritarian rule under Ahidjo, Biya turned the corner in the mid-1980s. His creation of the CPDM, then its New Deal and Communal Liberalism phases, educated a population requiring tutelage and constituted an experimental phase of gradual, prudent democratization, completed by the "Liberty Laws" on the press and parties of late 1990. "Advanced Democracy" implemented further advances, especially through two episodes, 1991's restoration of the prime minister as head of government to invigorate the National Assembly and balance the power of the president as head of state, and 1992's multiparty elections for the National Assembly and presidency. The speaker then castigated two forces in particular, a hostile private press since 1990 and the 1991 direct action known as "villes mortes/ghost towns," for interrupting and compromising Advanced Democracy through lies and violence which cost Cameroon dearly and made both domestic and foreign waters hard for the CPDM to navigate. But, he concluded, the party persevered despite impatience and intolerance, and its struggle will prevail against the manifold crisis at this mid-1990s juncture. This was *étatisme* fully unfurled.

It was the starkest possible contrast with an encounter one week earlier, March 16, in a Bamenda taxi. A nursery school teacher recounted her son's troubles at the University of Dschang since the 1994–1995 academic year began, with few classes, scanty facilities, and press reports about phantom administration, corrupt finance, and the vice-chancellor's attempt to shift the entire operation to his own locale of origin near Bafoussam, forty miles away. The teacher's summation, embracing Cameroon's broader spectrum of experience this represented to her: "It is not poverty, it is wickedness!"[1]

She was next encountered four months later, July 12, 1995, at the SDF party's Ayaba Ward monthly meeting, where forty-five people, twenty men and twenty-five women, fifteen-twenty of them between the ages of eighteen and thirty, overflowed the meeting room of the host, Ayaba's quarter head in the customary local governance attached to the palace of the fon (chief) of the village of Mendankwe. The meeting comprehensively demonstrated civil society's local emergence, meshing formal political and informal cultural experience. It opened with Christian prayer, used Pidgin speech throughout, and closed with traditional Mendankwe food. It was directed by three women on the ward's executive board, two of them civil servants (a nurse and our teacher) who endured pay arrears and working conditions they considered the regime's fault, the third a farmer. As ward secretary, the teacher summarized resolutions from the SDF's recently concluded National Congress at Maroua, which argued the total bankruptcy of CPDM governance. All these features very closely matched the style of the region's myriad private "savings and loan" societies known as *njangi*, corresponding to the better known *tontine* among francophones, and the food recalled our earlier metaphoric framework. Gathering traditional village leadership, distressed adult civil servants and under- or unemployed young men and women in a familiar local setting in the SDF's core area, the meeting broadly reflected the party's gender, age, status, and occupation demography five years after its confrontation with the regime in 1990 triggered a resistance movement firmly entrenched by 1995. It also served our teacher as a political stage to counter CPDM wisdom like that on national radio months before.

With these diametrically opposite, hegemonic and counter-hegemonic versions of Cameroon's reality in place, one the product of Yaounde's

Crisis Years, 1991–1992

government mass communication monopoly, the other from Bamenda's mass political mobilization, we can begin our coverage of 1991–1992. The Yondo Black case, the SDF's bloody birth, and laws on the press and political parties began this new era of Cameroon's experience in 1990. Much of this chapter will focus on parties and their popular manifestations, then on institutional and electoral matters through 1992. But perhaps the best synoptic introduction to these years is through the less obvious medium of journalism.

As we saw, the Freedom of Mass Communication Law No. 90/052 of December 19, 1990, enacted the same day as Law No. 90/056 enabling party formation, made it easier to start and publish newspapers but still bound journalism to many Ministry of Territorial Administration (MINAT) powers. A test case appeared within days, and demonstrated the intricate bonds between the press and politics as challenges to the regime mounted. Pius Njawe, publisher of *Le Messager*, the most critical paper throughout the 1980s, printed the writer-economist Célestin Monga's broadly accusatory "Open Letter" to Paul Biya, December 27, 1990. Both were charged with bringing Biya into disrepute and disturbing public order. Their January trial, ending in fines and suspended jail terms, spawned defence committees, brought thousands of protesters and troops using guns into the streets as far away as Garoua, and became a key democratization rallying point.[2]

This episode established print journalism's place in the protest repertoire. The regime added newspapers to its staple *Cameroon Tribune* but independent newspapers proliferated.[3] The same impulse and rhythm animated them as political parties, and they came fundamentally to record, define, even play a role in creating the fault lines between state and civil society, and defining democratization, in the 1990s. Since they provide much of our evidential base for those years, and as prelude, then, to extensive political analysis, we take some care to establish the character and value of independent newspapers, and their significance in public affairs, before moving to a more formal and conventional study of politics.[4]

The Press

Few Cameroonians in 1989, a year before Law No. 90/052 took effect, could have predicted what the press would be like in late 1991, a year thereafter. The government's *Cameroon Tribune* was in 1989 not just dominant; this core platform for state and party was practically monopolistic, if not statutorily so like CRTV's place in *non*-print communication. Six times weekly in French, twice in English, with a weekend features edition, it dwarfed nongovernmental papers like *L'Effort Camerounais* (from the Catholic Church, since 1955), *Cameroon Post* (since 1969), *Le Combat-*

tant (since 1976), *Le Messager* (since 1979), and a few other less well established private newspapers, some appearing regularly, some not.

Cameroon Tribune was a carefully crafted work of persuasion, meant to impart to Cameroonians a sense of consistency, indeed serenity, in high places. The standard front page used small photographs of Biya and brief texts from his writings or the public record of his governance and larger photographs bearing on public persons or events. These were often of Biya himself with some foreign or local dignitary in the Etoudi Palace reception room with its sofa and chairs, carpet, table, vase and flowers, not opulent by most Franco-African standards, indeed sobered by Biya's uniformly "western" dress. Print complemented photography at or near the front of *Cameroon Tribune*, with greetings sent to or received from other heads of state for some "national day" or its equivalent, in ritual, repetitive language concluded with the words "renewed assurances of my highest esteem" or some close variant.

This scenario, a counterpart to Biya's display in pictures on walls and on robes for men and wrappers for women throughout the republic (the latter are produced by a parastatal; the most florid are called "Purple Pauls" by critics), cued the public expression of national unity. There is no need to elaborate here the hegemonic science and art of this kind of African "presidential monarchist" display.[5] Suffice it to place at the center of such efforts in Cameroon a ranking scholar and practitioner of semiotics, Jacques Fame Ndongo, returned from doctoral work in France c.1980 and posted in the next decade, often simultaneously, as Director of the Advanced School of Mass Communications, editor of *Cameroon Tribune*, and in key offices at the presidency. One clearly grasps his and the regime's view of authority and order near the conclusion of his major scholarly work, *Le Prince et le Scribe* (1988, surely styled as a modern Machiavellian guide for Africa):

> Un consensus est nécessaire: que le scribe accepte la nécessité de l'Etat et que le Prince favorise le débat d'idées, source de progrès et rempart contre la sclérose de l'Etat. . . . C'est la condition *sine qua non* d'un progrès authentique et durable. Ceci ne signifie pas que la fonction du scribe soit, par essence, de s'opposer ou de critiquer/ A consensus is necessary: let the scribe accept the necessity of the State and let the Prince encourage the debate of ideas, as source of progress and bulwark against sclerosois of the State. . . . It is the condition *sine qua non* of authentic and lasting progress. This does not mean that the scribe's function ought, by nature, to be oppositional or critical.[6]

Cameroon Tribune faithfully reproduced Fame Ndongo's creed. Despite *Le Messager* and its independent counterparts, it basically commanded this field until (we will now see) the ground was contested in the wake of Law No. 90/052, despite the restrictions it imposed.

A year after the press law, at least sixty more newspapers from the private sector challenged *Cameroon Tribune*.[7] If they did not drown it out, they much reduced the impact of its previously central role in news production and distribution. One prime example made the point, five months into the new era, when *Cameroon Post*, May 30, 1991, cited a Paris source, *Jeune Afrique Plus*, to the effect that *Le Messager* now printed 90,000 copies a week, while *Cameroon Tribune* sold just 20,000 copies of its French edition daily. These two were now about equivalent in weekly production. If one adds the impact of the dozens more independent papers by then operating, even allowing for limited reading and buying habits, the general point is clear: there was a substantial transformation in the character and circulation of news in print, which questioned official sources and created an alternative "public record."[8] By February 1995, *Cameroon Tribune* fell from 1990's standard eight appearances a week with 120 pages in the two official languages to just five times weekly with forty pages. By contrast, *Le Messager* retained its once a week twelve page format of 1990, increased it to sixteen pages for a time, and intermittently printed an English language edition from 1991.

Independent papers waxed and waned in the years after 1990, as grinding effort, competition, and economic hardship took their tolls. But a dozen or so appeared consistently once or twice a week when not banned (*The Herald*, a 1992 entry, published three times weekly by 1997), and were decently patronized or at least scanned if not bought at large city kiosks. One key consequence of opening the press to private enterprise was the large Bamileke presence in critical journalism as entrepreneurs and writers, from an ethnic group harboring special grievances against Biya and the Beti over exclusion from state opportunity channels since Ahidjo's eclipse.[9]

Génération, February 8–14, 1995, in its twenty-fourth weekly appearance, showed how deeply the surface of the official journalism dominant in the 1980s was now scored by determined new ventures and their investigative press policies and techniques. Five pages broke the news of the impending liquidation of La Société des Transports Urbains du Cameroun (SOTUC), the parastatal which had since 1973 supplied public bus transport in Douala and Yaounde, but now operated with under 20 percent of its peak fleet, clientele, and staff, and irretrievably in debt. Documentation for the lead story about "l'acte d'euthanasie" to be performed two weeks later was from Auxence MEKONGO, an eponymous alias *Génération* adopted for such sources, here identified as follows:

> Sous ce pseudonyme se cache un haut responsable du ministère des Transports qui, pour des raisons évidentes, a requis l'anonymat/Under this pseudonym is to be found a high Ministry of Transport official who, for obvious reasons, has required anonymity.

The news staff's own four page autopsy followed, with seven reporters covering finances, personnel, inventory, projected impact, and comparative perspective based on other countries' experience. SOTUC's demise did come to pass; instructively, it was announced for February 22 but busses actually shut down February 20, thwarting possible last day protests from sacked workers and travellers facing one more cut in living standards (surviving busses were then offered to waiting entrepreneurs, but their condition and the disrepair of roads made even their deflated prices unattractive).

Serving to exemplify both the "leakage" which made previously secret or privileged governmental activity public and the investigative reportage of the 1990s, the SOTUC story reflected Cameroon's sea change in information access and use, no small factor in the experience this chapter documents, with its heavy reliance on precisely such private press sources which critics labelled politically partisan and sensational, hinting at opposition party origins.[10] But they missed the fundamental point of independent journalism's key "fourth estate" role. Government by secrecy and privileged knowledge became difficult to sustain; it is easy to believe what was rumored after 1990, that voices replaced paper to a significant degree for communication in high regime echelons, to foil the press.

It should be noted that the private press also tracked the opposition parties and personalities critically, even as it framed the critique of the regime and especially savaged Biya and the CPDM in language and visual art.[11] Regime and opposition alike became targets in both the written and cartoon or caricature formats that spread widely; satiric visuals were used by all major papers and especially in *Le Messager's* supplement *Popoli*, emulating France's *Le Canard Enchaîné*. The peculiarity in the opposition leader Samuel Eboua's spoken use of French was fair game, and John Fru Ndi's quiet dinner at the end of 1994 with the French ambassador, against the grain of his own populism and of many of his supporters' sensibilities, became a lead cartoon in *Popoli* January 26, 1995. A more serious example was the column by the versatile, influential Herbert Boh, a key SDF stalwart and its first press secretary, in *La Nouvelle Expression* March 8–10, 1995, on the theme that

> l'opposition n'a brillé que par son manque d'unité d'action et de stratégie, et par son désir de courir (en rangs dispersés)/ the opposition has shone only

through its lack of unified action and strategy, and by its desire to run (in scattered ranks)

for four years, permitting Biya's survival: "Paul Biya ne peut mieux faire/Paul Biya could not do better." The anglophone private press equally scorned the opposition's weak effort to mobilize boycotts and demonstrations early that year; witness *The Herald*'s March 13–15 editorial, "Opposition at siesta?."

Another crucial part of the press environment needing attention was the use of pre- and postpublication censorship which Law 90/052 (despite its "Freedom of Mass Communication" title) kept in place at need.[12] The "big five" independents were banned at a crucial period late in 1991 before the regime called for the "Tripartite" conference on the political crisis, helping it forestall opposition demands for *its* choice of forum, a Sovereign National Conference. *Le Messager*, their senior partner, was the special target, not just for censorship but for attacks on its Douala office, which for a time went unrepaired, so that passersby witnessed the impact of direct regime suppression. These were the consequences of pioneer work by its editor Pius Njawe, from his publication of Monga's open letter in 1990, then a June 12, 1991, opinion poll which first informed Cameroonians that support for both the SDF and a second new opposition party, the National Union for Democracy and Progress (NUDP) surpassed the CPDM's, and many provocations ever since. Over the next few years *Le Messager* skirted bans by changing its name while keeping a familiar masthead and critical content, as did *Galaxie* (*L'ami du peuple, Galaxie Media*) and *La Nouvelle Expression* (*Expression Nouvelle, Expression*) for temporary respite. Such work won Njawe 1993's Golden Pen award from La Fédération Internationale des Editeurs de Journaux for upholding press freedoms under pressure.

This was *not* Fame Ndongo's Cameroon. Retaliation earned Njawe and others time in jail and an unreliable passport.[13] Another way to counter the critical private press was to add new papers like *La Caravane* and *Le Courrier* to previous regime organs like *Le Patriote*, so superior in production values as to suggest money available far beyond the resources of opposition papers. The practice of buying favorable journalism abroad continued.[14]

Frequent reversion to prior censorship and banning orders or postpublication seizures from kiosks took place when the regime caused or faced crises and when standard criticism was deemed offensive. *Galaxie*, for instance, at a time of no great trouble, had five consecutive May–June 1995 issues seized from vendors, thus incurring the fixed cost of printing without recouping the point of sale revenue which was virtually its only income source. Less obvious pressures were at work, like

Newspaper Kiosk, Bamenda, 1995. Photo by the authors.

the risk of advertising in the critical press for firms seeking government contracts, the technological and fiscal dominance of the state printer, with few modern press options of satisfactory quality and price, especially from 1994 when CFA devaluation priced Nigerian alternatives out of reach, and the fall of Cameroonians' disposable incomes. *Le Messager* March 13, 1995, acknowledged that its print run, formerly 80,000–120,000, was now in the 20,000–25,000 range. Newspapers that year appeared in some kiosks stapled shut, so that readers could not browse and put them back; some vendors also adopted a "rental" fee of CFA 100 francs as against the prevailing CFA 250–300 francs sale price.

Independent papers did compete with each other, not always on friendly terms. Disputes at *Le Messager* led to spin-offs by its former staff, *The Messenger* in English and *Dikalo* (Bassa for "messenger") in French, and it was common to hear charges that newspapers raised money by selling their pages to political parties or to fuel personal and political rivalries.[15] But they could also join forces. A prominent independent journalist, Benjamin Zebaze of *Challenge Hebdo*, started Rotoprint as his own and other papers' alternative to the state's publishing house, SOPECAM. L'Organisation Camerounaise pour la Liberté de la Presse (OCALIP) began under Njawe's auspices in 1993 to pool protests and print occasional "solidarity" issues for its half-dozen French language members, and a parallel venture linked *Génération* and *La Sentinelle* with the English language *The Herald* in mid-1995. Earlier that year, OCALIP appealed for the ministry of finance to lower import charges on newsprint as its world price rose, in line with UNESCO's Florence Convention protecting cultural enterprise, and ran cartoons asking Cameroonians whether CFA 300 francs was better spent on a newspaper or a beer.

Both these campaigns, as the papers scrambled for survival in their challenges to the state and appeals to the wider society, accurately reflected the tension and improvised search for recovery in Cameroon's public life. We can say, as we move now from print journalism toward the parties and the political experience minutely charted across the spectrum from *Cameroon Tribune* to *Le Messager*, that the papers, the parties, and the politics were subtly aligned. The subsidized *Cameroon Tribune*, despite its loss of readers and pages, clung to its appearance and routines, like the CPDM. Conversely, unsubsidized newspapers were by mid-decade, like opposition parties and their constituencies, much at risk but collectively resilient, a condition the rest of this chapter amply demonstrates for Camerooon at large in 1991–1992.

Opposition Challenge and Regime Response to Mid-1991

Subject like journalism to registration and discretionary sanction by MINAT, political parties proliferated following Law 90/056 of December 19, 1990. The reinstated UPC was the first to appear and dozens more registered by mid-1991.[16] Some were ephemeral or so localized and clearly opportunistic as to count for little beyond the registry paper, others resulted from the CPDM fishing the multiparty waters to create satellites and thus complicate politics to its own benefit, still others emerged from disputes the new parties could not contain. As it was for newspapers, half a dozen or so became significant forces, either national enough in their appeal or locally important enough to be reckoned in any national calculations. The collective party action will command recurrent attention below.

Effective collaboration appeared by April 1991 in the National Coordination of Opposition Parties and Associations (NCOPA). It linked the new parties with human and civil rights groups formed during the Yondo Black trial a year before. Its key demand became a Sovereign National Conference (SNC), drawing on the model and tradition of the 1789 Etats-Généraux and on the very recent experience of Mali and Benin in francophone Africa.[17] It invited CPDM participation, to conduct interim governance and frame constitutional and electoral reform, as the necessary political prelude to any fundamental attack on the broad socio-economic crisis. The argument and conclusions were clear: pluralistic political structures based on democratic initiatives were preconditions for the success of any efforts directed to policy matters, including the economy. Those in power at home, those abroad interested in Cameroon, should recognize Europe since 1989 and Africa since Mandela's release in 1990 as the backdrop for Cameroon's mobilizing democratization.

The first serious popular manifestations following the Njawe-Monga trial were in April 1991. Students at the University of Yaounde were at the time Cameroon's most volatile mass of potential resistance.[18] The university was swollen to three or four times its planned enrollment, badly managed, and scarcely maintained, and students living off-campus were demographically entwined with Yaounde's population. Although the city had a strong Beti ethnic affiliation with the regime, it was by no means uniform. Migration patterns over decades created a diverse population in its civil service, petty commerce, and labor ranks, rendered potentially volatile as the economic crisis cut employment and services among salary and wage earners as well as those chronically worse off.

During Biya's trip to France the first week of April, to bracket himself and Mitterrand as democrats and to seek funds, student demonstrations criticizing university life began on campus and threatened to support and join democratic openings in the city. Troops went in. Minister of Information Augustin Kontchou Kouomegni began a swift rise to prominence in the regime and notoriety in the opposition, and gave the latter its first memorable slogan, by claiming "Zéro Mort/None Dead" (which a government commission of enquiry repeated), but students added six more named youth fatalities to Bamenda's a year before and rumors circulated of dozens of bodies in nearby lakes.[19] Classes were suspended and students dispersed for weeks, not to the regime's advantage in cases like one we will cover in Chapter 7 below: the return of 300 students to Kumbo-Nso, North West Province, where they joined forces with their elders and brought such a pervasive resistance to bear on all local governmental institutions that an anthropologist greatly experienced there has written that the area has in many ways by the mid-1990s "withdrawn from the state."[20] Classes resumed sporadically in May, then came exams which the government forced on ill-prepared students driven by batons and tear gas to the halls, so as not to "lose" the academic year and its own credibility.[21]

Mobilization of the campus and its occupation by security forces were the first episodes which translated Bamenda's protests of May 1990 from a vicarious, isolated experience to a reality which significant numbers of Cameroonians elsewhere could understand. Not just lawyers like Yondo Black and aspiring party politicians like John Fru Ndi were now of national interest; student leaders like Senfo Tonkam and the eponymous Schwarzkopf and Mandela were detained, trumpeted defiance, made headlines, threatened a resistance not to be contained by conventional means among easily managed social sectors. Tests of nerves ensued. The opposition through frequent NCOPA meetings added the rhetoric of tax refusal and other civil disobedience measures to Sovereign National Conference demands in its successive Plans of Action, while the regime countered, for instance, with rumors of a plot to kill Biya linking Fru Ndi and the senior anglophone army officer, General James Tataw.[22]

The regime recognized the danger signs. A special session of the National Assembly late in April restored the office of the prime minister, seven years inactive, and added the title head of government to the appointee, Sadou Hayatou, so as to create the sense of broad reform through a shared executive and stronger legislature. The move was additionally conciliatory because he was a northerner not a Beti, a party technocrat not a regime ideologue. But with the constitutional carrot came the security stick. By May 17 a military "Operational Command" was in place under revamped loyalist leadership, to parallel and at need to displace civilian authority in three clusters (defined by key road links) covering seven provinces considered potentially disruptive: Far North-North-Adamawa; West-North West; Littoral-South West.[23] This was a significant juncture, for it meant that less than half a year after the December 19 pluralist press and party legislation, Biya and the CPDM considered only the other three provinces, Center, South, and East, to be secure.

One month later, at the end of June, came the episode which truly mobilized Cameroon's city streets and village markets beyond the Bamenda and Yaounde outbursts. Biya's foreign travels ceased and the crisis was faced in a long anticipated address to the National Assembly at the end of its budget session, June 27, carried on state radio and television.[24] It halted all other public life in Cameroon that morning. Speculation had built amidst hints of more concessions like April's appointment of Hayatou, who had toned down at least the image of an executive monolith by the "consultative" tone of his public appearances and statements. Would the regime dig further in? Would it offer dialogue and concessions, perhaps nudged by foreign creditors and donors who were starting to pay attention, if only to hold whatever ground it still controlled? Would it more generously follow the trends in Mali, Benin, and Zambia toward democratic transition through a national conference or electoral process, and genuinely reconstruct governance?

The choice revealed June 27 was for "law and order" in a text prescribing the *status quo* as of 1990's legislation, until a National Assembly election now moved forward to late 1991, on the basis of a revised electoral code for its conduct. Dominating this prospect, however, was the memorably obdurate heart of the speech itself:

Je l'ai dit et je le maintiens: la Conférence Nationale est sans objet pour le Cameroun.... Seules les urnes parleront/I have said it and I maintain it: The National Conference has no purpose for Cameroon.... Only the ballot boxes will speak.[25]

Douala, June 27, 1991: The Response to President Paul Biya's "Sans Objet" Speech. Photo by Omer Songwe, used with permission.

The public response was swift. To "Zéro Mort" was added a second powerful slogan of scorn and rage for 1991 as Biya's now indelible nickname "Monsieur Sans Objet" keyed the resistance and "went national" in a way he did not intend (or else precisely calculated to provoke a crisis the regime welcomed; its paralysis in the next few weeks, documented below, suggested otherwise). As the speech ended barricades went up and fires were set in Douala. By nightfall one gendarme and three civilians were dead and the house of the mayor of New Bell Quarter was burned down. This historically most radical and street-politicized city in Cameroon, where workers' closure of the port had weeks before built a fundamental threat to the state's revenue base which (alongside foreign infusions) the regime depended on, now gave Bamenda its first vanguard support in francophone Cameroon. Other cities and towns in four provinces (North West, South West, Littoral, West) quickly followed, as direct action previously confined to Douala known as "villes mortes/ghost towns" gathered wider momentum.

Villes Mortes/Ghost Towns, Mid-Late 1991

This episode in certain locales approached the intensity of a general strike, although it lacked the workers' organizational base found in the classic general strike literature and experience from industrial societies. Any account of "villes mortes" (the most generic term used in Cameroon) needs its physical, emotional, and rhetorical context established and coupled with those slogans introduced above, "Zéro Mort" and "Sans Objet." Consider two accounts from the independent press in mid-July, pointing out that not just the state apparatus but also the opposition parties which translated the strike blueprint of a young Douala militant, Mboua Massock, into action fell behind the momentum of forces and events.[26] Célestin Monga, since his trial an eloquent journalistic and political voice, and regime target, described for Douala in *Challenge Hebdo*, July 10, "une rapide sociologie des bandes de gamins qui dressent des barricades dans les rues/an escalating sociology of bands of kids who build barricades in the streets" and take Sylvester Stallone-Rambo, Chuck Norris and Arnold Schwarzenegger as their models for liberty, rather than Aimé Césaire and Cheikh Anta Diop.[27] A week later, *Le Messager* on July 18 turned such observations into analysis, asking in the course of an attack on Biya whether he *or* the opposition was in touch with the experience developing:

> qu'il comprenne, lui et ses appuis occidentaux, que ce qui se passe au Cameroun aujourd'hui est une révolte populaire que les partis d'opposition s'efforcent simplement de contenir/let him and his western backers understand that what is happening in Cameroon today is a popular revolt that the opposition parties are trying hard simply to contain.

Here, among what *Aurore Plus* August 19 called "les enfants de l'Intifada," were the earlier symptoms of anomie and morbidity noted by Bayart and Mbembe, moved from the shadows of hegemony into the arena of debate and action by the chronic poor and recently dispossessed, the migrants becoming vagrants, whom politicians would now have to reckon with.

The rhetorical force behind "villes mortes" can also be gauged in the recurrent press and popular play with acronyms. CRTV became "Centre de Réetablissement Total de Voleurs/Centre for the Complete Resettlement of Thieves." The SDF proclaimed "Suffer Don Finish" on the party's green and white visors, billed caps, and T-shirts which flourished on the streets and helped build its finances.

Confrontations spread rapidly from Douala after Biya's June 27 speech and two more aggressive regime decisions: security personnel prevented a rally planned July 5 in front of Yaounde's presidential palace; six human rights groups were banned the next week (the NCOPA was already outlawed, but defied the order). Spontaneous street demonstrations, market and other economic boycotts became trenchant, fully organized civil disobedience which for three to six months, depending on the locale, made massive rallies and other highly visible challenges to state authority Cameroon's dominant features. They spread from provincial capitals of North West, South West, West, and Littoral Provinces, and also from smaller centers like Kumbo-Nso (North West), where we noted returned

Yaounde students active, Kumba (South West), and Mbouda (West). Their grievances were old and new, including traditions and grudges dating back to precolonial, colonial and *maquisard* times, and more recent language, ethnic, and other forms of marginalization. Rural people either joined actively or were drawn into the movement, some reluctantly by threats. "Villes mortes" held firm in all parts of these four provinces for at least ten weeks, longer but with some slippage in Douala, until Christmas in Bamenda and Bafoussam.

Street demonstrations were the most dramatic features. The first major episode in Bamenda, a city where perhaps 150,000 people live, market, or trade regularly, called to protest the ban on Yaounde's rally, brought at least 20,000 toward the governor's mansion July 5. Officials met them for a half hour at the foot of the cliff beneath, then turned them away unsatisfied. Some stayed to taunt and challenge troops blocking the road uphill. They included female *takumbeng* elders who displayed traditional paraphernalia and postures of defiance to excessive authority, including disrobing, against very uncomfortable security forces, not for the most part familiar at first hand with local customs but clearly aware that this was open, if unarmed, defiance.[28] Others moved through the city and burned tires. Security responded with fifty-sixty tear gas canisters from ground forces and Bamenda's state security helicopter. One person died and property damage from looting by both sides was substantial.

After this first outburst, the streets became far more disciplined arenas of protest. Bamenda, the SDF's terrain, in fact became the unofficial opposition capital, with its rallies by all reports serving as models for NCOPA actions everywhere the disaffection spread. Representative, then, if more frequent and intense than elsewhere, Bamenda's rallies of three-four hours, three times a week, mobilized 8,000–10,000 people, 30,000 on special occasions, and 50,000 at a crisis march of defiance October 2 (we describe it below). Separate parties marched with their colors and converged at a point renamed "Liberty Square" to honor the 1990 martyrs. There, smaller party leaders spoke from a truck bed before Fru Ndi appeared, usually in customary Grassfields dress, speaking predominantly and powerfully in Pidgin for a Sovereign National Conference remedy as his central theme. Ahidjo effigies were common, backdrop features.

These were almost literally, except for religious worship, the only weekday gatherings in central Bamenda for five months. Most schools, in recess when "villes mortes" began, stayed shut. All four-wheeled inter- and intracity vehicle transport, banks, shops except pharmacies on a rotating schedule, and all but the most minimal curbside markets, closed Monday-Friday, as they did wherever "ville mortes" took hold. Postal and health services continued, but the public sector otherwise closed

John Fru Ndi, Chairman, Social Democratic Front. Photo by Omer Songwe, used with permission.

down. Most of the water, electricity, and telecommunications staff went on strike and the rest stayed away, for their offices were the targets for non-payment campaigns, subject to deterrence attack and risky to work at.[29] As normal state revenue from utility rates fell, tax collection also suffered when people turned to clandestine Nigerian vehicle fuel.

Virtually all Bamenda observed the strike, whether from enthusiasm or fear of reprisal. Only newspaper and small food sales continued. Visible large scale business and people using cars instead of scooters, bicycles, or their feet for leisure, risked sanctions. The best example of the city's determination to nullify the regime and valorize local society was the beer truck fleet operated by the Brasseries du Cameroun, a regime and (it was rumored) a Biya family fiscal mainstay. It was warned off the road weekdays

A Bamenda Rally Begins, July 1991. Photo by the authors.

and complied after one fiery lesson, while cheaper palm wine ("white mimbo") was permitted entry by motor scooters from nearby villages. Weekends took up the slack. Markets opened Saturday; so did banks, subject to chaotic "runs" on cash, which was curtailed by capital flight abroad and removal of funds from the public to the private domestic economy. Sunday was for church and visits, including the region's *njangi* meetings for private credit and loan purposes, the local economy's retreat.

Cities competed to be known as "hardest" and most self-sufficient and denying, mixing the release of frustrations and the euphoria of resistance with the nihilism Monga noted, revealing the riskier and unsettling side of the strike experience. Located midway between two obvious contenders, Bamenda and Bafoussam, Mbouda (with its strong 1960s *maquisard* memory) may have been the strictest of all, on its own people and outsiders. Regime supporters, profiteers, or people simply trying to maintain livelihoods by dodging the food supply blockade of Yaounde—for "starving" the capital and regime heartland was one key but never successful element in "villes mortes"—gave Mbouda a wide berth and operated only at night.

In these ways, taking Bamenda's lead or following their own trajectories, up to 2,000,000 unarmed Cameroonians directly or indirectly linked to "villes mortes" worked three striking changes in its core area's political culture in mid- to late 1991, at considerable risk and cost to many of them. They created the courage and discipline of a largely nonviolent resistance movement, drastically reduced the 1991 state revenue base, and stretched security forces, whose units at times protected civilians in the opposition they were called on to punish, and even turned on each other.[30]

Facing pockets of insurrection, the state for weeks after June 27 shut down all but basic security responses throughout much of the four province heartland of the strike. The NCOPA and its constituent parties rallied there continuously, insisting on the Sovereign National Conference and rejecting Biya's calls for one-on-one meetings with their leaders. He was confined well into August to his capital and village residences, Etoudi Palace in Yaounde and Mvomeka'a (South Province, where a recently built golf course fed the fury of critics). Opposition leaders, by contrast, moved freely despite harassment, and set Cameroon's political agenda and pace. Moving far beyond the regime's 1990 concessions to liberalization, popular forces operating both in disciplined and spontaneous, disorderly ways gave Cameroon's civil society a force and a presence which the new political parties, "political society," were hard pressed to channel. But institutional politics soon emerged, including firmer party organizations, competition among them, and electoral politics. The first phase of democratization, in the streets, was over; harder, more complex work to unseat the regime lay ahead.

Political Parties and Mounting Tensions

The SDF was most prominent in the opposition, with the better part of a working year's head start and the martyr's legacy, even though it was like the others legalized only in early 1991. A smartly produced *SDF Echo* party newspaper, position statements on policy issues, a party constitution, and a detailed party history with founding documents appeared in print by July, to accompany its rallies in most southern cities and the sale of party cards for CFA 500 francs (less than US$2).[31] Notable facets from the start were its inclusive and equal membership rather than the CPDM's distinct wings for women and youths, in reality separate and unequal, and a clear reliance on wards of 100 people as the base structures where all elections up the party echelons started. The SDF's popular impact registered with core supporters in North West Province; long used to the derisive "No Way!" attached by others to the NW initials of their automobile licence plates, they now declared "Na We!" (Pidgin for "It's Us!" or "Our Turn!").

The restored UPC joined the SDF for a time in the opposition forefront. Most of its veterans had stayed clear of CPDM links, but the history of competing internal and exile wings compromised its prospects even before 1991 and leadership rivalry between three old hands as the year unfolded intensified its problems. Its nominal leader, Douala's Prince Dika Akwa, had in the past spent years in jail and exile, then settled into a notable scholarly life; his style remained mercurial. A returned exile closely linked with the party's syndicalist and Marxist past, Woungly Massaga, now aligned his new Parti de la Solidarité du Peuple in the regime's camp; it "displayed" him in Yaounde's fashionable Hôtel des Députés.[32] Another veteran UPC hero long in British exile, the anglophone Ndeh Ntumazah, returned home late in 1991 to join its leadership race. UPC disarray was obvious. The NCOPA in fact suspended the party in September when private talks between Dika Akwa and Biya became known.

Very substantial regime links, far more so than Fru Ndi's until 1988, marked the leadership contestants in another opposition party, formed in 1990 and legalized in 1991. The National Union for Democracy and Progress (NUDP) was the base, first, of Samuel Eboua, a southern francophone and formerly Ahidjo's secretary in the presidency, then by late 1991, after his return from Nigerian exile, the Muslim northerner and Biya's first prime minister, Maïgari Bello Bouba. Their regime backgrounds made the NUDP's genesis ambiguous and their leadership rivalry which Bello Bouba won early in 1992 made its subsequent history contentious.[33]

One more key party and leader appeared, less encumbered with such baggage. Unchallenged at the helm of the Cameroon Democratic Union (CDU), Adamou Ndam Njoya built on his reformist work and reputation as minister of education late in Ahidjo's presidency and his absence from domestic politics after 1982. His royal blood as a scion of the sultanate of Foumban added credibility and strategic location in the disaffected West Province to his prospects, as did his work and stature abroad as an international civil servant and an ecumenical Muslim leader. What happened in Foumban after the CDU's creation in April showed the fluidity of politics by mid-1991. The palace's and city's CPDM ranks were until then closely managed by a senior prince with continual and powerful cabinet posts, Ibrahim Mbombo Njoya (he became sultan a year later). But once the breach opened with Ndam Njoya, the elderly sultan pronounced his people free to choose their own party alignment and a junior prince resigned his CPDM divisional presidency and party membership, reportedly followed by 20,000 rank and file.[34]

Dozens more parties mushroomed alongside these four but were not historically or currently consequential. Many in fact quickly folded into the CPDM, giving credence to charges of its payroll at work against pluralism by splitting the opposition and bewildering the public into passivity. But CPDM splintering itself continued. Most notably, its vastly ex-

Maïgari Bello Bouba, President, National Union for Democracy and Progress. Photo by the authors.

perienced strategist and elder statesman François Sengat Kuo, from Douala, started a reformist wing within the party, the Courant des Forces Progressistes, issuing manifestos which called for a crisis CPDM congress, the Sovereign National Conference, even Biya's resignation.[35] The younger Douala leader Jean-Jacques Ekindi, once in the UPC and imprisoned by Ahidjo, left the CPDM and created his own party, the Mouvement Progressiste. A prominent women's wing leader and parliamentarian, Victoria Tomedi Ndando, resigned and joined the SDF.[36]

Such defections made the CPDM resemble a calving iceberg while the opposition remained dynamic and unified. The NCOPA moved "villes mortes" forward. All its supporters would probably in mid-1991 have agreed in principle with a sign advocating "Elected Governors and Divi-

Adamou Ndam Njoya, President, Cameroon Democratic Union. Photo by Omer Songwe, used with permission.

sion Officers" displayed where rallies mustered on Sonac Street, Bamenda. More precise positions were still on the horizon, but the weightiest early leaders, Fru Ndi among anglophones and Ndam Njoya among francophones, appeared to have much in common during the third quarter of 1991. Both wanted executive appointments curtailed, and executive, legislative, and judicial powers separated and balanced.[37] On broader philosophy and policy issues, here was Fru Ndi's speech, May 26, 1991:

> in our manifesto and ... our Party's Constitution ... it becomes very clear that the SDF stands for welfare state liberalism ... fair play in politics ... fair deal in economic and social matters.thorough reform in the area of economy, finance, education, law, health.[38]

The CDU manifesto issued two months later advocated

> a humane market economy ... the State and the Civil Society together ... reinforce healthy competition ... individual responsibility and private initia-

tive [with] judicious and fair state intervention in order to improve social justice.[39]

This was a potentially consensual blend, ideologically fluid on a liberal to social democratic front, combining private enterprise *and* a strong public service sector, calling for "accountability." Ndam Njoya's character and reputation were patrician, but little separated him from the populist Fru Ndi in this excerpt from his July 13 speech at the CDU's first Bafoussam rally:

> Il n'y aura pas au Cameroun des producteurs d'un côté et des consommateurs de l'autre/There will not be producers on one side and consumers on the other in Cameroon.[40]

The first CDU rally in Bamenda drew close to 5,000 at Municipal Stadium August 17; Ndam Njoya eulogized the city's 1990 victims and called it "the democratic capital of Cameroon." This, while massive turnouts at *shared* rallies continued, was opposition politics at its highest level and potential, establishing common ground between the anglophone considered the heart and soul of the resistance and a francophone widely respected for his past record in CNU government and avoidance of CPDM politics.

Their alliance never took firmer hold.[41] They had to remain silent on one very delicate issue for the opposition, the future constitutional basis of the state, so as not to risk debate and possible schism between francophones used to a strong central authority and anglophones advocating federalism. A more *generic* problem faced these and all other opposition parties, which cooperative NCOPA work did not remove: they were still narrow in their interests and appeals. SDF roots in North West were spreading to West, Littoral and South West Provinces, but other parties remained more localized.[42] The CDU base was in fact very narrow; Ndam Njoya proved unable to develop a significant following outside Foumban and remained in basic respects an enclaved leader. UPC fortunes by now limited its appeal to Douala and an arc of territory inland, by no means unchallenged there, with a few spurs among Bassa people who kept residual ties to its powerful history alive. The NUDP was the major emerging voice in the provinces of Far North, North and Adamawa. But the rivalry to claim Ahidjo's legacy already noted between Bello Bouba and Eboua, as well as historic tension between the North's minority but unified Fulbe-Muslim constituency (Ahidjo's and now Bello Bouba's base) and the majority but scattered non-Muslim ethnicities, limited NUDP hopes to develop within that region, let alone to spread its appeal elsewhere. Finally, no party claimed any firm base among the Bamileke, the largest ethnicity, surely the wealthiest in private sector terms, the dominant press influence, and probably the prime target population of all parties' recruitment efforts beyond their core areas.

But reference to Bamileke ethnicity raises a complication for Cameroon's opposition parties *within* their local catchment areas. Reference to regional or provincial bases, as if they were sufficiently coherent to provide firm springboards for further party launches, misleads. There were good historic reasons why no "Bamileke" party on its own surfaced, which writings on both its (more properly, *their*) earlier history and more contemporary political analyses by Bayart and Mbembe readily explain.[43] The reality of internally patchwork ethnic populations within these areas made politics a constant search for local as well as broader affiliations and alliances, with as many centrifugal as centripedal forces at work. These were more than "nuances." Each party to some degree reflected the "all politics is local" dictum, and had local flanks to protect against rivals at the same time it tried to expand its base; this was true even of SDF experience among anglophones, much firmer in the Grassfields of North West than in coastal South West Province.[44]

The NCOPA kept driving wedges into the CPDM throughout 1991, and sent delegates to Europe and North America to lobby the opposition case. But if the CPDM resembled a calving iceberg, the specific gravity and critical mass conditions on which the opposition could unite, beyond its desire to bring down the regime, were far from settled, at local, regional, or national levels. And despite the solidarity we have noted at the rallies, nothing precluded NCOPA policy disputes or competition for allied parties' members.[45]

One policy issue above all demonstrated how hard it was to turn even the most promising in principle of all the opposition affiliations, between the SDF and CDU, into wider, firmer practice.[46] The NCOPA, forty strong, met in Bamenda September 20–22, 1991, to debate whether to boycott the 1991–1992 school year, as part of "villes mortes." No thornier issue could be raised, given on one hand the popular belief in education's value and on the other hand the multiple calculations of the boycott's likely impact. There were obvious class dimensions. It would harm well-off families who could afford schools and, with more to lose, wanted their children to "get on with it," but might relieve more marginal people whose income "villes mortes" stopped at just the peak time for earnings from crops and youth vacation work, needed for fees, books, and clothes. Further: if a boycott took place, but with different local levels of adherence, children in militant zones would suffer in academic competition, in an exam-driven system, with those elsewhere not complying. And the issue went beyond schools. If *they* opened, taxi and food vending sectors must do so to service them, and the strike would collapse. If they stayed shut, those sectors' loss of income would intensify, testing interests and loyalties severely in an economy already shrinking because of "villes mortes." There was another key variable, touching education's producers as well as consumers. Teachers were a potentially strong opposition constituency because of chronic salary arrears, vaulting class sizes, and failing infrastructure, and they in fact spent much mid-1991 vacation time building new professional associations. But exams they threatened to leave untouched *were* marked, disclosing gaps in militancy between elders with a conventional sense of calling and cadets, and between francophones and anglophones.

No easy reading of education's consumers and producers guided NCOPA; no ideology or strategy could straddle or resolve these bread and butter differences. In the event, the CDU and Ndam Njoya (a former minister of education, after all) tried "to persuade the opposition to leave the school palaver out of politics," but a boycott was narrowly approved on SDF advocacy. Results under such cross-cutting conditions were bound to be mixed. Successful for its first month in Bamenda (where one large school was torched when its proprietor started classes), it was ignored in Yaounde and found varied support elsewhere before losing its force. The first truly contested NCOPA policy vote split its ranks and compromised its coherence at precisely the time when it could least afford vulnerability.[47]

For the regime counterattacked after weeks of paralysis following June 27. August 21, five days after Bamenda first received Ndam Njoya, Bafoussam turned out for Fru Ndi. But he was also greeted by security forces, who wounded and warned him with a rubber bullet to the foot. A week later, at Maroua in Far North Province, Biya made his first stop in a six week "air lift" tour of all provincial capitals. Official texts of all ten speeches printed in *Cameroon Tribune* October 10 revealed a common format in quarter- to half-hour deliveries. He thanked and welcomed his hosts, presented himself as head of state rather than party leader in opposition strongholds, or in both roles where he was comfortable doing so. He cited in detail the government's local development projects and promised more, explained economic hardship as a result of commodity price falls abroad, not caused or soon reversible by domestic initiatives, deplored violence, and claimed (like the March 23, 1995, radio text) that "Advanced Democracy" CPDM policies, especially since the laws of late 1990, made *him* the source for dialogue and unity. Massive CRTV coverage favored his campaign, while the opposition was denied nonprint media access and was substantially silenced as five major newspapers were banned.

Responses to Biya varied according to local political views of the crisis and its dynamic. The regime's own *Cameroon Tribune* reported a crowd of just 4,000 at Bafoussam September 12, in a setting where presidential visits normally mean public holidays and massive crowds. Bamenda and Yaounde were, again, the polar contrasts. Heavy rain interrupted September 13's Bamenda rally, and the crowd was not much larger than what attends "palaver" for any Commercial Avenue road mishap. Both in speech and symbol, Biya was directly confronted. The urban delegate (mayor) appointed by the regime, Jomia Pefok, read an astonishingly defiant text about what happens to the grass when elephants fight, by no means a "neutral" message since it assumed the opposition's right to contest the field. The speech was reliably said to be an SDF product, replacing the text supplied Pefok by the regime, nullifying and ridiculing Biya's own rehearsed speech in response, for Pefok spoke in English and Biya could not rebuke or rebut him spontaneously.[48] And his gift from the area's traditional chiefs was a carved table, signifying support for the Sovereign National Conference. Quite different was the last stop, Yaounde October 4, with a massive turnout on Biya's home ground and the most belligerent gifts imaginable—drum, spear, machete—to which his response became his second most quoted line of 1991: "Tant que Yaounde respire, le Cameroun vit/As long as Yaounde breathes, Cameroon lives."[49]

As the calendar advanced and the tour unfolded, skirmishes between regime and opposition intensified. Reports by *La Vision Hebdo* August 21 and *Le Jeune Observateur* August 22 connected violence and many deaths since May at Meiganga, Adamawa Province, with new political tensions, whereas *Cameroon Tribune* July 22 and September 11 played down the violence and blamed old ethnic and land use conflict between Fulbe and Gbaya, now said to be reconciled. Difficult to evaluate, the news was significant for its glimpse of unrest in the seldom reported three northern provinces.[50] In Douala September 23–24, a NCOPA rally was broken up, with Ekindi (the city's as well as country's most prominent CPDM defector) and the elderly Eboua arrested and beaten badly enough for hospital treatment; foreign consuls intervened to secure their release.[51] This was three days after Biya's appearance there late in his tour, and his speech to a populace whose opposition militants wanted to deny him access; beginning it with the unscripted challenge (not in all published accounts) "Me voici donc à Douala!/So here I am in Douala!" showed, like Bamenda's and Yaounde's rallies, the fissures now opened in Cameroon by both sides.

They widened again October 2, as Biya's Yaounde rally was being prepared: Bamenda exploded.[52] A *modus operandi* worked out between locals and an anglophone governor experienced in the area had made recent rallies orderly by day and private patrols useful in keeping residential peace after dark, reducing tensions. But these bargained conditions fell apart when the senior divisional officer (SDO) refused a permit for October 2's routine rally; he was a Bassa francophone newcomer attached to MINAT, nominally inferior but elevated in the structure of authority now that there were military Operational Commands. This provoked Bamenda, and matters escalated when security's helicopter began dropping tear gas and concussion grenades on the city two hours earlier than rallies ever began.

It became the harshest episode since May 26, 1990, as 50,000 people moved toward a rally site never used before; with a new SDO, so many people, and no fixed line of march, "understandings" collapsed. The unarmed crowd surged toward half a dozen armed troops, who fortunately backed off, averting real carnage, and the rally went ahead. There, Fru Ndi's car was hit by grenades and bullets in a way that renders plausible the SDF's claim of an assassination attempt, whether planned or spontaneous. The car's bullet-pocked and charred hulk sits in his compound to this day, a reminder of the regime's violence and Bamenda's resistance. The crowd dispersed peacefully but was harassed. By day's end there were two people dead, five amputees from concussion grenades they were forced at gun point to pick up, and dozens more hospitalized. Security forces also destroyed property, to punish a population still observing "villes mortes" by withholding rate, tax, and licence payments from a local treasury (admittedly by now empty) and from other government offices and projects those revenues serviced. October 2, 1991, was in one more way a key episode, for Amnesty International and BBC television circulated evidence smuggled out of Bamenda, the first such documentary coverage abroad for Cameroon's militant democracy struggle.

The Balance of Forces, Late 1991

Contrary to the SDF credo, suffering clearly was *not* finished and another reading for its acronym emerged: "Suffer Dey (for) Front." The hostile SDO soon became the North West Province's governor, his name Bell Luc René transposed locally to Bell Luc Grenade. With this lesson dealt Bamenda, and Biya's belligerent message sent from Yaounde's rally two days later, a new phase of the crisis began. Those two cities served well to identify the fault lines exposed in Cameroon by late 1991: Bamenda as the peripheral countryside and its disaffected, in places insurrectionary civil society; Yaounde as the metropolis where the state and its leadership held firm against the pockets of dissent.[53]

While the state could still punish Bamenda, Yaounde was not immune from an independent civil society's encroachments. The capital's security forces, civil service apparatus, sustained road link to Douala, and its core Beti ethnicity kept "villes mortes" at bay in essentials. But its economy was pinched before "villes mortes" and further reduced by the strike, and the university's discontent before mid-year and some tentative, certainly risky, party organizing there after mid-year added to pressures on and in the city. The opposition never mobilized Yaounde with the discipline exercised in its heartland, nor did the indiscipline Monga found in Douala's underclass surface there. Its intelligentsia, however, responded independently to the crisis. Breaking away from the dominant pattern of what the dissident scholar-teacher Ambroise Kom called the "organic intelligentsia" and its support of Ahidjo's and Biya's statecraft through the 1980s, a critical writer's group published a white paper entitled *Changer le Cameroun: Pourquoi pas?* in October 1990. Its 397 pages constituted a blueprint, with details for each ministry, for the replacement of "une République des fonctionnaires/a Republic of bureaucrats" by a new politics capable of creating "un pays à la mesure de notre génie/a country commensurate with our genius."[54]

The opposition parties simultaneously worked as best they could to upset Yaounde's routines and link the dissenting political and civil society forces. July 4, 1991, for example, as the worlds of "sans objet" and "villes mortes" collided, party leaders met at Ndam Njoya's Yaounde compound to organize a march the next day. Security forces sealed them off, a search warrant when demanded was scribbled on paper a soldier's back supported, and the march was called off. But word of the initiative spread, both through the city's streets by "radio trattoir/pavement radio" and overseas by phone links (which FAX and E-mail soon superseded) to British and French radio services, which then fed the news CRTV was not reporting back to Cameroonians.[55]

Even in this most hostile environment, therefore, dissenting forces were in touch and mobilizing by 1991. They linked widely separated terrains, and began a process whereby lines of class, ethnicity, and religion presumptively dividing their leaders from leaders, followers from followers, and leaders from followers, were modified by a basic literacy and common experience in opposition politics. This apprenticeship was thorough, and the lessons of its successes and failures remain accessible in the long term. More immediately, what both the regime and opposition were now able and willing to bring to the struggle, and the limits on such capacity, was becoming clear. Which would prevail? The civil disobedience challenge and improvised unity of the NCOPA as the agent for civil society's diverse constituencies? CPDM fortress incumbency using the state apparatus to cut losses and to freeze the opposition in a war of attrition? Would compromise between them ensue? Six years later, these issues remain unsettled, but no one as this story advances can miss the complexity of the forces at work, especially the opposition's, or mistake how difficult it is to dislodge a determined African regime from power by peaceful means.

The balance of forces by October 1991 was delicate. The opposition was intact, but recently harassed by the regime as never before and subject to internal pressures now reaching the surface of the NCOPA both collectively and in its constituent parts. Its basic operation became more difficult to sustain. No NCOPA meeting took place in a crucial period September 22–October 17 starting with the crackdown in Douala, and the heavy use of press censorship during those weeks effectively cut the opposition's media access to its public. Conversely, the CRTV nonprint monopoly made all it could of Biya's tour through October 4, then on October 11 served as the channel for the regime's next major initiative.

Five days earlier, *Cameroon Tribune* reported a press conference with France's veteran Africa policy official, Michel Aurillac. He was long associated with Pierre Messmer, one of France's oldest "Cameroon hands" whose experience there began in 1935 and included the high commissioner's post in the late 1950s (prior to his term as French prime minister). Aurillac's text read extraordinarily like the core of Biya's speech televised October 11, again refusing a national conference but announcing two key steps forward.[56] The government offered for the first time to meet opposition leaders collectively rather than singly, with the prime minister presiding (*not* a Sovereign National Conference), and scheduled legislative (though not presidential) elections for February 16, 1992; it was classic due process negotiation following the recent use of force. These consultative and electoral initiatives, added to the presidential and party purses and the official media, blunted the opposition and tilted Cameroon's politics in the regime's favor. The speech completed Biya's, the state's, and the party's move from a passive to active role facing the crisis.

What was at stake was quickly, clearly recognized. *Cameroon Tribune*'s October 14 headline invitation "AUX URNES CITOYENS!/ CITIZENS TO THE POLLS!" was countered October 15 by *Challenge Hebdo:* "Cameroun, 16 février 1992: "LE PIEGE"/"THE TRAP." The latter's concern for the shift in momentum and the opposition's temptation was well founded. Prime Minister Hayatou convened a "Tripartite" conference in Yaounde October 30, assembling members of the government, leaders of opposition parties, and "notables" (most of these were CPDM partisans). It had the air and look though not the name of a nonsovereign national conference, on the regime's ground.

There were delays and walkouts until the talks closed, but an agreement known as the Yaounde Declaration reached November 13 was signed by forty of forty-seven parties thus far legalized, most notably the NUDP, CDU, and fragments of the UPC, but not the SDF. Fru Ndi's withdrawal from the meeting in midcourse signalled once again his party's militancy but it destroyed the NCOPA. *Cameroon Tribune* headlined this split in the opposition ranks and the regime had other good reasons for satisfaction. Key to its needs, signatories agreed to abandon "villes mortes" with its revenue haemmorhage and to defer the Sovereign National Conference demand pending the election. In exchange, the government announced a period of grace for those with tax and other arrears caused for whatever reason (as participants or bystanders) by the strike, lifted July's ban on human rights groups, ended Operational Command structures, freed people jailed during the strike, and opened the way for some exiles (or their remains) to return to Cameroon. It also agreed to further discussion of electoral and constitutional processes.[57]

A breathing space, and more, was immediately obvious. Biya was in Paris by November 18 for a meeting of French community states, and secured by month's end both an aid package of the kind denied earlier in the year and French fiscal backing for renewed IMF debt renegotiation. *Cameroon Tribune* November 29 could say that "the situation is improving . . . courageous measures have been taken by the President and his government." Indeed, with the Yaounde Declaration signed and French support resumed, with the National Assembly in a quiet regular session, and a legislative poll scheduled, the regime's recovery by the end of 1991 was indisputable. It now projected the appearance of "normalcy" within Cameroon and beyond its borders. By contrast, the opposition was split and the independent press was full of commentary on "trahi/betrayal."[58] In the most practical terms demonstrating the force of the agreement, "villes mortes" effectively ended except in Bamenda where it continued until year's end, and it was only rank and file militancy voiced by 15,000 at a four hour Municipal Stadium rally November 22 against SDF leaders' wishes that prolonged its flagging energy a few more weeks there.

Legislative Electoral Test, Early 1992

Substantially relocated from the streets to sites where more closed and conventional practices prevailed, Cameroon's politics shifted gears at the start of 1992. The regime's recovery was first of all compromised by problems in translating the Yaounde Declaration and its formalization in the National Assembly's Law No. 91/020, December 16, 1991, from paper to reality. Lax implementation of electoral procedure details, due to apparent CPDM bad faith, disappointed many signatories to the pact. The core opposition meanwhile sought rejuvenation at a Bamenda meeting January 5. It changed its name from NCOPA to Alliance for the Reconstruction of Cameroon-Sovereign National Conference (ARC-SNC), and adopted a New Plan of Action signed among others by Sengat Kuo, leader of the progressives in the CPDM, reminding Cameroon that regime politics also remained unstable. But neither the NUDP nor CDU were present at this Bamenda session, the UPC was there only in fragments, and the electoral campaign rather than the continued demand for a Sovereign National Conference and other ARC-SNC initiatives held the field of national politics.

One issue now stood out: could a fair and free National Assembly election be held? Critics on a series of issues confronted the CPDM on its ground staked out since the Yaounde Declaration. Only a comprehensive new electoral code and independent electoral commission could erase flaws inherent in a voting system still keyed to one-party rule. Awkwardly gerrymandered and demographically anomalous constituencies, needing reapportionment, remained in place. CPDM appointees through MINAT still controlled electoral lists and polling places. Nonprint media restrictions and press censorship continued to hamper opposition parties. These defects were cited not only by domestic critics but also by agencies from abroad, which opted out as election observers and thus denied the regime the look of legitimacy it sought.[59] The result was an election boycott movement, spreading from the core opposition to its fringes. The SDF led twenty-two ARC-SNC allies to a boycott decision January 5, and a smaller bloc gathered by the CDU in Yaounde followed suit January 21, with significant impact on the poll results, discussed below.[60] Elements of the UPC boycotted. Ultimately thirty-five of sixty-nine registered parties did so, but not the NUDP which Bello Bouba wrested from Samuel Eboua's leadership January 4–5, or Augustin Frédéric Kodock's UPC bloc.

The campaign lurched through February as the election was postponed two weeks until March 1, amid unsettling news from Bamenda of growing anglophone frustrations the city vented. The constitutional issues Fru Ndi and Ndam Njoya skirted six months earlier were no longer "off limits" since their split at the Tripartite talks, which their separate boycott decisions did nothing to heal. A militant sector of Fru Ndi's following pressed constitutional issues in a lengthy and eloquent pamphlet from Bamenda's "Cameroon Anglophone Movement (A Social Cultural Association)" (CAM). This first of many such manifestations to surface was excerpted by *Cameroon Post* January 15, with a sampling of its 5,000 signatures. It demanded far more now than "Elected Governors and Division Officers" six months before:

> Now, in 1992, the moment of decision has come . . . [It] is no longer between Federalism and Unitarism. It is between *Federation* and *Separation*. No to Federation = Yes to Separation. If we cannot return to the 1961 federal structure . . . we should liquidate the union peacefully and go our separate ways.

Display of the Federation Flag, Bamenda, 1992. Photo by Omer Songwe, used with permission.

Bellicose ethnic and linguistic as well as constitutional code words laced the document, which defiantly abandoned the anonymity of these 5,000 people willing to consider a secession option which went far beyond the SDF's stated posture. This option went more public on the streets of Bamenda February 11, an emotionally charged day because its originally festive character, based on "reunification" when anglophones in 1961 voted to join francophones, was soon gutted when Ahidjo made it "Youth Day" to efface the memory of the distinctive spheres. So, February 11, 1992, a Bamenda rally displayed the two star federation version of Cameroon's flag discarded in 1984; demonstrations continued for weeks despite efforts to suppress them, causing wounds though not deaths.[61]

Such activity added to the campaign's surreal character, initiated by the boycott. Voting under these conditions was bound to be inconclusive, because of the low turnout, 65 percent, and the endless speculation about how votes *not* cast would have changed the count. *Cameroon Tribune*'s text of official results March 11, blemished by faulty figures for Haut-Nkam (West Province), showed roughly 2,500,000 votes cast. There were 234,000 spoiled ballots, a high number, raising suspicions that many of these as well as votes not cast registered a protest against the poll itself, since more than half came from opposition strongholds like Douala (53,000), Bafoussam (21,000), Nkongsamba (16,000), and Bamenda (11,000), plus

Yaounde (31,000) with a diverse population presumably including dissenters.[62] The CPDM elected eighty-eight members to the 180 seat National Assembly, just short of a majority, the first clear break in a quarter century pattern of virtually 100 percent regime support. Opposition was vested in two major blocs. First, the NUDP elected sixty-eight deputies from all but two provinces; it dominated in Bello Bouba's North and Adamawa bases, and in South West, and won a majority of the West Province seats. This national standing led Bello Bouba to proclaim the NUDP as both the true heir of Ahidjo's "Great North" political legacy based on Fulbe-Muslim leadership *and* as the future home for a restored North-South axis in Cameroon's politics, which Biya's more exclusive and partisan "southern strategy" Beti support base in the CPDM had weakened. But the victories were subject (like the CPDM count) to doubts surrounding the vote boycott's impact. Second, Kodock's UPC fragment took eighteen seats, restricted to Douala, Littoral Province and adjacent West, Center, and South West Province remnants of former strength. The remaining six seats fell to a Far North Province party of people neither Fulbe nor Muslim, the Movement for the Defence of the Republic (MDR), which the CPDM cultivated, so as to counter the NUDP.[63]

Ironic and highly salient features of the boycott created ghostly anomalies favoring the CPDM in the vote. It swept all twenty seats in the SDF heartland North West Province on a reported 28 percent participation, which the SDF called inflated. Similarly, in CDU core territory, Noun Division of West Province with Foumban city at its center, the CPDM won all five seats on a reported 15 percent poll. These twenty-five seats constituted 30 percent of the CPDM's seats nationally; had they been contested and won by the SDF and CDU, and had the SDF won seats beyond its base area in North West, as was likely, an opposition coalition would have had at least the numbers (coherence is another matter) to organize a National Assembly majority. Disputes ever since about the wisdom or folly of the SDF and CDU boycotts rage even hotter than speculation about how those parties the boycott benefitted would have fared in a larger voter turnout.

In the event, following a brief organizational session of the new legislature, Biya in early April revealed the strategy for Cameroon's first governance in a quarter century to include a substantially pluralist party politics in the legislature. A belly politics alliance of the MDR with the CPDM gave the former four of thirty-nine cabinet posts and the latter a slim working majority of ninety-four National Assembly seats. Responding to both the peril and the opportunity the SDF boycott presented, Biya replaced Sadou Hayatou as prime minister with a Bamenda CPDM veteran, Simon Achidi Achu, elected on a low constituency turnout, to contest Fru Ndi on their joint home territory (they are from the same village complex, Santa; Achidi Achu's CPDM list had defeated one including Fru Ndi in the 1988 National Assembly election).

The regular June session lasted past its normal termination into late July, as debates and amendments reflected the NUDP and UPC opposition presence. The budget and economy faced pressure from international creditors, such that the nature and impact of a new labor code were more than normally complex issues and took two weeks to pass. The Assembly debated at length the next stage in the normal electoral sequence, for urban and rural councils, and the relation of those to be elected to *appointed* urban delegates (mayors) at their head. The results disturbed those with pluralistic hopes, preserving the appointees' autonomy familiar since 1974 and the regime's right to suspend councillors. Uneasiness about the government also attended the passage of laws regulating the National Assembly's internal conduct. Its president and enlarged executive bureau, and the party leaders, obtained budget powers, perquisites, and immunities more appropriate for executive than legislative styles of government, and for countries less impoverished. Party leadership took decisive control of any disputes created by the resignation or ouster of party members in the Assembly, against the grain of constituent initiatives and interests. One truly invasive move against constituent-responsive traditions gave members a secret ballot on Assembly legislation; this was rescinded a year later.[64]

Presidential Electoral Test, Late 1992

If this mid-1992 National Assembly experience left room for doubt about the CPDM's commitment to a transition from monolithic and executive toward democratic governance with a significantly autonomous legislature, a presidential election crowded all issues of policy routines and procedures completely off Cameroon's political map for the rest of 1992. The decision to bring this poll forward from 1993 must have been taken directly upon late July's end of the National Assembly and despite the death of Biya's wife July 29. First it was made known August 18 that local elections scheduled for October were postponed eighteen months, then Biya went on CRTV August 25 to announce his candidacy for the presidency in an October 11 election.

The legislature began an extraordinary session September 8. One question mattered: would old electoral laws respond to the new conditions? Substantive and technical issues fused in one key debate, about whether to adopt, with more than seventy parties now in action, a "first past the post" once-only vote or a two-stage vote like France's.[65] The government argument for precedent and cost reduction narrowly prevailed, favoring the one ballot procedure, against objections that multiparty evolution (as elsewhere in Africa) required the second stage and a winning majority, not just a plurality. Key to the vote was jockeying within the already splintered UPC, whereby its leadership differed on this point and the largest fraction of its legislative caucus voted with the CPDM. Its leader, Kodock, embraced Biya's candidacy September 28 and later took cabinet posts for himself and colleagues, a pitiful end to the UPC's principled half-century resistance to Cameroon's regimes.

Legislation also clarified residential qualifications for returned exiles as presidential candidates, including Bello Bouba of the NUDP but *not* a UPC leader contesting Kodock, Henri Hogbe Nlend, known to oppose the once-only presidential ballot. Another key vote rejected a scheme to enable parties to sponsor candidates collectively, in coalition, which would have been at least potentially useful to the opposition.[66] Law No. 92/010 and Decree No. 92/194 of September 17 turned the election process over, as before, to MINAT. In fact, Biya was already on campaign tour and issuing decrees which realigned administrative units and their personnel in directions favorable to the CPDM. If democratic transition *does* require the independent conduct of elections, as most analysts and advocates agree, the CPDM violated this standard throughout these preparations.

There were also significant developments bearing on the campaign outside the National Assembly, where SDF energies since the boycott of the legislative poll were predicated on winning the presidency, in an election it knew the CPDM would not long defer. We have noted Fru Ndi's refusal of the Yaounde Declaration late in 1991; in 1992, as he was joined in the March vote boycott by many who had signed it but then recanted as they discovered its implementation flaws, the SDF could and did claim the mantle of both popular and constitutional, in short truly organized opposition. Its First National Convention, May 21–27, 1992, in Bamenda, denounced Biya's politics of the past year as manipulative, arguing that he gave no economic relief and that he neither placed in jeopardy nor offered alternatives to the regime's mainstays, the unitary state and the power of the presidency, or the control of the armed force which ultimately sustained those civilian bulwarks.[67] The SDF reversed its National Assembly election boycott and mobilized a formidable challenge on paper and in the field for the real power it correctly located in the presidency.

But a bewildering fragmentation of the opposition set in, reducing its chances for success. There were early prospects for a coherent challenge. Bello Bouba appeared at a Bamenda rally in August and called for one opposition candidate against Biya, and a meeting there August 27 brought together key figures from inside and outside the ARC-SNC coalition. Simultaneously, the unprecedented resignation from Biya's government of the minister of public function (civil service), Garga Haman Hadji, and his subsequent appearances in opposition settings, created new momentum for the challenge.[68] But contradictory incidents and episodes also developed, raising doubts and suspicions. Ndam Njoya convened a group different than Bamenda's in Yaounde early in September. Fru Ndi on his way to the ARC-SNC's formal support for the presidency, conferred September 8, was briefly challenged for the presidential nomination from within his own party by the "Douala ten" defense lawyer, Bernard Muna.[69] Disarray in the opposition alliance structure reduced the impact of huge crowds Fru Ndi pulled to rallies. All this campaign rivalry compromised the opposition's democratization struggle, which reverted to politicking among its leaders, destroying the unified front against Biya and deserting the ground of policy and principle.[70]

By the end of the campaign, then, electoral laws still predicated on one-party practice and favoring the regime remained in place, and the opposition proved incapable of settling on one candidate. Comparing the conditions of October 1991 and October 1992, the NUDP and the largest UPC fragment, in 1991 both still NCOPA coalition members, were now in the National Assembly, and that UPC bloc's leader supported Biya. The *extra*parliamentary opposition was not coherently prepared for the election, given the basic SDF-CDU split in the intervening 12 months. Fru Ndi assembled a successor to the NCOPA and ARC-SNC, the Union For Change (UFC), to support his candidacy, but there was a void where the NUDP, CDU and some UPC elements had been in the rallies and meetings a year before. Ultimately, Bello Bouba and Ndam Njoya at the head of their parties challenged Fru Ndi as well as Biya for the presidency. These were all measures of the opposition's problems over a year's time, but also—what price incumbency?—reflected the CPDM's tactical success. It had found partners in government to replace Sengat Kuo, Ekindi, and others who had deserted its own ranks (Ekindi now also ran for the presidency). By doing so it had weakened an opposition now divided, as *its* erstwhile allies jockeyed for comparative advantage now that prospects of high office were involved.

There was one more significant, external factor leading up to October 11, polling day. A year before, in September 1991, a delegation from the National Democratic Institute for International Affairs (NDI), based in Washington D.C. with a congressional mandate and budget to assist in democratic transitions, arrived in Cameroon just when the open conflict between the regime and opposition escalated. Three NDI visitors, two Americans including New York's secretary of state in charge of its elections and a Paraguayan jurist, were in Bamenda September 13, 1991; they met Fru Ndi at his compound and viewed on Commercial Avenue what we described above, Biya's coldest reception of his national tour, verbally and symbolically, by elites present as well as commoners absent. NDI was among the potential foreign observers which declined to monitor the March 1992 legislative poll, finding the process too hasty and confused to prepare, supervise, and assess. All others chose or were persuaded to stay away from the October presidential poll as well, but NDI persisted, perhaps alerted by the earlier trip to Cameroon. It sent thirteen people from nine countries as presidential poll monitors, and trained 175 locals as poll watchers. The visitors were present in all but one province on election day October 11 and one of them, Maine's former attorney general, stayed another eleven days to observe the postballot process and consult a variety of concerned parties.

Perhaps the best preface to the vote and its outcome are in fact the NDI reports, both the interim findings released in Cameroon October 28, then its final version early in 1993. This was the core of the October 28 text, with full independent press coverage (*Cameroon Post* November 4–11):

> NDI ... seriously calls into question, for any fair observer, the validity of the outcome. It would not be an exaggeration to suggest that this election system was designed to fail. While several parties were responsible for election irregularities, the overwhelming weight of responsibility for this failed process lies with the government and President Biya.

The final, official report, in fifty-five pages and twelve appendices, found "serious fault with the electoral process" which "simply does not make it possible to determine which candidate received the most votes or which candidate would have been the winner in a fair election."[71] It cited irregularities before, during, and after ballots were cast, from the laws which governed the vote to the reports which announced its result, not without criticizing the opposition's conduct but more heavily citing the regime's. The NDI report concluded (the furious regime claimed it *determined*) that this election, like that of March for the National Assembly, would continue to block rather than resolve Cameroon's crisis of governance.

Controversy still surrounds what really happened October 11 and the next few days. Official results published October 23 in *Cameroon Tribune*, certified by MINAT's officials and upheld by the Supreme Court against protests, gave Biya 39 percent, Fru Ndi 36 percent, Bello Bouba 19 percent, Ndam Njoya 4 percent. Among the protests Fru Ndi lodged were procedural criticisms anticipating the NDI's and, most directly, the discrepancy between reported and real tabulations, including the reversal of early CRTV vote count announcements, such that Fru Ndi's actual victory by 2 percent became Biya's, stolen then upheld, by 3 percent. It is tempting but futile to analyze the SDF tally, but we must note before a detailed look at the CPDM's, which prevailed, that the total *it* gave Fru Ndi and Ndam Njoya almost matched Biya's, and that their combined vote *plus* Bello Bouba's totaled 60 percent to Biya's 40 percent. The opposition's fragmentation and the once-only vote returned the incumbent to power, as in Kenya shortly thereafter among other parallels.

The regime's *own* data demonstrated, from a total of 2,843,000 votes, 243,000 more than the March 1 count for the National Assembly, that the SDF's 1,006,000 tally included significant incursions both in CPDM territory and in areas where other parties, especially the NUDP, benefitted from the SDF's boycott of the legislative poll. The critical terrain was a bulge in most of three provinces contiguous with or near the SDF's North West base, where close to 1,000,000 people voted. In these provinces, South West, Littoral, and West, Fru Ndi polled 52 percent, 68 percent and 68 percent respectively, placing their NUDP and UPC majorities in the National Assembly election in true perspective. Not just the raw vote was significant; two key patterns emerged. The 68 percent support in Littoral and West meant that Fru Ndi won on former UPC terrain and bridged part of Cameroon's historic language gulf. Equally compelling were his margins of victory in the major urban centers of all three of these new support-base provinces, where commerce and industry create a cosmopolitan mix of indigenous and migrant populations, incipiently class-constituted and ideology-conscious, less marked by "vertical cleavages" of language, religion, and ethnicity. Fru Ndi's victories in those key electoral districts included Fako (Limbe city) 77 percent, Wouri (Douala) 69 percent, Mifi (Bafoussam) 89 percent, and Mungo (Nkongsamba) 87 percent; the exception was Noun (Foumban), won by Ndam Njoya with 59 percent. In the fifteen *départements* of those three provinces, Fru Ndi won eight with majorities and two with pluralities. Such results were striking in socio-economic terms, and pointed toward a composite growth in civil society, created by "villes mortes," crystallized and tempered by the subsequent pressures, and not deterred by the opposition's campaign disarray.

Perhaps most startling elsewhere (unless one recalls the spoiled ballots March 1) was Fru Ndi's 39 percent tally in Mfoundi (Yaounde) against Biya's 52 percent, for Yaounde was partly in terms of ethnic identity, and precisely in terms of civil service employment, the heart of the natural CPDM constituency. This was, however, Fru Ndi's only show of strength beyond the four provinces he won. In the CPDM strongholds, East, South, and Center outside Yaounde, and in NUDP territory in Adamawa, North, and Far North where Islam, Fulbe direction of electoral politics, and quasi-feudal society maintain a more "vertical" social structure and patron-client public life, he won as much as 10 percent of the vote in just four of thirty *départements*.

One dramatic incident added tension and color after the votes were cast and before results were issued. The Governor of East Province and

Blockade Against Security Forces During Bamenda's State of Emergency, November 1992. Photo by Omer Songwe, used with permission.

brother to the prime minister, George Mofor Achu, resigned October 19. His letter to Biya, published immediately by the independent press, revealed instructions from MINAT to do "everything fair and foul to ensure at least a 60 percent victory for the CPDM party candidate."[72]

Many ("most" is tempting, but impressionistic) Cameroonians believe the SDF won the 1992 presidential election. Most certainly the vote demography and electoral landscape related above deflected political momentum from the CPDM. But any moral or other form of SDF victory was bittersweet, for the regime proved the power of incumbency, perhaps its primal nature ascribed especially by Fatton among our sources, in the hands of determined or desperate rulers. To the Supreme Court's October 23 verdict in Biya's favor was added a severe press ban and then a State of Emergency for all of North West Province, October 27–December 28, to quell public disturbances there (although turbulence elsewhere, some election-related, some not, escaped this measure).[73] The province suffered at least four deaths and 173 irregular prison detentions in that time, many involving severe beatings incurred by rank and file supporters and by such prominent victims as retired Supreme Court Chief Justice Nyo Wakai. Fru Ndi and 152 others were literally confined to his compound, although defended from state security forces by relays of thousands of Bamendans (including the *takumbeng* "Amazons") who prevented his removal to Yaounde, and visited in shows of solidarity and for additional cover by Archbishop Desmond Tutu and numerous ambassadors and other delegates from foreign governments. As in the 1991 Bamenda resistance, publicity abroad spread knowledge and afforded protection.

Thus Cameroon ended a third year of open crisis and high drama somberly. The two elections of 1992 were no easy road to democratization, or panacea for the country's cumulative crisis. Quite the opposite: the campaigns and the conduct and results of the ballots opened more wounds than they healed and very likely caused commoners with optimistic expectations about light at the tunnel's end to turn disillusioned and cynical. The state apparatus plugged its gaps and checked or reversed civil society's advances through the elections, but failed to restore its credibility abroad because of the flaws in the polls; its gains were thus minimal, though indispensable French support remained intact. Rather than any real democratic transition taking place, let alone consolidation, indecisive stalemate was the outcome. With the election cycle over, the regime still in place and a certain exhaustion of energy and effort inevitable, 1993 ushered in a contest of wills and struggle of attrition over the long haul, less dramatic but no less instructive about the ways of African governance in the contest between autocracy and democratic pluralism.

Notes

1. A month later *La Nouvelle Expression,* April 13–17, 1995, documented her tirade about the university. Here, and for much subsequent text covering 1991 and 1995, evidence includes direct authorial and indirect informants' witness.

2. Charles Manga Fombad, "Freedom of Expression in the Cameroonian Democratic Transition," *The Journal of Modern African Studies* 33,2 (1995), pp. 211–226 discusses the enforcement of press constraints, an analysis amplified here. We thank Célestin Monga, personal communication, January 8, 1996, for details here and below.

3. For a brief history of censorship since 1960 and many parallels to the coverage here, see Francis Beng Nyamnjoh, "Contrôle de l'information au Cameroun: Implication pour les recherches en communication," *Afrika Spectrum* 28,1 (1993), pp. 93–115. For newspapers at the inception of political change elsewhere in Africa, see Monique Pagès, "L'explosion de la presse en Afrique francophone au sud du Sahara," *Afrique Contemporaine* 159 (1991), pp. 77–82, and Carol Lancaster, "Democracy in Africa," *Foreign Policy* 85 (1991–1992), pp. 148–165, which records the hazards of challenging autocracies.

4. Against the standard cautions about how few people read the little news African papers produce, we found a more than casual clientele looking for information in southern Cameroonian cities where politics are active, and a press that worked hard in difficult circumstances to cover events and provide analysis. The press is worth the attention accorded here.

5. See Achille Mbembe, "Provisional Notes on the Postcolony," *Africa* 62,1 (1992), pp. 3–37 for a Cameroon slant, Egbomi Ayina, "Pagnes et politique," *Politique Africaine* 27 (1987), pp. 47–54 for Togo detail with Cameroonian applications, and Kathleen Bickford, "The A.B.C.s of Cloth and Politics in Côte d'Ivoire," *Africa Today* 4,2 (1994), pp. 5–19.

6. Jacques Fame Ndongo, *Le Prince et le Scribe* (Paris: Berger-Levrault, 1988), p. 320.

7. That was the number discovered in casual kiosk browsing in 1991. In 1995, the SDF Bamenda's secretariat stocked issues from eighty different papers.

8. The influence of the independent press also spread to the state monopoly *non*-print media in ways that mattered. When both *Cameroon Tribune* and CRTV reported the six Bamenda shooting deaths of May 26, 1990, as a case of people "piétinés" (crushed) as if the crowd rather than security forces were responsible, the anglophone newsreader whom Bamenda people say introduced TV coverage of the event with the phrase "Lies from Bamenda . . ." (not "Live . . .") had print media training, and in 1995 became editor of a major opposition paper. Herbert Boh and Ntemfac Ofege, *Prison Graduate: The Story of Cameroon Calling* (Yaounde: United News Service, 1991) documents this and other anglophone challenges within CRTV newsrooms at the time.

9. *Challenge Hebdo*, March 9, 1995, interviewed Séverin Tchounkeu, editor of *La Nouvelle Expression*, and tried to draw him out about connections to Bamileke politicians, finances and Laakam, an ethnic association, but he declined. These two and *Le Messager*, among the most prominent critics of the regime and most often in court or sanctioned, were published by Bamilekes. Independent papers' finances were beyond our capacity for systematic study, but a June 14, 1995, interview with the publisher of *Galaxie*, the well known writer and regime critic Patrice Ndedi Penda, at the time battling censors and the courts, yielded information likely to be representative for this critical cohort. With an annual CFA 22,000,000 francs ($50,000) income from sales, he and his business manager wife each month paid eleven staff a total of CFA 575,000 francs ($1,300), ran local bureaus for CFA 75,000 francs ($160), and paid CFA 165,000 francs ($370) for office utilities. Each week's print run cost CFA 350,000 francs ($750). Sympathetic lawyers offered *pro bono* defence when Ndedi Penda was on occasions since 1991 prosecuted.

10. For the tart analysis of a star TV journalist after he left Cameroon to join CNN and then the World Bank in media jobs, see Eric Chinje, "The Media in Emerging African Democracies: Power, Politics, and the Role of the Press," *The Fletcher Forum of World Affairs* 17,1 (1993), pp. 49–65. Some papers were frivolous or worse, but *Le Messager, La Nouvelle Expression, Galaxie* and *Génération* among the survivors would stand any professional scrutiny, especially given circumstances sketched in this section. The English language *Cameroon Post* and *The Herald*, despite lower production standards, some hyperbolic news writing, and inflated opposition crowd estimates, carried news and commentary it took equal courage to gather and print.

11. For the use of political cartoons, with many reproduced, see Achille Mbembe, "La 'chose' et ses doubles dans la caricature camerounaise," *Cahiers d'Etudes Africaines* 36, 141/142 (1996), pp. 143–170.

12. Manga Fombad, "Freedom of Expression," pp. 215–216 covers the working details of censorship.

13. *OCALIP Info* 1, September 1993, a publication Njawe started so as to focus attention on the condition of the critical press, and which continues to appear at intervals.

14. See *Cameroon Post*, August 19, 1991, *Cameroon Express*, August 23, 1991, and Beng Nyamajoh, 'Contrôle de l'information" for these topics, amplified in August 1995 interviews with the independent journalist Omer Songwe, who specified Fame Ndongo's *grande école* as the regime papers' source. *La Caravane*'s motto conveys its politics: "Le Chien Aboie, La Caravane Passe/The Dog Barks, The Caravan Moves On."

15. Interviews with Omer Songwe and charges about *Dikalo* from SDF leaders are, respectively, the 1995 sources here. If any opposition political party operated newspapers without identifying the party origin, we are not aware of it.

16. *Le Messager* published a twenty-four party roster June 12, and called the list incomplete; others were operating but not yet legal and some were legalized but not yet public enough to be reported. There were 109 registered by mid-1995.

17. See Fabien Eboussi Boulaga, *Les Conférences Nationales en Afrique Noire* (Paris: Karthala, 1993); John Heilbrunn, "Social Origins of National Conferences in Benin and Togo," *The Journal of Modern African Studies* 31,2 (1993), pp. 277–299; Jacques Nzouankeu, "The Role of the National Conference in the Transition to Democracy: The Cases of Benin and Mali," *Issue* 21/1–2 (1993), pp. 44–50; Pearl Robinson, "The National Conference Phenomenon in Francophone Africa," *Comparative Studies in Society and History* 36,3 (1994), pp. 575–610.

18. Syndicat National des Enseignants du Supérieur (SYNES), *The University in Cameroon: an institution in disarray* (Yaounde, 1992) best covers the background. SYNES was formed as a faculty union by critics, subject to repression including slash wounds in a late 1991 attack on its principal founder, the physicist Jongwane Dipoko.

19. The Cameroon Human Rights organization, one of many by now in the field of investigation and reportage, named fifty-eight students "reported missing and feared killed"; *Cameroon Post*, April 25–May 2, 1991.

20. Miriam Goheen, *Men Own the Fields, Women Own the Crops: Gender and Power in the Cameroon Grassfields* (Madison: The University of Wisconsin Press, 1996), xii.

21. As the campus seethed, Biya fed the fury with a trip late in April to receive an honorary degree at University of Maryland (Eastern Shore). It was his last such occasion in the U.S.A. Shaw University (North Carolina) offered him a degree in 1993, but withdrew it at the last moment to forestall protests.

22. *The Sketch*, June 19, 1991, recorded this among other rumors, which included secession plans between Fru Ndi and Nigeria. SDF colors, green and white like Nigeria's, proved inconvenient in this sense, but may in compensation have attracted Muslims.

23. *Le Messager* (English ed.), June 20, 1991.

24. One of the authors arrived for five months of research the day before. This section and the next three thus draw especially heavily on eye witness, a variety

of actors and observers whose accounts are weighed and cross-checked, or both. Crowd numbers are estimates, usually far lower than what the opposition parties and independent press claimed.

25. *Cameroon Tribune*, June 28, 1991. A process of "la sublimation de la légalité" emerged; Patrice Bigombe Logo and Hélène-Laure Menthong, "Crise de légitimité et évidence de la continuité politique," *Politique Africaine* 62 (1996), p. 20.

26. More than his leadership of a small Parti Social Liberal Démocratique, this "villes mortes" coordination kept Mboua Massock at peril over the next years. He was in and out of jail through 1995; the press made his difficulties known.

27. Monga's passport was in constant jeopardy until he left to write and teach abroad in 1992.

28. A key unpublished paper, Emmanuel Fru Doh, "Women, Events and the Revitalization of Culture: Takumbeng as a Mankon Cultural Phenomenon in the Bamenda Grassfields," University of Buea Conference on Cameroon Writing, November 30–December 4, 1994, relates the tradition to such actions. More details are deferred to Chapter 7; *takumbeng* persisted in years ahead, especially in defence of Fru Ndi's compound, and took on an "Amazon" aura.

29. Skirmishes recurred in Bamenda long after the strike's end. See *Cameroon Post*, March 20–27, 1992, for the story "SNEC, SONEL Carry Out Mass Disconnections in Bamenda, Agents Beaten Up" when water and electricity company officials attempted to collect from those refusing to pay bills.

30. In and near Bamenda in 1991, police with local origins, families and domiciles warned civilians against sweeps, some in reprisal, by gendarmes largely from elsewhere in Cameroon and lodged in barracks. Radio France International reported September 28 gunfire between local and national security forces near Foumban (West Province). *La Détente*, October 3, 1991, reported the refusal of an army colonel to attack Douala civilians, and his replacement. A key sub-text soon emerged, as gendarmes and soldiers became the only rank and file agents of the state sure of their pay in the following years.

31. The first *SDF Echo* appeared a year to the day following the May 26, 1990, party launching. Edwin Wongibe, *The Social Democratic Front and the Thorny Road to Social Justice*, published without imprint in Bamenda, provides the party's best early history and select documentation in print, covering the period to June 1991.

32. It is casually of interest, to give this time in Cameroon its full flavor, that the veteran African coup specialist Bob Denard was reported present at the same hotel by *Cameroon Express*, October 11, 1991.

33. An SDF inner circle member speaks of contacts with Bello Bouba in Nigeria before his return to Cameroon in August 1991, which any full history of Cameroon parties will need to explore.

34. *Cameroon Express*, July 9, 1991; that number of defectors seems high.

35. *L'Aurore*, October 2, 1991.

36. She gained special attention because her resignation from the CPDM in April led to her expulsion from the National Assembly, raising a key constitutional issue about the balance of rights and interests between party, deputy, and electorate; see Samuel Efoua Mbozo'o, *L'Assemblée Nationale du Cameroun*

(Yaounde: Hérodote, 1994), pp. 43–45. Ndando was later an unsuccessful candidate to be the SDF's Secretary General.

37. Closer analysis is reserved for the next chapter, for their constitutional proposals appeared only in 1993–1994.

38. *CamerooNow,* June 5–12, 1991.

39. The text is from the section on Economic Policy Design in Cameroon Democratic Union, *Manifesto* (n.d.), p. 4.

40. *Le Messager,* August 10, 1991.

41. Full discussion of its failure is deferred until the next chapter.

42. The SDF's (thus far only) first national vice-chairman from its start exemplifies its own and other parties' search for national standing. Souleymane Mahamat is a northern Muslim by birth, a fluent bilingual speaker, and was long employed by the petroleum parastatal in South West Province until his politics intervened.

43. Jean-Louis Dongmo, *Le dynamisme bamileke* (Yaounde: CEPER, 1981) is a key source historically.

44. Much more on this topic follows in the next chapter, along with coverage of SDF fault lines between its anglophones and francophones and its leadership rivalries; none of them surfaced in 1991, but all did so in 1992.

45. A mid-July 1991 incident in Bamenda demonstrated these difficulties, when anglophone SDF supporters from salaried professional ranks, gathered for a beer, were shown the "trial balloon" transitional government proposed by the oppositionist *Cameroon Express* newspaper from Bafoussam, July 9. Keeping Biya as a figurehead president, naming Ndam Njoya prime minister and Fru Ndi foreign minister, the slate was dismissed out of hand for its ethnic balance. Nine out of twenty-eight from West Province, seven of them Bamileke, were too many for these anglophones, despite West Province's strategic importance in the opposition movement. Their derisive attitude was as telling as their ethnic calculus. But their scorn was understandable on local grounds, because this Bafoussam French language paper's cabinet member list included just two anglophones.

46. Details are from *Cameroon Post,* September 27, 1991, and briefings with a member of the SDF National Executive.

47. The meeting also divided the SDF from all others in attendance on the issue of whether voter registration and electoral competition should be NCOPA priorities. Far more prepared for a campaign and ballot than its partners, it accepted their majority view to retain sole focus on the Sovereign National Conference, as the NCOPA's experience with choice and complexity intensified.

48. *The New Standard,* September 28, 1991, and SDF sources provide this account of Pefok's text. *Challenge Hebdo* October 15 printed a French translation with the caption "My God!"

49. *Cameroon Tribune,* October 7, 1991, translated Biya's thanks for the gifts as follows: "I will use the drum to invite Cameroonians to unity and concord. I hope I will not be required to use the spear and the matchet. But, as the elders say, 'If you want peace, prepare for war.'"

50. Subsequent research confirmed serious violence, recurrent into 1992, caused by all the factors mentioned; Philip Burnham, *The Politics of Cultural Dif-*

ference in Northern Cameroon (Washington: Smithsonian Institution Press, 1996), pp. 1–3, 133–138.

51. *Cameroon Tribune*, September 26, 1991, reporting three deaths in Douala and Fru Ndi's simultaneous arrest at Nkongsamba, and *La Détente*, October 3, 1991, reporting Italian and American consular involvement.

52. Eye witness, casualty lists, and interviews with participants are sources for these three paragraphs. *The New Standard*, October 12, 1991, carried a full account, and *Amnesty International Report, 1992* (London, 1992), pp. 79–81, is a serviceable summary.

53. By extraordinary coincidence, football teams from the two cities met September 24 to determine which would advance to the premier division. PWD Bamenda won, and its return home with a week day motor escort was the only true relaxation of Bamenda's strike in all those months.

54. Ambroise Kom, "Writing Under A Monocracy," *Research in African Literatures* 22,1 (1991), pp. 83–92; Collectif, *Changer le Cameroun: Pourquoi pas?* (Yaounde: C3, 1990), pp. 18, 393. Two more books from this anonymous but traceable source followed in 1992 and 1993, *Le Cameroun Eclaté?* on the theme of Cameroon's "explosion" by the manipulation of ethnicity issues to conceal more fundamental roots of the crisis, and *Le 11 octobre 1992* on that year's disputed presidential election.

55. *Challenge Hebdo*, July 10–17, 1991, and an interview with Ndam Njoya, May 12, 1995.

56. See *Cameroon Tribune*, October 6–7 and October 14, 1991, for these texts and their close concurrence.

57. Intense newspaper coverage mid-October to mid-November recounted the working sessions held at intervals, interviewed principals and observers, printed draft proposals, and should inform any scholarship more detailed than our summary here. *Cameroon Tribune*, November 14, 1991, is the single most comprehensive source to our knowledge, for it included the text of the agreement and facsimile signatures, but the independent press stories and commentaries added variety and criticism.

58. *Challenge Hebdo's* headline for November 20–27, referring to Yaounde signatories, was "ET POUR QUELQUES DOLLARS DE PLUS" and the editorial was on "Les prostitués."

59. *Le Messager* (English ed.), January 23, 1992; *Le Messager*, January 30, 1992; *Dikalo*, February 6, 1992; *Cameroon Post*, February 28–March 6, 1992.

60. *Le Messager* (English ed.), January 13, 1992; *Cameroon Post*, January 29–February 4, 1992.

61. *Cameroon Post*, February 19, 1992.

62. Disclaimers against the total count should be noted; *Cameroon Post*, March 27, 1992, stated that just 1,000,000 voted.

63. For the MDR's genesis and its leader Dakole Daïssala, see Kees Schilder, *Quest for self-esteem: State, Islam and Mundang ethnicity in northern Cameroon* (Aldershot: Avebury, 1994), pp. 224–226.

64. Efoua Mbozo'o, *L'Assemblée Nationale*, pp. 87–95.

65. Efoua Mbozo'o, *L'Assemblée Nationale*, pp. 96–103, provides working detail about this debate, essential background for the campaign itself.

66. Whether it could have made multiple sponsorship and coalition work is questionable, in light of campaign details below.

67. The inaugural special edition of *The Herald*, June 11, 1992, is the best source for the convention, alongside the party's own productions of texts and *SDF Echo*, and includes photographs of representatives in attendance from the embassies of the U.S.A., Great Britain, and others from abroad, increasingly concerned about their interests in Cameroon and the country's stability.

68. *Cameroon Post*, August 21–28, 1992; *Challenge Hebdo* (Special Edition), August 28, 1992.

69. *Cameroon Post*, September 4–11, 11–18, 1992, including the latter's subheadline "7 Candidates So Far!" story as a reflection of the mushrooming chase against Biya.

70. Comprehensive campaign coverage would also include the extraordinary photographs published in France of the massive crowds at Fru Ndi's rallies, in the hundreds of thousands at Douala and Yaounde. One in particular, in *Afrique Magazine* November 1992, shows him holding a pistol and describes it as having been passed up from a Yaounde crowd, taken from a suspected assassin whom the crowd killed on the spot despite Fru Ndi's plea not to do so.

71. National Democratic Institute for International Affairs, *An Assessment of the October 11, 1992 Election in Cameroon* (Washington, 1993), Executive Summary, p. 6.

72. See National Democratic Institute, *An Assessment*, pp. 109–111 for the text. Mofor Achu went into hiding and then left Cameroon.

73. Cameroon's independent press, much of it dodging censorship with name changes, and *Jeune Afrique Economie* (Paris) covered Bamenda in this period comprehensively, in words and photos. FAX capacity played a key role, as messages from Bamenda to embassies at home and human rights groups and others abroad documented its ordeal. One symptom of the postelectoral crisis was the removal of close to 100 non-Beti troops, mostly anglophones and Bamilekes, from Biya's elite security corps; *The Herald*, December 16, 1992. The SDF inner circle speaks of its own intelligence contacts within state security at the time, enabling the party to anticipate reprisals and reduce casualties in Bamenda.

6

Impasse, 1993–1997

Cameroon became more "normal" 1993–1997, in the sense anglophones use this word to register the *conventional* uncertainty of their condition, than immediately before. It entered a phase of its national experience where the constant clash of polar opposites gave way to more subdued routines: the force of the regime's incumbency and its effort to marginalize all challengers; prolonged attrition, stalemate, and the pedestrian and often indecisive grind of institutional rather than movement politics; the lures of office-holding; party efforts for comparative advantage in the search for larger constituencies, understood in "zero sum" terms as requiring attacks on rivals; wars of words between parties and among factions within them. Dramatic episodes flared, especially as anglophone politics emerged, but the story line from 1993 forward lacks the previous years' passions and immediacy.

Still, the period we now examine was no less important for an understanding of politics and the choices leading to losses or gains for the democratic process in Cameroon and Africa. Political systems which absorb shocks like Cameroon's during 1990–1992 risk lapsing into apathy or lurching into anarchy, and the African examples of such outcomes are sobering. If Cameroon's experience was less compelling, 1993–1997, the stakes remained high in the search for a practice of politics capable of peacefully incorporating the new interests and resolving the multidimensional crisis within sectors of the state system and between the state and civil society.

We thus turn to that period's specific details and general shape. The postelectoral institutional and party alignment entering 1993 is the first focus. Then comes attention to multivocal, subnational factors which disturbed the visible "politics as usual" profile and revealed the maintenance of crisis at the sharpest edges of regional and ethnic politics, especially the emerging anglophone sovereigntist movement and subtler but substantial currents from the North. We then address texts and actions related to the constitution, a controversial arena which helped define both the interest group politics of the mid-1990s and the longer term visions of Cameroon's future. The chapter concludes by returning to a survey of the parties, their leadership, conduct, and prospects as of 1997. Impasse is the major theme, but we will consider the chances for democratic transition and consolidation as we proceed.

Party and Institutional Politics: The Early 1993 Profile

The 1992 contest for the presidency gave three parties a clearly national standing. Three others maintained a bargaining capacity at the center, or identifiable local bases of support, or both. Others survived only as satellites or ciphers.

Both the CPDM and SDF polled in the 35–40 percent range October 11, 1992 (depending, we recall, on whose results are believed), followed by the NUDP's 19 percent; recognizing their core appeals in regions, these were clearly national parties. The CDU, MDR and UPC entered the reckoning with some tangible measure of support, or in alliance with more substantial forces. Foumban gave the CDU and Ndam Njoya 59 percent of its votes for the presidency, combining with his less tangible virtues to make Ndam Njoya a presence in all "high office" speculation and calculation. The MDR did not offer a presidential candidate in 1992, but its six seats in the legislature, four ministerial posts, and niche among non-Fulbe, non-Muslim people in Far North Province gave it a governmental presence in alliance with the CPDM. What remained of the UPC continued to squabble over leadership, but the prevailing faction after the 1992 elections maintained its role in the National Assembly, blunted a still militant wing (UPC-MANIDEM), and under Kodock accepted four ministries and yoked its future, like the MDR's, to the CPDM, yielding what remained of the UPC's militant following to the party's other factions it defeated and, more to the point, to the SDF. Some other leaders and parties of marginal consequence folded into the CPDM and took rewards, like Antar Gassagay (National Party for Progress) who became the deputy in the powerful ministry of territorial administration. Others stayed in opposition as part of various SDF-led coalitions in the following years, like Samuel Eboua and the Movement for Democracy and Progress which he formed with support around Nkongsamba, Littoral Province, and a scattering elsewhere after losing the NUDP presidency to Bello Bouba early in 1992.

The formal politics of 1993 and subsequent years in fact began November 27, 1992, with a close resemblance to Ahidjo's earlier cooptation and alliance strategy for the CNU. Four members of the UPC wing under Kodock, Kodock included as the minister of agriculture, joined Biya's government. Two NUDP notables, Hamadou Mustapha as both vice-prime minister and minister of urban affairs and housing, and Issa Tchiroma as minister of transport, also took office under Biya; Bella Bouba and the party denounced but did not expel them.[1] These appointments brought members of two *pre*presidential election parties of National Assembly *opposition* into the ranks of government. This realignment served the CPDM's need to present the face of a national unity coalition at home and abroad following the disputed poll, and significantly enhanced the patronage, perhaps the influence, but not the power, of UPC-Kodock and a segment of the NUDP. What it meant for the idea and practice of a more legislatively oriented Cameroonian governance, proclaimed as a goal by the CPDM since early 1991, and for an autonomous opposition *within* the state apparatus, was clearly enough negative.

The remaining *non*parliamentary opposition responded to these events early in 1993, and the SDF again took the lead. Very soon after the state of emergency was lifted and the postelectoral lull ended, John and Rose Fru Ndi attended President Clinton's inauguration, invited by the Inaugural Committee on behalf of the Congressional Black Caucus. It was Fru Ndi's chance to defy Biya's regime for its irregular conduct of the poll, the state of emergency, and his people's and his own suppression, as well as to cut a figure abroad. Widely circulated photographs of what was clearly projected as Cameroon's true "presidential couple" meeting the Clintons, Nelson Mandela, and others made their point through the independent press back home.[2]

Fru Ndi wasted no time resuming action after his return. The Union for Change (UFC) alliance issued a Bafoussam Plan of Action, and began a six week tour of provincial capitals with a massive Yaounde rally February 25. Bamenda's streets filled again March 25, with the display like a year before of not just SDF banners but also the two-star flag of the growing anglophone federalist movement which we will document at length below; military intervention this time was harsher, adding two deaths and twenty wounded to Bamenda's toll. Fru Ndi persisted within the limits set by regime violence, choosing the provocative date April 6, the ninth anniversary of the abortive coup against Biya, to issue a call as "Legitamate President-Elect" for the Sovereign National Conference which still primed SDF policy. No matter its quixotic sound and limited post-1992 prospects: it sustained the symbolic politics begun in Washington D.C. and it opened the next competitive cycle of Cameroon's politics.

Regional and Ethnic Politics, 1993–1997: Anglophones

But once politics resumed an active postelectoral course, activity mobilized *outside* the institutional and party fabric was likewise visible and remained significant for the next four years. Most strikingly, and concurrent with the SDF's return to action, masses of anglophones began to work beyond not just regime but also SDF discipline. Their efforts provided mid-decade highlights and headlines, and posed problems for both Cameroon's state and the SDF. Less urgent but still noticeable parallels among other subnational groupings likewise broke the political surface. We start with anglophone activities, the most dramatic and consequential, then cover other experiences which denied the regime's hopes for quiet, untroubled governance.

We have seen 1992 and 1993 Bamenda evidence of anglophone frustration and militancy, with casualties mounting as the federal flag appeared. These episodes reinforced Bamenda's determination and foreshadowed initiatives elsewhere, renewing specifically anglophone energies dormant since 1972 and complicating national politics. Their impact on the country's ideological repertoire and political dynamic was immediate and consequential.

Reference to the veteran anglophone militant Albert Mukong, briefly introduced in Chapter 4, best frames these developments. Born in 1933, seasoned by student nationalism of the 1950s in Cameroon and Nigeria, then by a rich brew of party, electoral, subministerial, and exile politics in the 1960s, his life in and out of *prison* commenced October 7, 1970, in a cell with Ernest Ouandié just before the latter's execution closed the armed phase of UPC resistance to Ahidjo. The legacy was passed along; Mukong spent half the intervening time until 1990 in Ahidjo's and Biya's jails. He became both the anglophone parallel to the UPC francophone martyrs in all but the ultimate sense, and Cameroon's best known prisoner of conscience. His attention turned by the early 1980s to specifically anglophone advocacy, with the lawyer Gorji Dinka his ally in a bid to enlarge the anglophone presence in the CPDM when it replaced the CNU at Bamenda in 1985. That failed, so the search for an autonomous anglophone politics began with pressure groups, often acting and publishing clandestinely, like the Cameroon Anglophone Movement (CAM) and the Free West Cameroon Movement at home and the Cameroon Federalist Committee abroad, all part of the prehistory of Mukong's much disputed but considerable role in the SDF's birth, 1989–1990.[3]

Freed April 1990 from (thus far) his last indictment and imprisonment, in the Yondo Black case, he left Cameroon a month before the bloody public launch of the SDF May 26, dissatisfied with the emergence in the party of people and policies he thought deviated from anglophone priorities. His purpose was to document and argue at the United Nations and elsewhere the case for the dissolution of Cameroon as presently constituted and the restoration of former British Southern Cameroons, not yet as a fully sovereign polity, but free to choose its own future in a way denied by the dereliction of undertakings given anglophones, and by their own leaders' faults, when it emerged from trusteeship thirty years before.

During Mukong's two years away, April 1990–April 1992, the SDF's growth gave Cameroon at large, not just anglophones, an institutional resistance to autocracy for the first time since the UPC's eclipse. When he returned April 12, 1992, the anglophone cause remained Mukong's focus, and he helped rally its forces principally outside the SDF, though retaining a more than honorific place in SDF circles.[4] The first organizational step was the open appearance of CAM at its First National Convention, Buea, July 4, 1992, with Mukong its secretary general. Further proof of his work and collaborations emerged in early April 1993 when Buea hosted the first All Anglophone Conference (AAC I), the weekend before Fru Ndi April 6 renewed his more inclusive all-Cameroon call for the Sovereign National Conference. AAC I was summoned to voice a more directly anglophone view of politics and the constitution than the SDF's, given that party's national policy orientation and substantial francophone constituency. The site for CAM and AAC meetings was not a random choice; Buea is the South West Province capital and also, far more significant for those gathered, was the seat of anglophone governance prior to reunification in 1961 and then in the federal era until 1972.

SDF members were in fact prominent among the 5,000 gathered at Buea, and one of the most complex and delicate Cameroon issues ever since is the connection within anglophone politics between the SDF, the AAC, and the latter's varied progeny. Questions about anglophone groups persist: how to separate AAC background pressure group politics from the SDF foreground of open vote and office seeking; how to place and understand their overlapping members and leaders, and the distinctively North West Province cast of SDF leadership against South West Province AAC direction; how to read their different stances regarding francophones and the Cameroon state; how to understand Fru Ndi's roles and motives as anglophone initiatives threatened to compromise his leadership of a national party but at the same time created anglophone solidarity the SDF found useful; whether in fact the SDF cultivated the AAC in order to make the SDF appeal, as comparatively moderate, to francophones and foreigners.[5] But there was no mistaking the determined tenor of the Buea meeting, calling for anglophone unity as a first principle of further action, or the momentum it built.

Mukong's two years consulting and lobbying abroad yielded one most vital argument placed in his background paper for Buea, "Where Things Went Wrong." This was its core: from the point in 1958 when the United Nations began to address Southern Cameroons' movement from Trust Territory status, through the Foumban Conference in 1961 which sealed the relationship with the already sovereign République du Cameroun, an agreement to give the anglophone people equal status in "complete independence" *prior* to any reunification was not honored. Southern Cameroons, the argument went, was thus abandoned to its own weak constitutional and political devices, from which flowed all the intervening grievances. United Nations, British, Cameroon anglophone and francophone mistakes, and perfidy were all responsible; whatever the nature and balance of the faults and blame, and constitutional niceties aside, here was the paper's decisive core:

> From the foregoing we may sum up that there is no agreement that today binds us to be in one state with the francophones. . . . If both parties are still interested in the union, we should assume our pre-October 1, 1961 status and then go into negotiations.[6]

Calling on the United Nations to resume its lapsed responsibility to anglophones, calling on *them* to execute a sovereign peoples' will, and incidentally citing contemporary Croatia and Somaliland as inspirations, Mukong laid the foundations for the "restorationist" policy the Buea Declaration endorsed April 3.

Its language was trenchant, enumerating cultural, economic, and political details embedded in the constitutional landmarks of 1972 and 1984, summarized in the vocabulary of "marginalisation" and "divide and rule." The "whereas" clauses led to one clearest article in the Declaration (fully capitalized in the original):

> all Cameroonians of anglophone heritage are committed to working for the restoration of a federal constitution and a federal form of government which take cognizance of the bicultural nature of Cameroon and under which citizens shall be protected against such violations as have hereabove been enumerated.[7]

Those gathered in Buea thus sought their *own* constitutional resolution of a frustrated history, and clearly chose a language driven, two state rather than a multiple federation. Cameroon's center of constitutional gravity was under siege.

The AAC's growing presence and impact on the SDF registered within four months, in the way the party debated Cameroon's constitution at its Second Congress, July 29–31 at Bafoussam. The francophone choice of venue, in a city which matched Bamenda's insurgency in 1991, and where the regime first attacked Fru Ndi bodily, signalled the party's drive to enlarge its *national* appeal. Fru Ndi's opening policy address very clearly avoided use of or reference to the federalism the AAC had specified at Buea; his language in fact was 1991's:

political, constitutional and administrative decentralisation, based on the democratic election of governors and other officials, within the context of a pluralist state

with details to be worked out "by the people themselves" was the policy he enunciated, close to our Chapter 5 measure, Sonac Street, Bamenda's "Elected Governors and Division Officers" sign two years before.[8]

But constitutional issues, thus far skirted rather than resolved in the SDF, went to a Bafoussam committee for debate. A paper from Mukong reflecting the anglophone current was circulated, as at Buea in April. Entitled "Let's Keep Cameroon One," it pushed the two state federal argument as prime party principle, arguing that the SDF had forsaken its 1989–1990 origins among anglophones in a search for national support which must be spurious so long as the flawed, illegal constitutional features from 1961 forward remained. In that process, Mukong argued, a francophone membership majority of 60 percent emerging in the SDF now led it astray. To get back on course, the SDF must incorporate AAC I's federalism adopted in April and the subsequent declarations for a specifically two state federation, based on the anglophone-francophone distinction, which North West and South West Province SDF Conferences adopted in June and July respectively and brought to Bafoussam.[9] Now was the time, Mukong and these texts argued, and the last time, for the party to return to its foundations, so as not to abandon anglophones and be abandoned by them in turn.

The argument achieved part of its purpose. A constitutional subcommittee recommended a policy which led delegates to resolve at the end of the convention that in this "multi-ethnic, pluri-cultural and officially bilingual Nation" they

> Envisage the putting in place of a federal republic ... the number of states being left to be determined by the Sovereign National Conference according to certain objective criteria such as geographic contiguity, ethnicity, cultural heritage, economic viability, etc.[10]

The compromise, the skill needed to forge it, *and* the distance still remaining between the national constituency the SDF had cultivated since 1990 and the regional and local anglophone pressures built since 1992, were clear. The SDF for the first time committed itself to Cameroon's future in a federal constitution, but avoided prescribing the number of components and the clarity of the two state anglophone-francophone design emerging from Buea. The Sovereign National Conference remained its preferred vehicle for Cameroon's democratic transition in constitutional as in other terms. This federalism in principle was given firmer shape in August 1994 by the National Executive Committee, advocating

> une Fédération de quatre Etats, plus le Territoire de la Capitale Fédérale dont le statut sera défini par une loi du Parlement Fédérale/a Federation of four States, plus the Federal Capital Territory, whose status will be defined by a law of the Federal Parliament.[11]

The SDF's next Congress at Maroua in May 1995 confirmed in principle "the option of Federalism adopted at the Bafoussam Convention" and continued to repeat this call for four states and a federal capital, without specifying *their* boundaries or *its* site.[12] These are matters of utmost delicacy, posing the choices not only of a (perhaps new) capital, but between South West Province's precolonial affinity with coastal peoples of Littoral and its colonial and later connections with anglophones in North West, an issue lying beneath every surface of anglophone and national politics.

Anglophone militants were not satisfied with the SDF's move to a looser federalism than the AAC favored. Mukong remained active outside SDF ranks, then in 1995 abandoned direct political action so as to concentrate on nonpartisan human rights work. But other anglophones pushed the AAC cause forward. Its second meeting in Bamenda, April 29–May 1 1994, proved dramatic in its conduct and the policy text it issued. Once again determined to suppress Bamenda and its politics, the regime reinforced security, preventing assembly Friday on the planned site, the Catholic cathedral grounds. Dispersing to private houses for secret working sessions, delegates gathered undetected Sunday at the Presbyterian Church Center, responding defiantly both to Cameroon politics on the anglophone question at large and the weekend's show of force. They offered "Francophone Cameroun" negotiations on the basis of a federation, specifically to include an anglophone state. But anticipating the regime's rejection, the Bamenda Proclamation (like the Buea Declaration's key passages, fully capitalized in the original) then stated that the Anglophone Council it now created would

> thereupon, proclaim the revival of the independence and sovereignty of the anglophone territory of the Southern Cameroons and take all measures necessary to secure, defend and preserve the independence, sovereignty and integrity of the said territory

and then proclaim itself a constituent assembly to write and adopt a constitution.[13]

Anglophone lawyers and legal scholars, trained and experienced in Anglo-Saxon, common law traditions, whom we will meet again when this chapter analyzes constitutional proposals, were by now emerging in the AAC. Younger than Mukong, they became the movement's strategic and programmatic voices, marking its transition from the heroic leadership style of Mukong in the wilderness to their more institutionally directed and enabling skills. Chairing AAC II at Bamenda was Sam Ekontang Elad from South West Province, the leader of the Liberal Democratic Alliance based there, which for reasons of autonomy rooted in historic North West-South West differences among anglophones as well as leadership and policy concerns had never joined the SDF's coalitions. He admitted when interviewed after AAC II the danger that "the multiplicity of Anglophone pressure groups" was eroding support for Fru Ndi, but that they were justified by implacable francophone hostility within Cameroon as presently framed.[14] His colleague Simon Munzu, also distanced from the SDF, actually read the Bamenda Proclamation at the secret Sunday meeting and soon became the movement's most effective public voice as well as ranking legal scholar. A third prominent figure at Buea in 1993, Carlson Anyangwe, had by the time of AAC II left for Zambia; as the only one of these three who was from North West Province and close to the SDF, in fact a party Founding Father, his departure widened the SDF-AAC gap and thus moved the constitutional center of gravity among anglophones as well as for Cameroon at large.[15]

The AAC later in 1994 took the new name Southern Cameroons Peoples' Conference (SCPC), with the Southern Cameroons National Council (SCNC) as its executive arm; there was now a roster of articulate and insistent voices to anchor the anglophone federalist cause on the basis, ever more specific now, of *two* states. When Biya made no effort to start the negotiations called for at Bamenda, anglophones took their next initiative. The name "AAC III" does not fit what took place May-August 1995, but that was very much the spirit as the movement expanded, working abroad and staging open demonstrations at home. With the striking leadership of eighty-four year old John Ngu Foncha and seventy-eight year old Solomon Muna, both Ahidjo's vice-presidents, once rivals but now reconciled elders, eleven anglophones chaired by Elad with Munzu as their spokesman spent a month visiting United Kingdom, United States, and United Nations politicians and officials, as well as supporters and publicists for their cause. They basically renewed Mukong's restorationist strategy at a higher level and with broader support and fuller documentation than he could claim, 1990-1992.

While abroad through June, they were front page news only for independent anglophone journalists back home; francophone counterparts covered them sporadically and *Cameroon Tribune* hardly at all. Their homecoming was another matter, and became the stuff of legends. The mission landed in Douala June 28 and crossed the Mungo Bridge at the historic border between the "two Cameroons" the SCNC proclaimed. Led by its two elders on foot, it sung the movement's hymn, then reached Buea and displayed the United Nations flag. It was a well orchestrated, symbolically charged campaign, which Great Britain's ambassador joined the next day, without a public speech but with strong if silent impact. Advertised with broadsheets and welcomed with Mosaic and other biblical rhetoric, and large crowds, the group spent July and August touring the anglophone provinces amid uneasy, unpredictable but usually hostile responses to their appearances by the authorities, who no longer ignored the movement. The plan to rally in Bamenda July 3 encountered troops who used 75–100 tear gas canisters, but negotiations with the Governor (still Bell Luc René) led to a calm assembly July 4 with 15,000–20,000 on hand, twice the Municipal Stadium crowd welcoming the SDF home from its National Congress a month before. Munzu was arrested July 12 at Kumba, South West Province, but released in circumstances including the population's threat to burn the city's gendarmerie site. Meetings hastily convened by the regime enlisted the two provinces' "elites" against the SCNC, but Bamenda's July 11 was moved inside the Congress Hall to avoid popular hostility; Prime Minister Achidi Achu, presumptively on home ground, was ignored as his cavalcade moved down Commercial Avenue to the meeting, his car was stoned as he departed, and tear gas followed. The British ambassador, having joined the SCNC at Buea three weeks earlier, sent regrets when invited to the *regime's* July 15 South West Province meeting meant to counter the SCNC.[16]

No longer obscure on Cameroon's margins, militant anglophones thus commanded national and even some international attention by mid-1995. The contrast between this era's elite defection and mass mobilization and the 1960s, when Ahidjo brought anglophone leaders into the CNU before springing the 1972 referendum on them, with Foncha and Muna in turn his closest collaborators, could not be more stark. The AAC-SCNC anglophone wing has yet to attain its principal stated objective, United Nations action restoring the freedom to choose between options ranging from an acceptable federalism within Cameroon's current boundaries to, in the extreme case, secession. Nor did it deter the Commonwealth from admitting Cameroon in November 1995, despite arguments that it should wait until the regime's anglophone policy and human rights record improved (membership would balance Cameroon's place in "la francophonie"). And SCNC leadership weakened late in 1995 when Elad lost credibility by responding to a CPDM effort to make him appear to be discussing Cameroon's future with Biya's agents without SCNC colleagues' approval.

How well the SCNC would fare, not being a party, once Cameroon's electoral cycle resumed and the SDF ran candidates in a "Cameroon" which SCNC secessionists rejected, or how the latter would negotiate separation or proceed to what they called a "zero option" and abandon politics for armed struggle, as some by 1995 advocated, citing Eritrea, was unclear.[17] So was the SCNC's capacity as a movement, like the SDF's as a party, to resolve differences between North West and South West Province anglophones, if the entire AAC-SCNC phenomenon, as some argue, represented South West Province efforts to counter the North West Province origin and appeal of the SDF as *the* agency for anglophones.

SCNC work continues. A mid-1996 press release declared the intent "to convene a Constituent Assembly of the people of the Southern Cameroons and declare its independence"; a Unilateral Declaration could happen.[18] The SCNC's influence on national politics through its pressure on and connections with the SDF remains. It might claim a place if a comprehensive forum about Cameroon's future constitution should ever meet. As the long simmering, now heated product of regional and ethnic rage, the SCNC surely recalls Jacques Benjamin's 1972 text (from, remember, Québec) on the anglophone potential for disruption of the Cameroon state's hegemonic project, and the likely inadequacy of a federal approach to governance.[19] Cameroon's nation-state was by the mid-1990s rendered all the more problematic by the SCNC's mobilization of anglophone sovereigntists.

Regional and Ethnic Politics, 1993–1997: The North

The regional and ethnic dimensions of Cameroon's "Great North" politics in the provinces of Adamawa, North, and Far North added another case study to the sub-national profile in 1995, that of La Société de Développement du Coton (SODECOTON). It was in its own way, without the more openly dramatic flashpoints of anglophone politics, equally or more threatening to the state's and regime's long term integrity. SODECOTON is the parastatal responsible since 1974 for selling in raw or processed form the North's staple crop of most consequence, cotton, once it is gathered from the quarter million rural families whose small plots and dated tools and transport still dominate production at the agro-pole of this agro-industrial enterprise. Its commercial and industrial component, employing 2,500, was in fact pre–World War I German in origin, amplified by France's creation for West Africa of La Compagnie Française pour le Développement des Fibres Textiles (CFDT) after World War II. It was 70 percent nationalized by Decree 74/473 in 1974 creating SODECOTON as an integrated regional development scheme, leaving CFDT with a 30 percent holding.[20] The business was profitable in all but two of the next twenty years, fiscal 1991–1992 and 1992–1993; CFA devaluation in 1994 restored sales abroad, so that fiscal 1994–1995 looked lucrative when, March 29, 1995, SODECOTON's finances became a very political affair.

The process of privatizing state enterprises demanded since the late-1980s by Cameroon's large creditors abroad changed course that day with SODECOTON, one of the few attractive parastatals left. With ordinances and decrees specifying procedures and candidates in effect since 1990, it was added to Cameroon's list of eligible enterprises in 1994. But a March 29 press release from the ministry of the economy and finance announced suspension of its sale, and Decree 90/056 then moved SODECOTON's dossier from an interministerial council in charge of privatizations to the presidency, "sur hautes instructions du président de la République." That intriguing language mobilized the independent press to cover a region without its own secular private press, in other ways isolated, and thus seldom reported; this journalism was the one print source available as a guide to current events or at least as a barometer of opinion, given the silence of government and other principals in a case so sensitive. Stories, speculations, and editorials during the next weeks examined the North's political economy and the geopolitics linking it abroad, and augmented recent scholarship, in ways very germane to our coverage of regional politics here.

Six months before, a series of purchases had begun to transfer the assets of two other parastatals still intact and a third in receivership, between them holding 48 percent of SODECOTON's shares, to a new enterprise, the Société Mobilière d'Investissement du Cameroun (SMIC), in order for SMIC to become SODECOTON's largest though still minority owner, with Cameroon's treasury maintaining its 22 percent and CFDT its 30 percent. One key element in every nongovernmental version of the story was the evasion, both practised and permitted, of the guidelines about competitive bidding and of the safeguards for strategic sectors of the national economy specified in the privatization codes. The purchase prices thus far transacted for SODECOTON varied between 15–25 percent of most estimates of fair market value since its postdevaluation recovery. Cheap prices might suit failed or failing ventures (like SOTUC for urban busses, noted above) but not one of Cameroon's few major enterprises still viable in commercial terms, which was furthermore a cornerstone of the remaining economy, public service structure, and private income in the North, one of Cameroon's most populous but least developed areas. Further, with evidence of the prime minister's and the minister of industrial and commercial development's signatures on documents approving SMIC's purchase widely spread through government, thus far unopposed, most accounts squared with what the paper with the fullest coverage, *La Nouvelle Expression*, stated April 4–10 about "vraiment le plus grand scandale politico-financier de l'année":

à n'en point douter que toute la classe dirigeante camerounaise était au courant des transactions menée avec la SMIC/there is no doubt that the entire Cameroonian ruling class knew about the transactions conducted with SMIC

now subject to the March 29 reversal.

More intriguing still was SMIC's northern leadership. According to *Perspectives Hebdo,* April 11–17, it crossed political party lines to link the North's key elites, directing affairs for 200 shareholders nation-wide. From the CPDM came Biya's 1991–1992 prime minister Sadou Hayatou, the current National Assembly speaker Cavaye Yegue Djibril, a party vice-president Sadou Daoudou, and the single most powerful northern potentate by most reckonings, the lamido of Rey Bouba (on whose lands by one report, *The Herald,* April 3–5, 40 percent of SODECOTON's product is grown). From the NUDP came its treasurer Haddabi Barkindo, and Hamadou Mustapha, since 1992 simultaneously one of the nation's two vice-prime ministers and minister of urban affairs and housing in Biya's government, although he was expelled from the NUDP in January 1995 (i.e., in the advanced stage of SMIC's SODECOTON purchase) for having taken the ministries without the party's clearance. Two lines of questions followed from these names, once disclosed. Most obviously, what derailed SMIC when so many key CPDM figures had a stake in it? Less obviously, but more crucially here, what brought the North's leaders together in SMIC *across* the party divide?

Answers on both fronts emerged, mingled, touched nerve centers of national governance, and involved subnational politics in complex but vital ways. Consensus on the answer to the first question and its *political* rationale surfaced quickly in the investigative press, initially in *Le Messager* April 3, then elsewhere: succession to the presidency was at stake. With the 1972 constitution amended in 1984 silent about a "dauphin" (Article 7.2 vested interim power with the president of the National Assembly, leading to a new election), a political struggle therefore pitted two likely candidates, Hayatou and the minister whose press conference March 29 ground Hayatou's SMIC venture to a halt, Justin Ndioro. Both were primarily economic managers and technocrats rather than ethnic politicians in their CPDM party roles and loyalties; Ndioro in particular was from an ethnicity of little political consequence in Mbam, Center Province, where neither regime nor opposition claims "natural" ethnic constituencies but the CPDM still wins elections. They became by the mid-1990s the most attractive among the CPDM's own aspirants for Biya's office, given domestic realities disqualifying anglophones and Bamilekes, and foreign preferences for experience in economic portfolios and against Betis. Hayatou directed agriculture, economic planning, then the economy and finance before becoming prime minister, 1991–1992, then was made director of the region's transnational Banque des Etats d'Afrique Centrale. He in fact introduced Ndioro to government in the economy and finance post he left to become prime minister in 1991, attracted by Ndioro's years of experience with the electricity parastatal and with Cameroon's aluminum plant that is part of France's powerful Péchiney combine. On this reading of the political calculus, both Biya and Ndioro benefitted by blocking SMIC, given a presidential succession scenario which foreigners influence, France perhaps in decisive fashion. Biya, if a non-Beti must succeed, would prefer a politician like Ndioro with a small ethnic base and thus more pliable, than one like Hayatou with a strong northern constituency. Ndioro would benefit from SMIC's eclipse by cutting Hayatou out of vast political leverage and fiscal resources which a successful bid for SODECOTON would provide him.

The *fiscal* rationale within the CPDM for Ndioro's and Biya's choice to derail SMIC was to protect the government's shaky standing with French and international finance. As the press corps made known, France's minister of cooperation (i.e., for Africa) Bernard Debré came to Yaounde a week before the announcement, then March 28 came Jean-Christophe Mitterrand, the president's son and African factotum, presumably to press the case for durable subregional French interests working through familiar state structures. They were heard. France, according to *Perspectives Hebdo* April 11–17, opposed SMIC so as to protect CFDT, because the latter's 30 percent SODECOTON stake takes 70 percent of its earnings, by controlling invoicing and sales abroad. The French may also have sought a way to add the Cameroon treasury's 22 percent stake to CFDT's 30 percent, for full control of SODECOTON, perhaps to hold off the World Bank, deemed by France to be the U.S.A.'s stalking horse. The World Bank's Cameroon Director Joseph Ingram, the only foreigner to speak for publication at the time, in *Dikalo* April 10—"le Ministre Ndioro a pris en compte la position de la Banque Mondiale/Minister Ndioro has taken the World Bank's position into account"—voiced his concern about the evasions and low prices in the SMIC transaction. However arranged, on whomever's behalf, the reversal would divert the privatization from SMIC in a case where present and projected earnings were lucrative enough to warrant foreigners' special attention, and of course serve their interests by transferring more of Cameroon's economy (not just its debt) into one or other of their hands.[21]

Speculation about Hayatou, cut adrift by his junior protégé's denial of SMIC's bid for SODECOTON, was complex and more interesting for our purposes. A conventional reading from Cameroon's political party history would cite his concern, nominally since 1992 from outside the government but still with influence, to serve primarily national CPDM and also, as a by-product, northern interests by creating, in SMIC, an ethnocapitalist alliance of 200 elites from both the CPDM and NUDP which could shore up the region's politics, society, and finances.[22] The strategy might alienate foreigners with capital to invest and loans to protect, if SMIC kept them in a minority posture. But it could pay off for the CPDM and Hayatou if it transferred immediate political weight in the North to them from the rival NUDP, making the former more and the latter less stable, and thus restored the basic North-South ruling coalition, supported by the French in Ahidjo's time. This axis was now eroded by Biya's ever narrower southern base, which clearly troubled all foreigners with a stake in Cameroon, France included. Hayatou could, through SMIC as an ethnocapitalist hub to anchor a more northern oriented CPDM and to weaken Bello Bouba and the NUDP there, also strengthen his claim to the presidential succession. Foreigners would tolerate such an arrangement, even with cotton as a SMIC enterprise in nominally indigenous rather than outside control, if it stabilized Cameroon's polity and economy.

But there was more than CPDM party and succession politics to Hayatou's and his collaborators' strategy, if one pursues another line of analysis, focussed more fundamentally on the precarious nature of Cameroon's state than merely on the CPDM's survival or the Biya presidency's transfer. In this view, which better accounts for its NUDP component, SMIC drew northern leaders from across party lines for *sub*national, not national purposes, and Hayatou's real challenge was not to Ndioro for short term possession of the CPDM and Cameroon's presidency, but to Bello Bouba and the NUDP for more enduring control of Ahidjo's legacy and the North, a surer base for high power, even if it might be deferred.

Cameroon's basic political economy, not just the surface of its politics, was the scaffold for this analysis. Timber in East Province, cocoa in South and Center Provinces, many tropical forest products in South West Province, and coffee in a number of provinces are key resource bases. Cotton is the resource counterpart in the North. Ahidjo's statecraft assembled directors of these resources, customary rulers, and entrepreneurs in commerce and industry as a national ruling class alliance, with the Cameroon National Union as its political framework; he added petroleum but attached it to the central government's fiscal domain.

Cotton as the North's largest contribution to and stake in this structure was, and is, a complex matter. Recent publications on local northern societies by Kees Schilder and Philip Burnham, which add ethnographic and political substance to the journalism, agree that the region's Fulbe-Muslim precolonial and the French colonial authorities, which in their turn dominated the North to 1960, were differently disposed toward local populations, and Schilder shows that cotton production and the CFDT were French initiatives challenging the previous Fulbe-ethnic, Muslim-religious direction of politics and business.[23] But Ahidjo after 1960 gathered together all the strands of polity and economy with French approval, cotton included. The 1974 creation of SODECOTON became the parastatal *Cameroonian* vehicle for the North, just as Garoua with its massive projects, as Ahidjo's base, became a "second capital." "Fulbeization" (the term Schilder and Burnham use) became a regional "hegemonic project" with a national link, in Bayart's full sense, although Ahidjo, its dominant politician, was himself not fully its product by birth or culture.

The North's fortunes, however, declined after Biya put down the 1984 coup. Both Schilder and Burnham trace the generically southern and specifically Beti incursion on locals' direction of the region, reflected throughout its governance and designed not just to advance outsiders but also to profit from historical tensions between Fulbe-Muslim rulers and their local subjects.[24] For the former were never the monolithic power that Ahidjo's policy projected. They comprise just 30 percent of the North's total population, and their domination or mediation of the politics of the ethnically mixed 70 percent majority they call "Kirdi" (connoting "heathen") has been contentious, under both CNU and early CPDM auspices.

The journalism and scholarship cited here lead from these circumstances to a likely scenario for the contest over SODECOTON. It locates Biya and Ndioro as agents to maintain both southern Cameroonian and French influence in the North. It places Hayatou through SMIC as the agent of a *Northern* alliance crossing the two major parties active there, challenging as the priority for his own politics both the non-Northerners and the Bello Bouba-NUDP network for renewal of "Fulbeization" and Ahidjo's full regional legacy, including the cotton sector. As southern Beti powers wane, the argument goes, a northern backlash or vengeance appears, taking advantage of the loss of foreign and domestic confidence in Cameroon as presently ruled. In this reading, SMIC's "baronial" interests under Hayatou would restore SODECOTON to a reconstructed, more autonomous northern elite, to direct the region's affairs like before 1984.

Perspectives Hebdo and *Génération* followed this line most thoroughly as the SMIC-SODECOTON story peaked. The former was extraordinarily bold April 11–17, 1995. Its headline suggested a link between the northern plot of 1984 and this 1995 episode, and its cartoon showed whites stepping off a Cameroon Airlines jet, greeted by a tarmac sale of parastatals, with Biya's Etoudi Palace "à privatiser" in the background. This was its argument:

> beaucoup estiment que trop c'est trop et que le fruit est suffisament mûr pour tenter un véritable coup de force en direction d'Etoudi. . . . Le lobby

nordiste vient encore une fois de plus/many believe that enough is enough and that the fruit is sufficiently ripe to attempt a real test of strength against Etoudi.... The northern lobby comes once again.

Not stopping in its provocation there but moving to the national level, it pointed to the consequences if other regions with still viable agro-industrial bases were to follow suit:

> Si jamais les nordistes réussissent leur entreprise de mainmise sur la SODECOTON, les anglophones du Sud-Ouest vont tout aussi réclamer la reprise du contröle de la CDC/If northerners ever succeed in seizing SODECOTON, South West anglophones are going also to ask for the recovery of control over CDC

referring to the Cameroon Development Corporation, with a South West Province history like SODECOTON's, which was also a clear target for lucrative privatization, whether into domestic or foreign hands.[25]

Then came *Génération*'s turn April 12–18 in its "Dossier" section. Since March 29

> France et la Banque Mondiale ont réussi a court-circuiter cette "revanche" des hommes d'affaires camerounaises fussent-ils du Nord ... [et] la classe-appui "lamidale" ... Aussi va la processus d'une recolonisation des secteurs les plus performants de l'économie nationale par les néocolons de Paris et leurs valets de Yaoundé ... /France and the World Bank have succeeded in short-circuiting this "revenge" conducted by North Cameroonian businessmen ... [and] the auxiliary class of lamidos.... Thus goes the process of re-colonization of the most productive sectors of the national economy by the neocolons of Paris and their Yaounde valets ...

But this analysis pushed deeper still into the contest over agrarian capitalism in the North between overlapping regional, national, and international political economies:

> elle est factionelle et pose ... le problème de l'accès des factions camerounaise et française ... a la propriété des *nouveaux* moyens de production et d'échange qu'offre aujourd'hui la SODECOTON/it is factional and poses ... the problem of access for Cameroonian and French factions to ownership of the *new* means of production and exchange that SODECOTON now offers (emphasis added).

Which party, *Génération* asked, will prevail "dans un contexte d'ajustement qui promeut une 'crisocratie' (phénoméne d'enrichissement par la crise et l'ajustement)/in a context of adjustment which promotes a 'crisocracy' (a product of enrichment through crisis and adjustment)" where "le désengagement de l'Etat" takes place?

"Disengagement from the State": in the context of future stakes in a changing global economy, this is for our purposes the key factor in these two sources' analyses of SODECOTON's privatization. *Génération* warned toward the end of its April 12–18 text of "une économie morcelée/an economy broken up" between different tribal groups, and elaborated this theme when its Dossier section again covered SODECOTON June 7–13, under the very explicit heading "La Revanche des Communautés sur l'Etat/The Revenge of the Communities on the State." It contrasted the all too familiar Cameroon experience with politics of the belly among elites "appelés à défendre les intérêts du RDPC [CPDM] dans leurs régions d'origine" with the new dimension SODECOTON introduced:

> Mais aujourd'hui, le phénomène a débordé cette étape . . . à la naissance des lobbies qui ignorent les colorations politiques, l'important étant la langue et le terroir/But today, the phenomenon has gone past this stage . . . to the birth of lobbies which ignore political shadings, the crux of the matter being language and soil.

Such regional lobbies crossing party lines like Hayatou's for SMIC "confirment la faillite de l'Etat-nation/confirm the bankruptcy of the Nation-state" and raise the following questions about Cameroon:

> à l'aube de la régionalisation survivra-t-il des lobbies communautaristes? . . . ne risque-t-on demain de voir les conseils régionaux, la crise aidant, se transformer en groupes sécessionistes?/at the dawn of regionalization will it survive communal lobbies? . . . do we risk seeing regional councils, using the crisis, transform themselves into secessionist groups?

The risk referred to, resembling what the regime claims anglophones posed, justifying *their* marginalization and suppression in the 1990s, was, and is, posed by SODECOTON specifically *within* the mainstream body politic by key CPDM people, members of the NUDP's National Assembly caucus, and one of the two renegade NUDP ministers in Biya's government at the time, all SMIC principals.

The same risk can be perceived throughout Cameroon, wherever the condition of rural capitalism resembles SODECOTON's. Parastatals depend on external sources for the capital and technology which add the value and profit returned largely to foreigners. Burnham provides Adamawa Province examples of a ranching project created with World Bank aid in the 1980s, and more recent grazing land grants to a French firm to settle an overdue government debt.[26] The composite political economy of Cameroon's resources remains vulnerable to competition and schism among domestic actors for the limited spoils they enjoy, or for "gatekeeper" roles with foreigners, and therefore to the "restorationist" North politics examined here, which parallels the anglophone sovereigntist and secessionist tangents in the case previously covered.

La Nouvelle Expression made the same point, April 18–21, in "Privatisations: Les risques de l'ethno-capitalisme." What would prevent the exercise releasing public sector agro-industrial economies to "liberal market" forces from lapsing into a scramble for exclusive political and economic control by the dominant ethnic group in each region, gutting any chance of recovery for the discredited national regime? There was clearly the danger of a rippling belly politics moving from Yaounde to the provinces, further entrenching regional ethnocapitalism. Burnham suggested similar forces, with broad consequences:

> This more whole-hearted interest of elites in playing an active role in their home regions is . . . no doubt an expression of the pervasive discontent with the Biya regime. . . . The Cameroonian state may have marked out the playing field but locally specific rules control much of the game . . . the state is finding it increasingly difficult to exercise effective political controls on a routinised administrative basis.[27]

In many dimensions of this mid-1990s profile, Hayatou occupied a "straddling" tactical position regarding the national and subnational politics we survey here. Bello Bouba as the NUDP's head must have similarly surveyed and calculated national and subnational options; his northern role was strengthened early in 1996 by the emergence of Mahmadou Ahidjo Badjika, Ahmadu Ahidjo's son, as the party's political secretary and Garoua Urban Council's elected mayor.[28] Factoring in the pressures the SODECOTON episode raises, political leaders, current and would-be, may leave future options open, including party affiliations and possible alliances at the national *and* subnational levels. Will Yaounde and the foreign interests it still attracts as a gateway, Ndioro's and Biya's chosen center of gravity for SODECOTON since March 29, 1995, "sur hautes instructions du président de la République," hold the balance against potentially strong centrifugal forces we have located earlier in Bamenda and Buea, and now in Garoua, or in other possible regionally directed scenarios not yet visible? Will Hayatou, Fru Ndi, and others who may emerge be nationally focussed politicians, even statesmen, regionally specific ethnic power brokers, or tactical straddlers?

These are serious long term questions about political leadership and rivalry, and will persist. They touch the composition of the state Ahidjo built and bequeathed, the recomposition Biya directed after 1984, and the possible decomposition in the 1990s now disclosed. Let us recall how, before anglophones and SODECOTON became such public issues, Achille Mbembe in 1993 formulated the problem: "l'équilibre régional" might give way under pressures since 1990 to what he termed "la déliquescence de l'Etat/the decay of the State."[29] Ahidjo's creation remained largely intact for the first five years of Biya's CPDM, but when put to the test of 1990's laws and 1992's two multiparty elections, "son rétrécissement et son repli sur des bases régionales a commencé/its shrinkage and withdrawal to regional bases began."[30] Mbembe shattered not simply the *composite* profile of regions once drawn together in "l'alliance hégémonique" by Ahidjo, but which was now compromised (his choice of the next term is striking) by "Cette fronde, d'abord sourde/This rebellion, at first silent" but now visible and multiform.[31] He also conducted historical and political surgery on many regions separately, to make the point that, like the nation-state Cameroon, so too for its constituent parts, unequal components in unstable equilibrium are the present reality.

Here was the general scenario Mbembe projected from the many variables he discerned:

> Tous ces facteurs structeront les évolutions possibles dans le court, le moyen et le long terme. Et pour plusieurs années encore, les conflits locaux anciens et nouveaux, ainsi que les noeuds locaux d'obligations et de réciprocités renforceront ou amoindriront la capacité des partis politiques et des mouvements sociaux à coaliser ou à fractionner/All these factors will fashion a variety of movements in the short, medium and long term. And for several years more, old and new local conflicts, just like local bonds of obligations and reciprocities will reinforce or reduce the capacity of political parties and social movements to coalesce or to split up.[32]

Such fluidity seems the essence of what the previous anglophone and this North subregional coverage demonstrate.

One more dimension of this key 1995 episode needs attention, which Mbembe anticipated and *Génération* in fact traced, most forcefully in April 12–18 text addressing "cette entreprise de prolétarisation paysanne/this venture in peasant proletarianization" and on how the entire affair unfolded

> laissant aux abois les petits cultivateurs de coton du Nord qui ne peuvent plus aujourd'hui acheter une bouteille d'huile de coton parce que trop chère pour leur salaire ... /leaving in the lurch the small northern cultivators of cotton who can now no longer buy a bottle of cotton oil because [it is] too expensive for their pay ...

None of the high politics contenders for SODECOTON are likely to have been as concerned with the populations whose lives are directed and limited by the wages and farm gate cotton prices paid producers, as with the capital and technology which add the values as agriculture creates industry. That neglect would extend to the consequences of privatization under *any* auspices for those producers' schools, dispensaries, and the entire, basically underserviced life of the North at large.

The producers of cotton and other commercial or subsistence crops are not entirely silent, or at least have some voices raised for them recently, even if nothing like the mass mobilization of anglophones has yet stirred the North. There are now ethnic and cultural associations, even political parties, not yoked to "Fulbeization," the CDPM, or the NUDP in the North, though not necessarily hostile, as Schilder for the Mundang and Burnham for the Gbaya make clear. The MDR party with its six National Assembly seats from the Maroua area is specifically directed by a Kirdi fraction, the Tupuri, against Bello Bouba's NUDP Fulbe-Muslim interests, in restless alliance with the CPDM as the more distant, less domineering force, and there is a "pan-Gbaya" formation, MOINAM, started in Yaounde in 1993. Both scholars stress the particularistic, protectionist, small ethnic group agency character of these formations, and Burnham excludes the SDF as a factor in Mbere Division and (at least by inference from his silence) in the North at large.[33] But the SDF's inclusionary features may have some potential in the region's scheme of politics, although elections thus far show no such capacity. As noted in more detail later in this chapter, it staged a National Congress in Maroua in May 1995, partly to draw attention to and take advantage of the region's neglect under a decade of CPDM rule and the suspicions of the NUDP outside Fulbe-Muslim ranks. Maroua's streets produced some raised fists and "Power!" salutes, new to the politics of the North. Seeking a place there under the leadership of another Garoua political family's scion, Seidou Maidadi, the SDF used the SODECOTON case to argue that neither the CPDM since the parastatal began in 1974 nor any new configuration like SMIC or the NUDP, nearer governmental ranks than the SDF itself, have taken the majority of northern peoples' basic needs seriously even though they draw its votes, or seek to.[34] Quoting again the last lines of Mbembe's 1993 analysis:

> sans un changement fondamental, le Cameroun poursuit lentement sa course folle vers des ruptures qui ... risquent à la fin d'être particulièrement sanglantes/without a fundamental change, Cameroon slowly pursues its mad course toward ruptures which ... risk in the end being particularly bloody

surely brings social forces of change far beyond the SMIC-SODECOTON elite range of concern as well as political control of the *status quo* to mind, in the North and elsewhere.[35]

Such were some of the key features of the SODECOTON case, with its multiple dimensions and possible consequences yet to be determined. The public soundings would clearly register for a long time ahead. When Biya set foot in the North for the first time since 1992, April 18, 1995, only three weeks after the story broke, his announced four day schedule was cut short to less than two, with local reference made to danger to him from those the CPDM have disfavored in a Chad-border area dispute between Showa Arabs and Kotoko. Intriguing, too, was Hayatou's choice of Biya's time in the North for a journey to Mecca, a *very* Fulbe-Muslim choice of response. The National Assembly in June 1995 established an investigative committee for SODECOTON, but it has never reported, nor was the case resolved when a Yaounde tribunal reinstated SMIC's claim to SODECOTON, March 19, 1996.[36] Justin Ndioro's demotion in September 1996 to head the ministry of industrial and commercial development is also of interest. The company's disposition remains uncertain in mid-1997. "L'affaire SODECOTON" shows for Cameroon's North, on grounds of ethnicity and religion (and perhaps incipiently of class) how contentious, even fragile, the country's politics have become in the 1990s.

Regional and Ethnic Politics Miscellany, 1993–1997

Other subnational interests surfaced by 1995, not so prominently or so close to party politics as the anglophones' or the North's, but visibly enough to claim attention. Most important were the Bamilekes. Their numbers, their lack of a single party presence, their demographic dispersal, and the way their diaspora meshes their lives with indigenes, often competitively, are factors that make their cultural associations influential wherever they live and settle. One in particular, Laakam, represents much of the organized Bamileke elite, with branches pursuing their interests throughout Cameroon and in France. Pronouncements in the press periodically address public affairs, and note threats to Bamilekes from outside their ranks. Laakam pays close attention to politics, as the closest communal approximation to a Bamileke party. It has memorialized the French presidency and published in Cameroon its support for either sweeping decentralization or multistate federalism of the SDF type.[37] It is reasonable to judge from inference rather than what one would prefer, hard evidence from Laakam texts, that large Bamileke numbers in the SDF reflect Laakam preferences, in the absence of any political party with a specific Bamileke base. Laakam faces Bamileke competitors for ethno-communal support, a striking example of the variations and tensions within nominally homogeneous communities, which were sketched in Chapter 5 as hurdles to opposition alliances but, of course, work in other ways. In this case, the CPDM in 1994 countered Laakam's oppositionist politics by forming the Cercle de Réflexion et d'Action pour le Triomphe du Renouveau (CRATRE) in West Province, utilizing a familiar, historically crucial link of Bamileke wealth like Victor Fotso's and Françoise Foning's in finance, industry and patronage to the party and state. This led to frequent and heated intra-Bamileke exchanges both in the press and through verbal and physical clashes on the ground.[38] Intra-Bamileke friction is a special case because Bamileke wealth and numbers make a large impact beyond regional into national politics. Indeed, those beyond their ranks who find Bamilekes too active, invasive, or evasive in Cameroon's very rough and tumble ethnic competition target Laakam as a major source of the national crisis.

This is especially so in a running public skirmish between Laakam and another strong ethnic voice since political crisis began, the regime core Beti's Essingan, apparently formed by Joseph Owona (fiercely partisan, he becomes more familiar directly below), but like Laakam a recent addition to a persistent line of ethnic associations.[39] Terms of the controversy clearly demarcate Laakam's opposition to and Essingan's support for Biya's form and conduct of the Cameroon state. The private opposition press publicizes Essingan as the bedrock of Beti ethnic elites and their regime monopoly, dominating strategic civilian and military offices and broader resource allocation, while the regime press defends it. A prototypical headline from *Nouvelles du Cameroun*, September 13, 1991, "Laakam-Essingan: LA GUERRE?/WAR?" reduced Cameroon's complex crisis to a narrowly ethnic competition for state domination, mistaking a part of the problem for the whole. The international Rosicrucian Order, brought to Cameroon, has the same character as Essingan, for Biya, Titus Edzoa, the Beti secretary general in the presidency, 1993–1995, and other top echelon CPDM figures are members, wearing its pin of identity and otherwise making no secret of its presence and influence. There are Cameroonians who believe that some Beti CPDM leaders, facing the mid-1990s challenge from the party's northerners covered above, want it to become an exclusively Beti, Center, and South Province party. Further, they see Essingan as conspiratorially secessionist, linked across borders to a project of union with Gabon and Equatorial Guinea, homes to their ethnic "brothers" the Fang, duplicating what some ascribe to North West and South West Province anglophones and extend to Nigeria.[40] Thus, there arose in 1995 a long, loud controversy over the route chosen for a petroleum pipeline from Chad to Kribi, Owona's political base in South Province, believed by regime critics to be inferior to a more "natural" terminus at the anglophone port of Limbe, and believed by still more suspicious observers to provide Beti irredentism with the major port needed for secession to be economically viable.

Such views of the Beti, as for anglophones, northerners and Bamilekes already surveyed, overlook their differences. "Beti" embraces many significant ethnic groups. Biya himself is Bulu and Owona is Batanga. And their regime's low reputation by 1996 persuaded peoples on their northern fringe like the Eton and Ewondo, content in happier times to be known as Beti, to distance themselves by insisting on the more precise ethnic vocabulary. A salient political measure of this move surfaced after local elections in January 1996, covered in more detail below, when the leader of the Eton's Parti de l'Alliance Libérale, allied with the CPDM, Célestin Bedzigui, nullified the pact, declaring that *his* party now had majority Eton support, not the CPDM.[41] Among the coastal Duala at Cameroon's economic hub, similar tensions surface as Ngondo, by reputation a "traditional" leaders' cultural association but in fact only half a century old, drawn toward the opposition in the early 1990s, is contested by a regime element at mid-decade. Ralph Austen's recent work on the Duala especially well reveals the "invented tradition" and internal cleavage features which are generic to ethnic politics in Cameroon.[42]

By nature of the close ranks they keep and limited information they disseminate, these ethnic and cultural formations are difficult to assess. The literature on "tribalism" ranging from primordial ethnic sentiment to more instrumental and politically calculating elite ethnic politics in modern states enjoins further caution, and there are differences among Cameroon scholars between Bayart's preference for the generic "politics of the belly" and Mbembe's more ethnically specific explanations and more serious attention to them. However they are understood, such political dynamics claim a place in any current study of Cameroon, particularly if a strong populist movement like the anglophones' or determined elites as in SMIC, Laakam, and Essingan are at work, or where political parties are primarily ethnic formations.[43] They will surely maintain their complicated influence on the mainstream of Cameroon's public life, where we now return for a detailed look at constitutional proposals which periodically engaged the participants in politics, 1993–1997.

Constitutional Debate and Action

Constitutions and the laws they frame receive little attention in contemporary African studies. There used to be good reasons for this, with their domains and the force of their writs so fundamentally circumscribed by state autocracy. Since 1990, however, this is not a technical or esoteric topic, any more, say, than in Québec since 1970. Debates about unitary and federal government are now vital. The Cameroonian John Mukum Mbaku makes the point well: autocratic governance in Africa was organized to keep disparate people together through unitary state structures in an "uncompetitive constitutional environment"; if pluralism and democratization are to develop, "appropriate instruments to effectively constrain government-as-leviathan at all levels" are their natural channels.[44] Both where national conferences have addressed and even changed the state's central character and in countries like Cameroon where no such channel for debate and action emerged, constitutional proposals have become public issues and provide a useful focus for political studies.

Cameroon's Tripartite Conference late in 1991 created a Technical Committee on Constitutional Matters.[45] It was ignored as elections dominated 1992, but the end of that electoral cycle led to the intermittent appearance of draft constitutional texts, 1993–1995, and the promulgation of a revised constitution for Cameroon, January 18, 1996. Some extraordinary effort went into the texts and debates about them. So we now address the quieter but strategic constitutional front, analyzing and comparing the major proposals and indicating where they lead. This material provides one of the best litmus tests for judging those active in politics who know what is at stake over the constitutional intricacies of the state apparatus: where better to build, and conceal, a hegemonic project?

March 23, 1993, sixteen months after the Tripartite Conference, the government announced a "Grand Débat" about a constitution to revise or replace 1972's, often amended since then, in order to forestall Fru Ndi's call for a Sovereign National Conference. This led May 17 to the appearance of a "Preliminary Draft Constitution," identified with the Tripartite Technical Committee's chairman, at the time secretary general at the presidency, the CPDM's legal scholar of choice Joseph Owona, so that the document became known as the "Owona Constitution." Its timing and text raised controversy. Beginning May 19, three anglophone members we met earlier, who had gravitated to the AAC and surfaced at Buea seven weeks before as *its* legal specialists, Sam Ekontang Elad, Simon Munzu, and Carlson Anyangwe, made their grievances public. They were altogether differently oriented than the majority francophones on the eleven member committee who were trained in France and active in either or both the University of Yaounde and the inner circles of government. The anglophones charged that the committee had not met for a year, and that the text was Owona's own, unmediated by any consultation. Munzu used a CRTV interview to criticize both Owona's neglect of the committee and the document itself, charging bad faith and incompetence, then Anyangwe and Elad joined Munzu to present a common minority front.[46] Their critique brought the federalist force to bear as the debate on Cameroon's constitution was defined and joined by all camps. The debate's main lines and the outcome as of 1997 follow, traced through the adoption of Cameroon's current constitution, Law 96/06 of January 18, 1996.[47]

Article 1 of the Owona Constitution termed Cameroon a "unitary decentralized state [with] local autonomy through an extensive administrative decentralization"; this modified the terms of the constitution in force since Law 84/01 defined Cameroon in 1984 as a "Unitary State ... one and indivisible." Deep in the text, however, against the clearest opposition voice we noted on Bamenda's Sonac Street, 1991, Articles 152 and 156 kept *appointed* governors for local regions, dominating elected councils the president could still dissolve, and specified that governors "shall control and coordinate the action of the State's decentralized services." There were conciliatory features like Article 5 where "The one party system is forbidden" and Article 142 where "The opposition shall have a natural and legitimate status" in the National Assembly. Especially notable in light of 1992's election of Biya was Article 36 limiting presidents to two five year terms, chosen in two stages if there were no first ballot majority, rather than a "first past the post" procedure which had enabled Biya's election without a majority of votes cast in 1992's multiparty balloting. But the fundamental powers of the executive branch over the legislature, and of appointive over elective offices, were retained in most matters of governance personnel and procedures, as in the use of ordinances touching budgets (Articles 40, 41, 55, 74, 78, 82) and for referendum initiatives (Articles 90, 161). A partially appointive Senate without specified powers in relation to the National Assembly, be they positive, negative, or balancing, appeared as a vague collaborative appendage (Articles 57–61). There was, despite the face of the passages declaring the principles of multipartism and the opposition's status, only perfunctory attention to the principle or practice of multiparty politics.

How likely was this initiative to close Cameroon's gaps in governance and encourage consensual debate, in light of the concurrent Buea initiative and of further alternatives we now examine, proposed by the other major political parties? The NUDP issued a proposal early in 1992 as part of its legislative election platform, then followed with another more detailed text late in 1995 from its National Assembly caucus as it prepared for a parliamentary session the CPDM announced as a constitutional debate forum. Its texts provide our first points for comparison with Owona's, useful because they differed less substantially than others from the CPDM's but still demonstrated some distance between the regime and the opposition party closest to it. More attention is then given CDU and SDF texts, which began in 1993 to offer constitutional proposals enabling firmer comparisons and sharper contrasts.

The 1992 NUDP text appeared in a February 20 Special Number of *NUDP News,* printing Bello Bouba's speech to the Garoua Congress which made him the party's president January 4–5. It conveyed the first solid sense of the party's stance on Cameroon's constitution. A new document was called for, not a revision of the current text. It prescribed "la séparation effective et l'équilibre réel des pouvoirs, en reconnaissant des prérogatives importantes au Chef de l'Exécutif/the effective separation and real balance of powers, while recognizing prerogatives important to the Head of the Executive." Only two brief clauses touched legislative matters. A prime minister would be named by the president and responsible before the parliament, in ways unspecified. And there would be "une seconde Chambre Consultative, représentant les différentes catégories socio-professionelles," no more specific about the character and powers of a Senate than Owona's text a year later would be. The judiciary's autonomy was more clearly spelled out and an allied clause called for a National Charter of Liberties. Governance routines under the still acknowledged primacy of the executive should be revised "par une double opération de Déconcentration et de Décentralisation." The former term stood for enhanced ministerial competence; the latter, against Owona's grain a year later, entailed an ambitious scheme for elected and sovereign local councils and elected mayors, as well as provincial committees to enact and administer economic plans.

The NUDP in these details lodged substantial changes against CPDM practice at this early 1992 juncture. But in some essential terms, there was congruent ambiguity between these parties. "Provinces autonomes, jouissant de pouvoirs réels/Autonomous provinces, exercising real powers" remained in place, with no further reference to their articulation with the central structure of Cameroon's state. Nor was anything said about the crucial question of the governor's source of authority and role in the provinces. There was the same silence about an electoral code or commission. The NUDP's depth of criticism and concern on constitutional and electoral affairs in 1992 was hard to assess. That year's elections were still ahead and the party meant to contest both of them, so that these matters were not the priority they later became.

The 1995 NUDP text, following those elections, then three more years of experience ranging between attrition and periodically bitter conflict in Cameroon, distanced the NUDP further from the CPDM than before on some points. This 1995 "Proposition de Loi Portant Revision de la Constitution" was from its "Groupe Parlementaire," technically not a draft constitution from the party executive, but surely a carefully considered text prepared for November's National Assembly session and for public use. In a setting far more complicated by emerging anglophone issues and activism than in 1992, the NUDP specified (like the constitution in force) "une République une et indivisible" with only the confirmation of two official languages to recognize the fact of anglophone Cameroon; there was literally not another word in its fifty-seven articles directed to anglophones as the distinct constituency they had become.[48] It was, as in

1992, not stated in Article 50's cursory glance at subnational structures whether governors of provinces were to be elected or appointed, and there was no text anticipating an electoral code or commission separate from a Constitutional Council empowered to regulate elections and review disputes *after* their appearance. Owona (in 1993) and the NUDP's Article 18 agreed on limiting presidents to two five year terms.

There *were* NUDP departures from the Owona text, our guide to CPDM policy. Article 40 specified that the presidentially appointed prime minister and cabinet were subject to the National Assembly's majority and its vote of confidence, and gave a solid blueprint for a six year Senate capable of initiating laws. The NUDP thus advocated a prime minister more firmly grounded than Owona's in the National Assembly in case of "cohabitation" if different parties (as in France, a reference point, of course) controlled the executive and legislative branches, and a bicameral legislature with more autonomy than Owona provided for. It also revamped the National Assembly's internal governance, for Article 10 created officers, bureaus, and committees for its full five year term rather than annually. This would reduce the executive's capacity to manipulate and control work there by rewarding or punishing key legislative operatives through a continuous influence at each year's elections. A balanced view of this text would recognize its stronger position than Owona's on the legislative presence, but also its noncommittal approach on subnational governance. It reflected the NUDP's satisfaction with the National Assembly role its sixty-eight seats gave the party since 1992, and its desire to consolidate that role, without offering those Cameroonians still not represented in formal institutions any added role or significantly greater electoral safeguards.

There was, by contrast, no doubt about the clear intentions of the CDU and SDF to reconstruct Cameroon's constitution in their proposals, 1993–1996, even if they led in somewhat different directions. The CDU's basic text held more surprises than the SDF's proposal, and we will therefore deal with it first, comparing it with Owona's text and, where appropriate, the NUDP's. The "Avant-Projet de Constitution du Cameroun" (Draft Constitution) of October 29, 1993, covered thirty-two full newspaper-sized pages with 298 minutely detailed articles. The first impression was misleading, because the language approximated Owona's. Ndam Njoya and his colleagues left no doubt about having vacated 1991's common ground with the SDF on constitutional approaches. This text's Article 1, introducing the topic of sovereignty, stipulated Cameroon as a "REPUBLIQUE UNIE" rather than federal, reversing Biya's 1984 language of simply "La république" (which Owona retained) but keeping Ahidjo's of 1972. Article 2's elaboration of Cameroon as "un Etat unitaire décentralisé" exactly followed Owona's Article 1. Thus far there was common CPDM and CDU ground on the words and ideas conveyed to the public.

But differences began to develop as CDU details accumulated. Article 6.1 on party pluralism was stated positively—"La Constitution consacre le multipartisme"—whereas Owona's draft text merely forbade the "one party system." And as the CDU specified what "unitary" and "decentralized" meant, it challenged the CPDM in basic ways. One difference was most striking. Owona virtually ignored emergent anglophone federalism; his 167 articles protected anglophones within the unitary structure only in Article 2 keeping official bilingualism and in Article 167 providing for the corollary, a bilingual official gazette for the constitution. Owona's Articles 97 and 100 assured no less, but also no more, than four among twelve to twenty-one Supreme Court justice appointments from the presidency to those of both "Anglo-Saxon" and francophone background (that asymmetric language use aroused anglophones). Article 148 on the shape of "Decentralized Territorial Administrative Units" also specified that "no division . . . may take place whose effect would be the fusion of Anglophone and Francophone areas"; this would prevent formal assimilation, but in no way satisfied mainstream let alone militant anglophones in 1993.

By contrast, the CDU draft, although rejecting federalism, offered firm anglophone protections. After repeating Owona's recognition of bilingualism in Article 3, it turned to anglophones comprehensively in Article 43.4, introducing the section on the organization and form of the state:

> Chacune des deux aires linguistiques officielles aura ses collectivités territoriales distinctes et aucune de celles-ci ne pourra être à cheval sur les deux aires/Each of the two official linguistic areas will have distinct territorial identities and neither will be able to sit astride the other.

Providing a positive basis for distinctiveness rather than merely a negative protection against assimilation, this text addressed anglophone concerns in ways detailed and augmented as the text unfolded. Article 62.1, for instance, provided Cameroon both a president and vice-president, one a francophone, the other an anglophone, and Article 65 then established the vice-president's succession to the presidency in case of death, resignation, or impeachment, with the right to run for that office's next term. Article 151.3.c among others applied the same principle, sharing the offices of president and vice-president for a key instrument Owona (and the NUDP in 1995) failed to provide, a National Electoral Commission, between the language communities. Another most crucial point for anglophones, given their 1972 experience, was the procedure for constitutional revision. The CDU's Article 293 required amendments proposed in parliament and regional councils to be approved by at least two-thirds of those voting in *both* francophone and anglophone units of governance made autonomous on language grounds in Article 43.4 (and specified immediately below). In a referendum, revisions to the constitution would need a majority in, again, *both* language communities' regions. This is not to say that the CDU text on offices and constitutional revision satisfied all anglophones in the 1993 context, or later, but it arguably established and safeguarded their constitutional position in a "unitary decentralized" state, unlike Owona's text, which few anglophones outside the CPDM would have found a basis for serious discussion.

Another clear distinction marked the CDU's sense of the words "unitary" and "decentralized" when compared to the CPDM's in Owona's text and to the second NUDP draft. The CDU clearly separated executive and legislative powers, at the state's center *and* down through its regional and local structures, where a framework parallel to the national presidency and parliament shaped governance so as to share it between executive members and legislators; institutional cohabitation was articulated in principle and practice. The structure unfolded in sequences of articles too detailed to summarize, but a few key examples suffice. A fundamental innovation in Article 59.1 gave regions (the new names adopted for provinces) the power to levy taxes for many new legislative and fiscal competencies lodged there. Article 78.1 made their governors elective, unlike Owona's appointees (recall here, also, the NUDP's 1995 silence on the point), and Article 79.1 made them like the president of the republic (Article 65.1) subject to impeachment by the legislature they shared the region's power with. This *was* Bamenda's Sonac Street, 1991, taken seriously.

The CDU draft contained other features responding to changes in Cameroon's politics by 1993, which Owona's tended either to ignore or reject and the NUDP's to skirt, although all three texts agreed on one crucial point, for the CDU's Article 62.2 also limited presidents to two five year terms. Its Article 37.1 lowered the voting age from twenty-one to eighteen (the NUDP in 1995, Article 2, made it twenty), to enfranchise those mobilized since 1990 and more likely to support opposition parties. Article 128.5 removed a president's power to sustain a prime minister in office following National Assembly defeat in a vote of confidence (on this point the NUDP and CDU agreed). A key section, 15 percent of the entire CDU text, Articles 150–186, provided an electoral code which addressed the shambles of 1992's process, a feature Owona ignored and the UNDP subsumed within the Constitutional Council it limited to *ex post facto* review of disputes.

The document fundamentally challenged Owona's. Starting from the similar rhetoric and model of a "unitary decentralized" state, the CDU included rather than excluded the possibility of debate and consensus with those anglophones it did not on first glimpse favor, and in fact much of its detail resembled the federal constitutions of Nigeria and the U.S.A. Whereas Owona's proposal, and later CPDM efforts, drew fire for never advancing past the limits of what Achille Mbembe has called "multipartisme administratif" designed to avoid change and perpetuate regime power, the CDU text offered both sweeping and specific lines of constitutional evolution toward a pluralist Cameroon state.[49] As we now move to consider SDF constitutional proposals, we will find sharp contrasts with the CDU's, but there were commonalities which might in future stand the pressure of negotiation, despite rhetorical and substantive differences in these texts and the break between the parties on other fundamental grounds.

The SDF's "Proposed Constitution for a New Social Order in Cameroon," dated December 16, 1994, with its distinctive title and 238 articles as detailed as the CDU's, was among its most intricate, absorbing efforts.[50] We need to frame it here both to the other parties' texts and to the surge of anglophone energy leading from AAC I in Buea, 1993, to AAC II and the SCNC in Bamenda, 1994. The SDF's Bafoussam Congress in 1993, we recall, struck a delicate balance between anglophones and francophones, advocating federalism but not Buea's two state structure, then the party executive a year later adopted a four state model with a federal capital. This late 1994 document expanded on previous SDF texts and covered terrain the other parties left unexplored.

An Introduction stated the SDF's view of the current constitution's basic, inherent weakness: the parliament is a "mere extension of the Chief Executive ... [who] may rule by ordinances and decrees.... We have a Presidential Monarchy." By the end of its five pages, this section outlined the SDF course of remedies which later details spelled out. Some echoed other parties' plans, like the separation of powers, a bicameral legislature, and a five year presidential term renewable only once. But innovations were far more striking. There was the clear rhetorical force, with obvious resonance abroad, of a National Assembly renamed a "Congress of the People" to add to the more substantive provisions for a "Federal Republic." A Senate held the right to confirm all major appointments by the president, who was not only limited by Article 98 to a five year term, once renewable, but was stated to be "just first among equals." The next section of the text, the Preamble, declared "We the people ... do hereby make, enact and give to ourselves and posterity this constitution" which Article 1.1 then called "the supreme law" with "binding force on all authorities and persons exercising power on its behalf." This language clearly separated the SDF's constitutional approach from all others in circulation, and aligned it with popular sovereignty traditions elsewhere.

Human rights figured early and often. Article 32.1 added the United Nations Charter, the Universal Declaration of Human Rights, and similar documents Cameroon had signed to this constitution, and the next twenty-four articles specified safeguards in Bill of Rights form, ranging from life itself through work, strike, and petition to religious belief, which Cameroonians possessed by virtue of this text, their own creation. The number and variety of these rights were extraordinary.

The changes in governance structure augmented the legislature, as noted above, and altered the executive. Article 96.2 (following the CDU) divided the presidency and vice-presidency between an anglophone and francophone. Article 103.2 moved this principle into the cabinet, designating four "sovereignty" ministries (internal affairs, external relations, defence, finance/economic planning) and lodging two with each language group, against historical practice which had completely excluded anglophones from these posts for thirty-five years. Central government's powers, by contrast with the Owona and NUDP texts but like the CDU's, were curtailed sharply by the impact especially of articles 11, 218, and 223–234, dealing with the four federated states and local governance. Their composite features gave each state its own constitution with executive, legislative, and judicial authorities and institutions parallel to the federation's. *States* (there is a constitutional significance to this word's choice rather than *provinces* and *regions*) replaced the central authority as the sources of local government areas and agencies within their boundaries. All state and local offices including governors, which were each state's creation, were stipulated as elective not appointive, except in the judiciary. Perhaps the *most* sweeping change beyond CDU proposals created a revenue sharing process by providing a "State Special Fund" drawn from the federation, in order to finance a broad range of services and commissions which Yaounde dominated before 1972 and monopolized thereafter, from education and environment to police and prisons. The importance of these features of the SDF text includes their connection with the sections earlier in this chapter on sub-national initiatives and Mbembe's perception of the state's decay; this would be precisely the constitutional apparatus to animate the principle of decentralization by returning public revenue to elected representatives closer to the populations which were its source.

Nowhere was the SDF's elemental sense of these matters, in symbol and substance, matching its specific history since 1990 with the broader scope of resentment from the deeper past and with the wider population, more clear than in Article 236.5: "The National Gendarmerie is hereby proscribed." This provision suffices, against the temptation to add more detail about the SDF's determination to deconstruct the past and reconstruct the present and future governance of Cameroon.

By mid-decade, then, all major parties and the anglophone movement had draft proposals for a new constitution in place, although knowledge of them was confined to the politically active and journalistically informed minority, rather than widely shared among Cameroon's people. The 1972 constitution, periodically amended through 1991, was still in place, and the issue of whether and how to reconcile all these documents remained. Such were the terms of constitutional reference and debate; the closest approach to *dialogue* between the March 1993 call for "Grand Débat" and late 1995 was a sudden invitation from Biya to an all-party Constitutional Consultative Committee late in 1994, with Prime Minister Achidi Achu in the chair and a short reportage schedule; the entire process resembled the Yaounde Tripartite effort of 1991.

The presidency issued a text revising Owona's, now eighteen months old, as the centerpiece for discussion. These December 15, 1994, "Proposals of the President of the Republic for Constitutional Reform" (thus, "Biya's" below) were largely unresponsive on points the other parties had addressed, and took a harder line than Owona's text on key issues. It did clarify the composition and role of the Senate first proposed by Owona (Articles 14, 27). But it dropped Owona's preamble language (deceptive as it was) eighteen months before about decentralization, and increased Owona's protection of the key powers of the presidency; Article 7 made the term of the president renewable indefinitely instead of once only, and dropped the two-ballot presidential election provision which would require a majority vote by the end of balloting. Appointed governors and senior divisional officers kept their control over all locally elected councils (Articles 48–52, 58). Constitutional amendment, compared to Owona's text, was made easier from the presidency but harder from the legislature in Article 55, and in fact reflected Biya's desire to restore the presidential capacity to manage amendments through the nominally more autonomous parliament he had delegated some powers to when he restored the prime minister's office in 1991. Biya's Article 2 also drew a line in the political sand Owona omitted in a way certain to provoke, by lowering the voting age from twenty-one only to twenty, against the CDU and SDF provisions for the age of eighteen, *their* way to enfranchise younger post-1990 militants. The next to last Article 61 repeated language from 1972, that legislation by "the Federal State of Cameroon and in the Federated States . . . shall remain in force insofar as it is not repugnant to this Constitution," a most anachronistic and unconvincing (indeed, repugnant) passage, given the nature of the integralist state ever since.

If Article 61 was meant to conciliate anglophones, or if the text at large was designed to be consensual nationally, the effect was quite different. Fru Ndi refused the invitation to this meeting, through a press release claiming it was issued verbally and on very short notice, mocked any sense of protocol, and that no such regime committee could undertake work requiring a full constitutional conference. Sessions began in Yaounde but nothing was achieved, since most anglophones of substance, notably Christian Cardinal Tumi, Archbishop of Douala, immediately or soon walked out, joined by francophones including Bello Bouba and Ndam Njoya.

Frequent speculation and invective on the constitution continued through much of 1995, but neither a divided opposition nor parties beyond Cameroon's borders, who preferred at least the appearance if not the reality of open rather than closed door governance by the regime, could force any action. Then, late in the year, constitutional debate resumed with the announcement that November's National Assembly session would be the venue, with another CPDM text the focus. The CDU and SDF would not of course be there, but the NUDP, as we saw, was ready with a new text from its caucus. Press coverage was heavy, covering reservations among and amendments from CPDM, NUDP, UPC-Kodock and MDR members—debate was serious—and critiques rained down from those not present, including the SCNC leader Elad's view in *The Herald*, December 4–6, that the text under debate "would not be binding on Anglophones . . . and may even precipitate the early declaration of the independence of the Southern Cameroons."

We can bypass the details of the debate, including disputes on whether the text simply revised 1972's or was a new constitution, and summarize what came into force as Law No. 96/06 of January 18, 1996. The preamble began with the SDF language of "We, the people of Cameroon" and turned immediately to a twenty-five clause Bill of Rights, but the third item compromised the entire preamble's intent: freedom and security shall be guaranteed each individual, subject to respect for the rights of others *"and the higher interests of the state"* (emphasis added).[51] Part I immediately carried that latter sense forward, for under the rubric "The State and Sovereignty" followed language in Article 1(1)–(2) which incorporated 1984's "Republic of Cameroon" with Owona's 1993 "decentralized unitary state" further defined as "one and indivisible." Certainly familiar CPDM language, this text led to Part II on the executive stipulating in Articles 5 and 6 a high power, high profile presidency with a *seven* year, once renewable term, with none of the CDU or SDF features like a vice-president to share the executive and ensure succession to the presidency across the language boundary or a prime minister (like the NUDP's as well) with a solid foundation in a legislative majority. As there was no reference to retroactivity, the document enabled a Biya presidency through the year 2010. Nothing was said about one or two stage presidential elections, nor did the text specify a balance of executive, legislative, and judicial powers, although Article 37(2) assured the judiciary's independence of the others. Articles 20–24 added a Senate to the legislature, not much more defined by functions than Owona's and oddly constructed, 70 percent "by indirect universal suffrage" exercised in regions (the document's new term for provinces), 30 percent by presidential appointment.

Perhaps this Senate was meant to address the issue of clearly separated and balanced powers; in any event it veered sharply from CDU and SDF texts, and offered little consensual potential. Another such factor was Part VII, creating a Constitutional Council with vast powers, not only to adjudicate constitutional issues but also to conduct and supervise all elections and referendums. This ran against the grain of a debate since 1991 about the need for *separate* agencies for these functions, and especially the issue of their independence, for this text made three of the council's eleven members the president's appointees and reserved a place *ex officio* for all Cameroon's former presidents. Such features ran contrary to a few more promising elements, like judicial independence, protection for ethnic minorities, and (a trickier issue) indigenes under pressure from migrants, as well as Article 66 requiring officials to disclose assets and property at the start and finish of their tenures, from the president down through parastatal directors and tax officials.[52]

It was difficult on grounds thus far covered to imagine broad enthusiasm beyond the regime channels where it was created for this constitution as it defined *central* governance, and the text's later *sub*national sections on regions and localities surely widened the gaps. Article 55(2) gave subnational governance "administrative and fiscal autonomy in the management of regional and local interests ... freely administered by councils elected under conditions laid down by law." But then, whereas the CDU and SDF built in powerful, autonomous, elective local governance structures, and the SDF added revenue sharing to make such provisions work, *this* text called for elective councils which would elect a president (an indirect adaptation of Sonac Street's "Elected Governors"), but then in Article 58(1)-(2) negated the force of these apparent CPDM innovations:

> A delegate, appointed by the President of the Republic, shall represent the state in the Region. In this capacity, he shall be responsible for national interests ... as well as maintaining law and order. ... He shall exercise the supervisory authority of the State over the Region.

With the appointed delegate in place to check the elected regional president, the text then moved in Articles 59–60 to broadly written language enabling the president to suspend or dismiss these elected subnational office holders, and in Article 61 lodged powers to change the boundaries and add to the number of regions with, again, the president in Yaounde, suggesting a free hand to manipulate elections by realigning their map, against the SDF text removing such rights from the central government.

By the time the next to last Article 68 repeated the Owona text's gratuitous reference to the long defunct federal constitution's remaining force "where not repugnant" to this document, the "transparency" which all post-1990 rhetoric on constitutional issues called for was hard to discern, what was offered with the left hand was reclaimed with the right, and the new constitutional center of gravity in force fell far short of a consensual base which could move Cameroon beyond impasse. The document was not, as many called for, submitted to referendum after its passage in a National Assembly where key parties were absent, and it gave Paul Biya a clear opportunity to retain the presidency another fourteen years.

Marcelin Nguele Abada's close scrutiny of Law 96/06 concluded that "le jeu politique reste confisqué/the political game remains forfeit" with presidential prerogatives in full force. "Seul capitaine/Sole captain" of that game, Biya had nullified four years of pressure for constitutional change.[53] Constitutional debate and action thus typified the shadow world emerging and dissolving whenever by mid-decade the notables within and beyond the regime orbit sporadically mingled but hardly touched. The constitutional nadir was as obvious as the poverty of institutional, electoral, and party politics, where we now return.

Institutional, Electoral, and Party Politics, 1995–1997

Cameroon's formal politics responded to the December 1990 multiparty law, took its first recognizable shape under new conditions for the electoral contests of 1992, and was then refined though not transformed by the alliance structure of the National Assembly in 1993. Since then, through the subnational and constitutional episodes we have traced, and many others of interest but more ephemeral, survival has been the CPDM's objective, inertia its policy, impasse its achievement. The national political agenda, which it still in the formal sense directs, thus neutralizes constitutional debate, postpones or manipulates elections, and (above all?) prevents any free rein for the SDF. No fully credible appraisal of all the forces at work is therefore possible. So many stories, so hard to plot their direction; comprehensive patterns await their definition and determination. The historian struggles to align political analysis to a broad view of Cameroon's condition and needs, and to find a consistent line through all this.

One that is useful appeared in March 1994, when postcrisis routines were in place, and a lull if not a calm characterized public life. France once again backed Cameroon sufficiently for IMF debt payments to be renegotiated. Here was Biya's response to the IMF, by radio broadcast to his people March 23, in English translation: "The international finance community's attitude clearly shows the confidence it places in our country."[54] On the other hand *Le Monde* that very day, surveying francophone Africa, wrote (also translated) of "a situation routinely painful. A caricature example, ghostly Cameroon—the government hardly meets any longer—experiences the greatest hardship paying its civil servants." The "ghostly Cameroon" language rings more consistently and truer than Biya's.

Other signs of vacuity or morbidity appeared. Biya spent roughly 100 days in 1994 out of the country, local elections were put off by decree, and beyond the two regular sessions of the National Assembly only December's abortive Consultative Constitutional Committee briefly rippled the institutional surface.[55] "Après moi le déluge": Nicolas van de Walle's usage in a 1994 publication on Cameroon seemed apt.[56] So did one of the most intriguing domestic journalism efforts of 1995, *La Nouvelle Expression*'s April 18–21 evocation of Louis XV as "le roi fainéant/the idler king" toward the end of France's monarchy, and the current parallel case of Paul Biya and Cameroon. That very week Father Engelbert Mveng, arguably the country's most versatile and eminent scholar-artist of the generation born between World Wars I and II, was brutally murdered at his residence near Yaounde. A month later, the records and accounts storage sections of the National Assembly building were gutted by fire, very much like Ibadan's Cocoa House a decade before, with all the suspicion attaching to the loss of evidence which any prosecution for financial misconduct would require. In neither the murder nor the fire case, thirty months later, has an investigative report appeared. At a one day October 7, 1995, CPDM Extraordinary Congress, ten days after the signature of a letter of intent continuing Cameroon's lifeline IMF loan assistance another six months, Article 50.1 of the party constitution was amended so that Biya's CPDM presidency could be continued another five years, the meeting's most consequential actions in this sixth year of Cameroon's governance crisis.[57]

Yet, mid-1997, Biya remained safe in state and party presidencies, just having completed a term as chairman of the Organization of African Unity, 1996–1997. One of us wrote mistakenly in 1995, referring to the crisis years of 1991–1992 and their aftermath, and citing the state's revenue collapse and the regime's growing isolation both at home and abroad, that "the capacity of Cameroonians to starve the state was as striking as the regime's capacity to punish the people, and the presumed powers of incumbency looked more like burdens."[58] Eight months in Cameroon later that year accorded greater powers to incumbency than that passage did. And the observations on the radio review "Cameroon Calling"

Attack on John Fru Ndi, Yaounde, November 3, 1993. Photo by Kuitché-Dzomaffo, used with permission.

March 26, 1995, that most people born 1950–1980 habitually identify party with state unless something dramatic intervenes, and that those born since the CPDM's decline and the SDF's birth, 1985–1990, will not vote until at least 2005, also demonstrate the difficulty facing opposition forces. The setting we now establish for Cameroon politics through 1997 bears this out.

Biya in the presidential palace at Etoudi (Yaounde) still sets Cameroon's formal political agenda and timetable, not Fru Ndi in his compound at Ntarinkon (Bamenda), or any combination of the opposition forces. The regime's strength through incumbency grew after 1992. It dug in for survival, and France kept vigil for their joint interests abroad, no matter how lonely. Various conventional and some less conventional practices can be cited.

In the conventional range, the CPDM timed events to its advantage, continued the practice of machine and constituency politics through powers of patronage and appointment no less well known in Cameroon than elsewhere, and periodically resorted to force and intimidation. Local elections due in 1993 were postponed eighteen months by decree so as to withdraw a stationary electoral target the opposition could prepare its campaign for, and especially to freeze the SDF, then were delayed a second time. As the date for a poll inevitably loomed in 1995, CPDM appointees replaced elected officials to manage it and electoral procedures, including registration, were kept in MINAT's domain.[59] Indeed, 1995 went by without a new electoral code, so that when local elections, the first since 1987, were finally announced for January 21, 1996, neither due process nor public opinion were primed to produce results likely to convince Cameroonians at large of their validity.

Polls were conducted in 336 urban and rural communes, with thirty-eight of 123 registered parties active, most very localized. The CPDM entered all 336, while a combination of disputed MINAT invalidations and less than fully national organizations held the NUDP to 182 and the SDF to 105 commune contests (SDF candidacies were voided in 138 cases, according to a BBC January 22 report). Cameroon's intricate "list" system gave each commune a certain number of seats, and parties drew up candidate lists. Since an absolute majority victory gave all seats to that party, but less than a 50 percent result for any party led to a division of seats among the frontrunners, and gave those named *first* the advantage in any close result over those who followed, the order of candidates reflected sharp competition for constituency level party selections.

As in 1992, disputes compromised the results, and entrenched uncertainty and impasse, even to the point of rival tallies, including a discrepancy between MINAT's published results and the corresponding local reporting agents' figures.[60] Disputes aside, a detailed *Le Messager* statistical review February 15, supplemented by other sources, revealed the poll's basic patterns. The CPDM, reflecting its national apparatus, won absolute majorities in some respects, with 219 of the 336 communes and 6,033 of the 9,986 councillors. But the demographic distribution of the vote was less favorable to the regime, and in fact closely resembled the 1992 presidential vote pattern. Its success was confined to Center, South, and East Provinces, and to rural areas elsewhere with salaried chiefs or government ministers heading many CPDM lists.[61] Of thirty-one mayors in Cameroon's *urban* councils, twelve were CPDM and twelve were SDF electees from the councils created January 21 (until two the SDF won in Yaounde were disqualified after the vote), and three were from the NUDP. The eleven *largest* of these urban councils elected six SDF, two CPDM and two NUDP mayors, leaving one post in dispute. The demographic reality was most sharply focussed by the SDF's victory in 40 percent of those eleven largest urban communes' seats, including virtually clean sweeps in the anglophone cities and in francophone Douala, Nkongsamba, and Bafoussam, and a (revoked) share of Yaounde's; the NUDP took Garoua and Maroua, the CPDM Yaounde and Ebolowa. By another key measure, the CPDM won communes generating only 25 percent of Cameroon's local revenue, which meant a distinct threat in future to the accustomed funnel of funds from local councils to the national treasury, and whatever ultimate state, party, and personal destinations followed. Results were consistent with 1992 in terms of opposing CPDM rural fiefdoms to SDF ("Great West")-NUDP ("Great North") urban blocs, and in the way those three parties dominated all others, taking 92 percent of all communes and scuttling any expectations that small ethnic bloc parties would hold their own. This factor shaped one very crucial case: the SDF polled successfully in anglophone Cameroon against listed candidates with views more sovereigntist or secessionist than its own, and although local politics rather than the constitution drove this poll, the SDF could claim that its brand of federalism was no hindrance.[62]

Voting was one thing, government action in its wake quite another. Within a month, all eleven largest urban communes had government-appointed urban delegates in place (effectively now mayors, replacing those elected), even though only two of them elected CPDM mayors January 21. Three people died when protest against the SDF's displacement turned violent the last weekend of February in Limbe, South West Province, and Garoua's response to the appointment of a former minister of defence included a declaration by Ahidjo's son Mohamadou Ahidjo Badjika, its elected NUDP mayor, that there would be fiscal disobedience and "Pas de cohabitation." Many ceremonies installing the urban delegates were boycotted.[63] "Plus ça change...": nothing obviously negated the commentary in *Le Messager* February 1 that, with its constitution having passed the National Assembly with no referendum *and* its urban delegates poised to nullify electoral defeat, CPDM governance by attrition and bad faith, and therefore Cameroon's impasse, would continue.

The conduct and outcome of this balloting nicely demonstrated the government's edge in making use, substantively and tactically, of its control of the political agenda and timetable. With electoral politics still manageable in the short term, conventional machine and constituency politics, as since 1992, kept the CPDM formally intact. We have seen the offers of government seats following 1992's elections accepted by UPC-Kodock and by individuals like Moustapha and Tchiroma of the NUDP, and the reduction or neutralization of opposition thus achieved. It also used high office and its perquisites and patronage to "shadow" opposition leaders on their own terrain. Thus, Achidi Achu as prime minister challenged his Santa village rival Fru Ndi among North West Province anglophones, as did the CPDM veteran Robert Mbella Mbappe, minister of national education, for Samuel Eboua and his Movement for Democracy and Progress in their shared ethnic and geographic domain of the Mbo near Nkongsamba. This strategy whereby appointed officials were chosen to offset electoral results applied with special vigor in these cases because the CPDM won only 4 percent of the 1992 presidential votes in Achidi Achu's electoral district, and 7 percent in Mbella Mbappe's.[64]

Another incumbency tactic, perhaps less conventional by some standards, the effort to destabilize other parties, commanded much attention and suspicion by 1995. The UPC's residual force was as early as 1991 further diluted by Woungly Massaga's defection on CPDM initiative, noted in Chapter 5, and the contest for the leadership of its fragments helped Kodock move his wing into the government alliance in 1992. The two NUDP ministers taken into government in 1992 stayed in the NUDP while it tried to sort out the anomaly for two years, until their expulsion in January 1995. They immediately formed an "authentic" wing, the NUDP-A then renamed the National *Alliance* for Democracy and Progress, and acquired legal status from MINAT while continuing to challenge Bello Bouba's hold on the parent NUDP in the courts. Something similar vexed the SDF across its divide from the institutional mainstream in 1994–1995, not always possible to distinguish from internal problems covered immediately below, but circumstantially blamed on CPDM "divide and rule" interference by the middle of 1995, when breakaway structures known as the Social Democratic *Party* and Social Democratic *Forum* were registered by MINAT. The net result of these and other developments was 123 parties by March 1996, no longer a bold multiparty expression, but more a confusing, exhausting, ultimately impossible political map for many Cameroonians to negotiate, much to the regime's advantage. These incumbency features enabled the CPDM to hold its disputed ground after 1992, even if it became more of a patronage agency with an exclusive clientele at home (the Beti channel and its auxiliaries) and abroad (the French and "Bretton Woods" finance) than a functioning political party, at least beyond its core support areas.

But whether or not one subscribes to the CPDM's minimilist, belly, incumbency politics, *or* believes in MINAT's organization and the Supreme Court's confirmation of Cameroon's electoral process, *or* believes that elections are a valid measure of Cameroonians' political will, "success" in regime terms was tangible in the legislative election of May 17, 1997. It used the time and experience of more than four years since the last national ballot and a year since the local elections to prepare its ground, and not just procedurally. Ministerial changes in September 1996 replaced Achidi Achu, who never proved effective against Fru Ndi on their common North West Province terrain, with a new prime minister, the South West Province's Peter Mafany Musonge. From the anglophone area least attached to the SDF, trained in the U.S.A. as a civil engineer, and with managerial experience during a decade's leadership of the Cameroon Development Corporation, he brought credibility, credentials and a potentially useful constituency to the post beyond Achidi Achu's capacities. The other important feature of the September 1996 shuffle was the higher allocation of ministries to North Province.

The controversy over voter registration, electoral boundary issues, and the mechanics of the May 17 vote itself duplicated previous experience, and the preliminary Commonwealth observer team's report, though more polite than the NDI's in 1992, was very critical.[65] Since the SDF contested *this* National Assembly election, comparisons between 1992 and 1997 are largely vacuous, but the 1997 result improved the CPDM's situation and prospects in conventional terms. It restored its majority, adding twenty-one seats for a total of 109 out of 180, and won 47 percent of the votes cast in a 75 percent turnout.[66] Official counts awaited the settlement of disputes by new elections August 3, but roughly twelve seats more than in 1992 in South West Province, six more in North Province and twelve more in the rest of the "Great North" keyed the CPDM's recovery, in precisely those regions the September ministerial realignment had favored. The SDF's popular vote was 24 percent and it won forty-three seats, sweeping all but one of twenty in North West Province and taking a large majority in West Province. Since the CDU took all five Foumban area West Province seats, it was clear that the twenty seats their 1992 boycotts gave the CPDM had "come home," but these CPDM losses were offset, and more, elsewhere. The SDF victory in only nine of thirty seats in 1997 in South West and Littoral, where its presidential tally in 1992 was so impressive in urban areas, was not conclusive, because the regime redrew their boundaries, but a forty three seat total was not impressive. The NUDP polled 14 percent and fell from sixty-eight seats to (again, pending review of disputes) fifteen, losing all but one of its thirty-six 1992 seats outside the North, and keeping a majority *there* only in Adamawa Province. On the margins, alongside the five CDU seats (one was Ndam Njoya's), only their leaders, Kodock and Daïssala, took seats for the UPC and MDR parties allied to the regime. With their virtual demise, splinter party avenues into electoral politics seemed an option, or threat, without the capacity for mischief some have feared. One final irony marked May 17, 1997; the twenty-one seats the CPDM gained were precisely the number a *combined* opposition vote, instead of divided candidacies, would have taken from the regime, and kept it in a minority.

Fru Ndi and the rest of the opposition denounced every aspect of the poll, but SDF members took their seats with others in the National Assembly in June. As a prelude to the contest for the presidency in October 1997, these results, the credibility given the election by SDF participation, and its own majority in a *functioning* National Assembly must have encouraged the CPDM. One final calculation from electoral data gave a reading of its task ahead. Biya led ministers to the presidential poll in 1992 whom (like any politician facing an electorate) he presumably considered capable of delivering a favorable CPDM vote, based on the electoral map's representation and patronage realities. In the thirty-seven electoral divisions his ministers represented, from Achidi Achu down the ranks, the CPDM won 50+ percent of the vote in just sixteen of them. Of the twenty-one others, it polled less than 30 percent of the votes in eighteen, and in nine of *those*, the CPDM vote was under 10 percent. This May

1997 election looked like an improved springboard for some restored semblance of Bayart's hegemonic project and Ngayap's ruling class, if the rules of the electoral game and condition of impasse remained in force for the presidential election later in the year.

What, in all this, of the SDF, the most active but least calculable factor in all political equations? There are only imperfect indices, including the May 1997 election, in the absence of fully reliable vote counts. But Fru Ndi has challenged Biya's authority consistently since 1990, while avoiding since 1993 the public clashes and the reprisals to SDF militants which Bamenda's deaths on many occasions proved they would provoke. Thus, party activity in recent years has shifted to his own well attended presidential style rallies with their speeches, motor cavalcades and market tours, more private but discreetly publicized palace and funeral appearances in areas all over Cameroon where Biya chose or dared not to go, the steady appearance of foreign dignitaries at his Bamenda compound which came to be known as "Ntarinkon Palace," and large rallies during election campaigns.

More quietly, in his own North West Province support base, he has pursued a substantive and symbolic politics to point the contrast with the CPDM's distance from most Cameroonians. Some 1995 examples make the point and demonstrate the effort, far from Yaounde, to build civil society's sinews. He led a SDF rural trek March 12–15 with the trappings, staff and all, of a Gandhi pilgrimage to Furu Awa, a remote subdivisional center with no roads for less than four wheel drive vehicles to reach it and virtually without government services, at a time when Achidi Achu as prime minister undertook brief "parachute" visits in a government helicopter to rural areas like this, promising projects their people had heard of before but never yet seen.[67] Soon after, the SDF sent out militants to clean streets and streams in Bamenda of garbage which the urban council with its CPDM members in place since 1987 left untouched: was it coincidence that there soon followed (like the PSU noted earlier in Yaounde and Douala) a Bamenda government presence in the form of two massive, freshly painted Dutch project garbage trucks?[68] And when violent land disputes flared near Bamenda, linked to officials in the presidency by circumstantial evidence too dense to report thoroughly, Fru Ndi visited the sites and undertook negotiations which local administrators seemed less interested in conducting.[69] The SDF approach was crystallized at the party's national convention at Maroua late in May (covered in more detail below), when Fru Ndi coined the phrase "up there" to describe Cameroon governance; no one *nationally* missed the reference to Biya's frequent flights to Europe, any more than North West Province people missed the contrast between Fru Ndi's feet and Achidi Achu's rotor blades.

John Fru Ndi Mediating Village Boundary Dispute, North West Province, mid-1995. Photo by Kuitché-Dzomaffo, used with permission.

But alongside this popularity and the exploitation of the CPDM's defects were more problematic dimensions to the SDF's post-1992 experience. Sorting out the constitutional tangle when anglophone two state federalism emerged in 1993 kept the party under constant pressure. The work to persuade francophones that Cameroon needed not just a decentralization of the unitary state but a specifically federal constitution went forward from the Bafoussam congress in 1993, but its tactics and timing were delicate matters. For every francophone convinced about the "soft" four state federal approach, there was the risk of losing anglophones who demanded the two state federation, with one state being anglophone, or a more sovereigntist or separatist solution to grievances which indisputably brought latent francophobia to the surface west of the Mungo River, with negative consequences for the SDF among francophones to its east.

The constitution and other policy issues in the SDF mingled with grittier, even seamier forces. The party's early leadership divided. We saw Bernard Muna, the son of Ahidjo's vice-president, who was the lead lawyer in Yondo Black's defence and then a high SDF official, challenge Fru Ndi as a candidate for Cameroon's presidency in 1992. He broke ranks further by deserting the Sovereign National Conference strategy in 1993 for Biya's "Grand Débat" and lost his party posts before resigning in 1995. A more fundamental separation was Siga Asanga's. He is Fru Ndi's uncle, co-signer with him of the SDF's first application for legal status, and the party's first secretary general. A literary scholar at the University of Yaounde with the SDF's highest academic profile, as close to an ideologue of the left as the public element of the party's inner circle has produced, Asanga became openly critical toward 1995. He charged the SDF with deserting its populist, social democratic character and national appeal in favor of a more regional anglophone alliance with traditional authorities and business elites, a retreat in class terms to Ahidjo's national bourgeoisie, with federalism as a constitutional camouflage and an exclusionary device against non-anglophone progressives.[70] This fed tensions between SDF anglophones and francophones, and led not only to Asanga's loss of party office and expulsion, but also to a number of Bamilekes and francophones with Yaounde and particularly university ties, in Asanga's circle, resigning or being dismissed from the party's National Executive Committee and other high offices in 1995.[71]

The SDF stood its ground by pointing to its compromise short of two state federalism, in a party with a majority of francophones and an appropriate number of Bamilekes in its leadership positions, evidence of symmetry between its policy and membership. But these disputes, acrimonious and the topic of front page headlines and inside page speculation in all newspapers, were not all about policy and principle. The party accused Asanga in particular of building a personal clientele within its ranks by appointing or "co-opting" high officials, now following him into dissent, in a party built quite differently, on the basis of local, ward level elections for its leaders, including Fru Ndi. It believed the dissidents to be motivated, at best, by too much taste for power and office, frustrated by the party's painful, patient, lonely witness outside Cameroon's institutional structure since 1990. Muna and Asanga in particular were charged with long standing personal ambitions to unseat Fru Ndi, based on their judgment of his career and character as too limited for the presidency of the party now and the nation in future. This was the "unqualified bookseller" critique the CPDM constantly used against Fru Ndi, which Muna's resignation letter quoted by *The Herald* December 25–31, 1995, in fact echoed: "it is evident that the SDF has no use for a man of my standing and experience." Those who resigned or were purged accused Fru Ndi and the SDF National Executive Committee of narrow vision and wide paranoia.

Such charges and rebuttals reflected severe SDF cleavages and made it plausible by 1995 to perceive the SDF caught between Yaounde's cosmopolitans and Bamenda's provincials. One group was from the intelligentsia and attuned to the political and ideological currents, debates, and compromises of the center. The other was still shaped by grass roots life at the marginalized periphery where violent repression remained more of a threat and militant defiance more the party's mainstream, intensified after 1993 by the AAC thrust. One side thought the other opportunist, the retort emphasized fundamentalism.

The SDF could hardly avoid such divisions between its political society and civil society components, and the speculations they raised, as the consequences of mounting years in the wilderness and the need to keep a hard faith without the rewards of electoral or other tangible success.[72] If, in crucial ways, Fru Ndi's Bamenda compound was its strength in solidarity, it was also a bunker, literally under siege during 1992's state of emergency, retaining that character politically and temperamentally thereafter. Despite its very open doors and many visitors, Ntarinkon in Bamenda displayed some of the hermetic character and rivalries it decried at Etoudi in Yaounde. In order to keep public trust in the democratization effort it led in 1990–1991 free from any hint of compromise or betrayal, the party considered it necessary and beneficial to abstain from political routines, except the 1992 presidential contest for real power, until the 1996–1997 electoral cycle. That concern for its legitimacy was fundamental, and correct, if one considers the consequences when perceptions of opportunism in the democratization struggle arise elsewhere in Africa.[73] But the isolation, disputes, and loss of leaders exacted a high organizational cost.

Denied a formal national role in politics for five years by the CPDM's inertial freeze and its choice of response, subject in its isolation to the external and internal pressures just cited, the SDF faced quite mundane problems as well. Delay of a scheduled party Congress at Maroua, 1994, meant to take the party banner north, raised questions about party finance: was it suffering in this sense too? Defections and dismissals were frequent enough by March 1995 for the broadly sympathetic independent press to speculate on Fru Ndi's leadership, even the SDF's survival as a viable political force.

The party in fact rallied. It replaced Asanga as secretary general in 1994 with a University of Yaounde faculty member and South West Province native, Tazoacha Asonganyi, from precisely the ranks its critics and defectors claimed it marginalized. It then went on the offensive by holding the Maroua Congress May 25–27, 1995, the fifth anniversary of its Bamenda launch, against heavy odds.[74] It was a gamble to convene in a part of Cameroon the CPDM and NUDP dominated, where Fru Ndi's 1992 presidential vote in the three northern provinces combined was 4.3 percent by the regime's count, 6 percent by the SDF's. Rank and file members travelled, most by road caravans, to Cameroon's northernmost provincial capital, adding 650 delegates from afar to 275 from Adamawa, North, and Far North. Maroua's people and others along the official party's route northward turned out by the thousands for public occasions mixing "meet the candidate" market tours and spontaneous stump speeches without any security presence or problems—we noted above the "Power" salutes, and should add from that previous text on the North

Impasse, 1993–1997 205

The SDF Cavalcade, Maroua Party Congress, May 1995. Photo by the authors.

the comparison with Biya's tour a month earlier, bristling with security but still cut short. Fru Ndi and other SDF leaders, with Maidadi (introduced above) their local anchor, sought and gained wide access and hospitality, and made their case, on visits within northern political circles not previously accessible. The Congress itself was held at a compound which on May 24 was totally unserviced and unequipped, save one hand-pumped water well, but which the delegates in thirty-six hours made into an adequate shelter and meeting place for the orderly conduct of their work. The Maroua Congress did not translate into support in the North's local polls eight months later or in the legislative election of 1997, for MINAT disqualified many candidates, and much of its northern leadership under Maidadi remains vested in "Great West" migrants rather than locals. But the SDF proved it could function self-sufficiently, in numbers, with energy, 1,200 road kilometers from its core area. It removed any doubts others raised in 1995 about its resilience and vitality, and by 1997 had a viable second echelon of leadership under Fru Ndi to replace the earlier losses.

Still, contrary currents within and around the SDF helped the CPDM to survive. The regime exploited remaining SDF differences, played on North West-South West Province rivalries among anglophones by Musonge's appointment, even enjoyed the rhetorical luxury of warning about threats to Cameroon's sovereignty because of the frontier dispute with Nigeria in the South West Province Bakassi peninsula area, close to parts of former Biafra. That spectre of secession and civil war was invoked to demonize anglophone links to Nigeria. In a situation already marked by inertia favoring incumbency, and controversy within the SDF, such factors further protected CPDM interests.

Likewise useful to the CPDM was the distance separating all the potentially united opposition forces from each other. We are persuaded by Mbembe's view that there is a link of historic legitimacy from the old UPC to the new SDF, and this text offers proofs for it. But there is also a plausible skepticism about the *entire* opposition's background and performance, expressed by Manga Kuoh as follows: "l'Opposition au Cameroun en 1990 est tout aussi l'héritière du système-Ahidjo que le Pouvoir/Cameroon's opposition in 1990 is as much the heir of Ahidjo's system as those in power."[75] Opportunism and disunity are his points, which can be measured by the reduced base of the original NCOPA coalition. It rallied powerfully in mid-1991, but within a year its early leaders Fru Ndi, Bello Bouba, and Ndam Njoya competed for the presidency and Kodock took the largest UPC faction into alliance with the CPDM. What remained of the NCOPA by 1995, the Front of Allies for Change (FAC), was a fitfully reactive group of roughly fifteen parties, none national in scope except the SDF. They were satellites in its orbit and the FAC scarcely stirred between irregular meetings which Fru Ndi by mid-1995 was as likely to miss as attend. The FAC effectively ceased functioning late in 1995 when the SDF suspended its participation, tired of contributing a lone national presence to its partners and of absorbing their sporadic criticism, arguing it was high time for each FAC party to prove its credentials in the January 1996 local elections; only the SDF did so.

Adding to the government's cushion created by opposition weakness was the role the NUDP under Bello Bouba played by 1997. Signing the Yaounde Declaration and participating in both the 1992 elections to stake a claim in Cameroon's formal politics, the party formed the official National Assembly opposition with sixty-eight seats and took 19 percent of the presidential vote. Bello Bouba, like Ndam Njoya well connected in France, has the Cameroon "Great North" base described earlier, even if it was weakened by electoral losses in 1997. His dismissal from Biya's prime minister's post he held in 1982–1983 and his Nigerian exile give him a regional "grievance" constituency like Fru Ndi's and Ndam Njoya's, and he plays an autonomous role, personally and as Ahidjo's plausible heir. He articulates this independence skillfully. On one hand is his consistent use of the term "responsible opposition" to establish his distance from Fru Ndi, tied to criticism of the SDF 1992 boycott when "active participation in the life of national institutions" would have challenged the regime with the strength of an opposition majority in the National Assembly.[76] At the same time, differences with the CPDM are made clear on occasions like the sentencing of eight NUDP members to eight-fifteen year prison terms after mid-1994 street clashes in Maroua and the imposition early in 1996 of CPDM urban delegates over locally elected NUDP mayors and councils in the North; the NUDP suspended its National Assembly participation and Bello Bouba threatened "ghost town" retaliation.[77]

The electoral calendar of 1997 brought the NUDP and SDF closer. Their leaders exchanged visits to each other's party congresses in late 1996 and early 1997, proclaimed a common front against Biya, and established a "harmonizing committee."[78] But Bello Bouba's political interest in preventing any advance by Fru Ndi must be noted. Assuming the maintenance of Cameroon's nation-state in its present form, and given the loss of NUDP seats in the 1997 National Assembly vote, Bello Bouba might choose to ally with Biya rather than Fru Ndi. He could claim a measure of power with the North as his base by reconstituting Ahidjo's North-South axis through a "union government" pact with the CPDM, whereas he could *not* so easily, perhaps never, challenge Fru Ndi with the latter in power, or dominate him in a coalition. In any event, no analysis about Cameroon in the 1990s can ignore Bello Bouba's autonomy and tactical strength in negotiations about alliances and power sharing. The real test of these matters will be whether Fru Ndi and Bello Bouba can agree on a single opposition candidate for the presidency prior to the October 1997 election, likely a third party, perhaps Garga Haman Hadji, who resigned as minister of the civil service in 1992. But no one should expect such an agreement, given the history of the opposition.

Its real missed opportunity, working to the CPDM's great advantage, was the failure to ally the SDF and CDU in 1991. Chapter 5 recounted their proximity, then their split at the Yaounde Tripartite. Documents at the time and later, and interviews with Fru Ndi and Ndam Njoya in 1995, disclose not just policy differences between the parties and their different constituencies but temperamental gaps and mistrust between their leaders.[79] The SDF deemed the Tripartite a litmus test to determine which forces in the opposition inclined to resistance and which to establishment politics. Ndam Njoya, by contrast, saw it as the chance to act as Cameroon's prudent patriot, bringing hostility under control by bridging the CPDM's politics of stalemate and the SDF's politics of the barricades. To the SDF's legacy of May 26, 1990, he hoped to counter with the signature date of the Yaounde Declaration, and (one infers) stake a legitimate claim to the presidency. Fru Ndi's path is the passionate populist's; he views Ndam Njoya as a leader with no domestic following outside the 4 percent of Cameroon's vote and five seats in the National Assembly

Foumban provided in the 1992 and 1997 elections, angling for the presidency more through personal and political ties in Europe than support in Cameroon. Ndam Njoya considers himself the rational and experienced statesman, and Fru Ndi in terms consistent with a press conference remark printed in *Cameroon Post*, December 30, 1991: "extremism has no place in the politics of today." Neither the affinities of their parties' constitutional proposals nor their considerable qualities and talents are likely to bridge these fundamental gaps.

Thus, the poverty of regime politics by 1997 is balanced by the opposition's largely misaligned parties and personalities. Domestic initiatives for crisis resolution in such conditions hold little immediate hope, and forces outside Cameroon have nudged more than budged its *status quo*. Executive powers and a manageable National Assembly allow the CPDM to sustain the residual single-party legacy, which has thus far proved impervious to demands for fundamental change. Appointed governors and local officials under MINAT discipline still operate forms of governance decades in the making, and remain the chief agents of political practice. And it may well be that elections which are prepared, conducted, counted, and confirmed under the disputed conditions of the 1990s have generated more frustration and indifference than confidence about the political process, prolonged rather than resolved the impasse in the five years separating the 1992 and 1997 polls, and stalled more than advanced democratic transition. There is surely, Africa-wide in the 1990s, more evidence of and correlation between flawed ballots leading to trouble, than for electoral activity by itself fostering democratic transition, and Cameroon's experience is a case in point. Richard Crook has shown how Côte d'Ivoire elections failed to break its regime in 1990 and 1995, in conditions uncannily like Cameroon's, with multipartism compromised by opposition politicians' rivalries, ethnic ruptures, and an inability to bring politics off the streets into a persuasive and positive program for change.[80] The parallel is clear, and discouraging.

Seven years of resistance to Cameroon's political order has rippled and sometimes riddled its surface, without uprooting its structures. It may well be that a reckoning still awaits the regime in late 1997's presidential election through a single opposition candidacy, or in some always possible, if sterile, scenario which the mid-1997 unrest in Kenya, Cameroon's close counterpart, reminds us could develop; nothing like Congo (Kinshasa) seems likely. But in the conventional measures this chapter covers, and despite the subnational evidence of trouble, Paul Biya has survived, if by default and French support, and has set a course for his presidency until 2010. There has been no genuine transition to democracy. Given the impasse, 1993–1997, it takes the long term optimist's leap of faith about the SDF's character and mobilization potential, and civil society's, over the short term realist's reading of the evidence to believe that Cameroon will move toward democratic transition and consolidation without a heavy, perhaps convulsive cost. Is any such optimism warranted? Chapter 7, as it develops, and concludes the book, will take up this question among others.

Notes

1. There was constant publicity for the next two years about this story, until their expulsion; some believed that Bello Bouba approved of their posts as antennae for the NUDP and good faith pledges for a potential NUDP-CPDM pact. Such were the suspicions swirling in post-electoral Cameroon.

2. It happened that Cameroon's Ambassador to the U.S.A., Paul Pondi, was then senior in the diplomatic corps there, so that *his* picture alone with Clinton at Georgetown University was part of the inaugural protocol and available for *Cameroon Tribune* to use against the Fru Ndi repertoire. When, as they do, Cameroonians on all political sides remark on Paul Biya's good luck, or "powers," this exemplifies what they mean.

3. Albert Mukong, *Prisoner Without a Crime* (Bamenda: Alfresco, 1985) takes his story to 1976: he brings it to 1991 in *My Stewardship in the Cameroon Struggle* (Enugu, Nigeria: Chuka Publishing, 1992). There is a valuable, often anonymous miscellany for use in any more complete account of anglophone circles during the 1980s and 1990s than we attempt here, some of it collected and published in the U.S.A. as Albert Mukong, ed., *The Case for the Southern Cameroons* (CAMFECO, 1990), with unsigned authorships it would be interesting to establish. We thank Mr. Mukong for providing many texts and interviews, which the following account of anglophone activity through 1997 places alongside other sources, to create at least the outline of a history which needs a more definitive version.

4. He was honored as an SDF Founding Father at the party's first Congress in May 1992, Bamenda, and was still so listed April 16, 1996, despite disputes traced below which distanced him from the SDF leadership by mid-decade.

5. This last convolution was suggested by a senior SDF advisor.

6. Albert Mukong, "Where Things Went Wrong": a paper presented to the Buea All Anglophone Conference of 2nd and 3rd April, 1993, pp. 13–14.

7. "The Buea Declaration," p. 10. The Declaration, Mukong's paper and much more from Buea appeared in print in a double-length special April 7, 1993, edition of *Cameroon Post*, which alongside *The Herald* became a key source and (unlike *The Herald*) publicist for anglophone affairs.

8. Second Ordinary Convention of the SDF: General Policy Address, p. 9.

9. "Let's Keep Cameroon One," especially p. 5 for reference to AAC two-state federalism adopted by the SDF provincial conferences in the weeks before Bafoussam. We are grateful to many in attendance at Bafoussam in 1993 for 1995 interviews on these very delicate issues.

10. SDF Second Ordinary Convention, Bafoussam, 29–31 July 1993: General Resolutions, pp. 3–4. Interviews with constitutional sub-committee members two years later yielded the assertion, disputed by others, that francophones there and

in the general assembly approved this broad rather than narrower AAC federalism. Whoever writes the SDF history will need to cover this episode carefully, for it remains contentious in light of party conflicts and defections we will note below.

11. National Executive Committee, Resolution 1. g), August 22, 1994.

12. Resolutions of the Third Ordinary Convention of the Social Democratic Front, May 26–29, 1995, p. 5.

13. The Bamenda Proclamation, p. 4. This theme recurs in anglophone circles; *The Herald*, February 2–4, 1996, reported the Cameroon Anglophone Movement's resolve to create a Constituent Assembly at its Bamenda executive meeting January 27.

14. For Elad's remark on the SDF and Fru Ndi, see *West Africa*, June 20–26, 1994, p. 1090.

15. A May 13, 1995, interview with Simon Munzu provide considerable context and detail here, and below, on constitutional and political issues. Audio tapes of him reading the Bamenda Proclamation at the church are prized in Bamenda.

16. *The Herald, Cameroon Post*, SCNC press releases and a number of broadsheets, especially "The New Nation" appearing every few days, carried details of this mid-1995 drama; personal observations supplement the Bamenda account.

17. This is a fascinating sub-theme we are not able to track; someone should. Journalists as far back as 1991 reported regime allegations about plots against it on Nigeria's behalf, fed from 1994 by the two countries' Bakassi Peninsula border skirmishes. The SDF categorically denies any interest in turning to arms Fru Ndi says in conversations it has been offered. A curious episode some years in the local background came to wider notice, *West Africa*, May 13–19, 1996, when Jeffrey Hughes, a 61 year old Australian "businessman and arms trafficker," was convicted in Queensland and jailed two years for "planning to overthrow the Cameroonian government or, failing that, to create an English-speaking state in the south of the country." It would be surprising, given Cameroon's course since 1990, if there were no such coup intrigues on the fringes of both its regime and opposition politics.

18. SCNC Press Release, July 9, 1996, and Piet Konings, "Le <problème anglophone> au Cameroun dans les années 1990," *Politique Africaine* 62 (1996), p. 32.

19. Jacques Benjamin, *Les Camerounais occidentaux: La Minorité dans un Etat bicommunitaire* (Montréal: Les presses de l'Université de Montréal, 1972), p. 149.

20. For the German background, see Harry Rudin, *Germans in the Cameroons, 1884–1914* (New Haven: Archon, 1968), pp. 269–271. Dossiers in March 31, 1995, editions of *Cameroon Tribune* and *La Nouvelle Expression*, then in *Génération*, April 12–18, and Paris' *Jeune Afrique*, May 11–17, pp. 34–37 (the best coverage overseas), described the French colonial and independent Cameroon eras from varied angles, and are substantially used in this account.

21. *Le Messager*, April 3 and *Dikalo*, April 10; "L'affaire SODECOTON" became *the* independent press story until June.

22. *Perspectives Hebdo*, April 11–17, was the source for this argument, drawing at length on Hayatou-Debre exchanges it quoted without attribution, with Hayatou arguing the case for northern indigenes at SODECOTON's helm.

23. Kees Schilder, *Quest for self-esteem: State, Islam and Mundang etnnicity in northern Cameroon* (Aldershot: Avebury, 1994), pp.141–144, 180, 212, on the Mundang of Kaele Division, Far North Province, and Philip Burnham, *The Politics of Cultural Difference in Northern Cameroon* (Washington: Smithsonian Institution Press, 1996), p. 170, on the Gbaya, Mbororo and Fulbe of Mbere Division, Adamawa Province, make these points.

24. Schilder, *Quest for self-esteem,* pp. 161ff., 172 (CPDM "party membership was generally regarded as a kind of tax" by the late 1980s); Burnham, *The Politics of Cultural Difference,* pp. 40–41, 93, 132–134.

25. *The Herald,* July 17–19, 1995, developed this parallel, quoting the Minister of Agriculture: "CDC Will Be Privatized." For a detailed analysis of CDC capital and labor, see Piet Konings, *Labour Resistance in Cameroon: managerial strategies and labour resistance in the agro-industrial plantations of the Cameroon Development Corporation* (London: James Currey; Portsmouth, N.H.: Heinemann, 1993).

26. Burnham, *The Politics of Cultural Difference,* pp. 131–132.

27. Burnham, *The Politics of Cultural Difference,* pp. 146–147.

28. *Le Quotidien,* March 14, 1996.

29. Mbembe, "Crise de légitimité, restauration autoritaire et déliquescence de l'Etat," in Peter Geschiere and Piet Konings, eds., *Pathways to Accumulation in Cameroon* (Paris: Karthala and Leiden: Afrika-Studiecentrum, 1993), p. 367.

30. Mbembe, "Crise de légitimité," p. 347.

31. Mbembe, "Crise de légitimité," p. 368.

32. Mbembe, "Crise de légitimité," p. 369.

33. Schilder, *Quest for self-esteem,* pp. 225–237; Burnham, *The Politics of Cultural Difference,* pp. 138–142.

34. Maidadi became a most interesting political figure in 1995 and is a person to watch in future. His father is the CPDM's minister in charge of National Assembly affairs, but the son is a tireless organizer who (with one author present) took Fru Ndi through streets and into palaces where the SDF had little or no previous northern access.

35. Mbembe, "Crise de légitimité," p. 374.

36. *Challenge Hebdo,* July 20, 1995; *La Nouvelle Expression,* April 12, 1996, with an added note from Paris sources that a court there was interested in the case: was someone invoking French law designed to indemnify companies like CFDT against losses from corrupt or hostile action abroad like SMIC's, the fascinating research domain opened by Pierre Péan *L'argent noir: corruption et sous-développement* (Paris: Fayard, 1988) and his later work?

37. See *Dikalo,* May 8, 1995, for its detailed February 1 advice to Mitterrand. Fru Ndi's membership in Laakam is public knowledge; both parties insist it is a bond of courtesy linking actors of substance, but enemies charge Laakam with "running" Fru Ndi.

38. For an example, significant again for its attention to opinion in France as well as Cameroon, see Laakam's December 9, 1994, press release attacking CRA-TRE, printed *Jeune Afrique,* February 13, 1995, p. 73, after its December 22, 1994, appearance in *Challenge Hebdo* (Douala).

39. The Owona-Essingan attribution is from *Cameroun Mon Pays*, June 29, 1993, a hostile source. Its publisher Théodore Ateba Yene, a dissident figure in Cameroon, died August 17, 1995, in circumstances deemed suspicious by *La Nouvelle Expression*, August 22–25, 1995.

40. *Galaxie*, February 6, 1995, provides a comprehensive dossier for this suspicion. Ethnic-specific material germane here, though not directed to this issue, informs Mario Azevedo, "Ethnicity and Democratization: Cameroon and Gabon," in Harvey Glickman, ed., *Ethnic Conflict and Democratization in Africa* (Atlanta: African Association Studies Press, 1995), pp. 255–288.

41. *Le Messager*, February 1, 1996.

42. *Challenge Nouveau*, February 22, 1996, described the Ngondo tensions. For Austen's fuller history, drawn on again in Chapter 7, see "Tradition, Invention and History: The Case of the Ngondo (Cameroon)," *Cahiers d'Etudes Africaines* XXXII,2 (1992), pp. 285–309, and *The Elusive Epic: Performance, Text and History in the Oral Narrative of Jeki la Njambè (Cameroon Coast)* (Atlanta: African Studies Association Press, 1995).

43. See the varied references, *Politique Africaine* 62 (1996), pp. 8, 19, 35, 61. By the mid-1990s in Cameroon, the "autochtone vs. allogène" vocabulary of tension, pitting natives and migrants, recurred in press reports from Yaounde, Ebolowa, Douala and Foumban.

44. John Mukum Mbaku, "Effective Constitutional Discourse as an Important First Step to Democratization in Africa," *Journal of Asian and African Studies* XXXI,1–2 (1996), pp. 48–49.

45. The issue's volatility registered when the committee's work surfaced in headlines like *Cameroon Post*'s, April 2–9, 1992: "*Row at Constitution Drafting Technical Committee:* Anglophone Proposal for Federation Rejected: Owona Threatens Beti Secession if Anglophones Insist on Federalism."

46. A fourth anglophone member, Biya's Minister of Tourism Benjamin Itoe, broke their ranks. A May 13, 1995, interview with Munzu and texts he supplied, including the trio's comprehensive "Press Statement" of May 19, 1993, are gratefully acknowledged as keys to much that follows immediately below.

47. For a precise and more technical coverage of what led to Law 96/06 but not the range of proposals discussed here, see Marcelin Nguele Abada, "Ruptures et continuités constitutionelles en république du Cameroun," *Revue Juridique et Politique, Indépendance et Coopération* 50,3 (1996), pp. 272–293.

48. An anglophone NUDP Vice President, Dr. Lawrence Formambuh, predicted in a conversation four months earlier (July 27, 1995) a federalist component in any future NUDP constitution; one wonders how he viewed this text when it surfaced.

49. Achille Mbembe, "La violence derrière le multipartisme," *Afrique Magazine*, November 1992, p. 56.

50. This was the work, following the 1993 Bafoussam Congress, of its Constitutional and Political Affairs Committee, whose chairman Nfor N. Nfor kindly provided and discussed the text in mid-1995, before its subsequent adoption by the party executive and publication.

51. This section enlarged the rights of citizens; the problems, of course, were twenty-four years of public life which compromised or contradicted the words on

those pages, and the CPDM's credibility when it used language like that emphasized here.

52. Much has been made of the Preamble's provision that "the State shall ensure the protection of minorities and shall preserve the rights of indigenous populations" as a pioneer effort in Cameroon's constitutional history. It remains to be seen whether this, in action, becomes a tool and rationale for ethnic divisiveness; see Nguele Abada, "Ruptures et Continuités," p. 293.

53. Nguele Abada, "Ruptures et continuités," pp. 286–287, 292.

54. *US Foreign Broadcast Service, Africa* (Washington, D.C.), March 24, 1994.

55. *The New Expression*, January 26, 1995, citing *Le Messager*, tallied Biya's 1994 time abroad, on trips usually reported by *Cameroon Tribune* or CRTV during or after the fact, not before.

56. Nicolas van de Walle, "Neopatrimonialism and Democracy in Africa, with an illustration from Cameroon," in Jennifer Widner, ed., *Economic Change and Political Liberalization in Sub-Saharan Africa* (Baltimore: Johns Hopkins University Press, 1994), p. 138.

57. *Galaxie*, October 9, 1995.

58. Milton Krieger, "Cameroon's Democratic Crossroads, 1990–4," *The Journal of Modern African Studies* 32,4 (1994), pp. 624–625.

59. The press kept this issue prominent; *La Nouvelle Expression*, August 8–11, 1995, gave the most detailed coverage of Law No. 92/002 of August 14, 1992, on electoral practice and its checkered career in the interval.

60. *Le Messager*, January 29, 1996. A Yaounde group, Conscience Africaine, with some funding and logistical support from outside Cameroon, was the major monitoring presence, with 189 observers in seven provinces, about the same coverage as the NDI achieved in 1992 but less experienced and financed, and unaccredited; an interview with its Secretary General, *Le Messager*, February 15, 1996, demonstrated MINAT's free procedural hand.

61. Chiefs were under heavy pressure to vacate the ostensible neutrality of their palaces for the balloting; those whose lists lost, and perhaps some who won, damaged their local standing, as Chapter 7 will point out. Speculation about their motives for running included expectation of reward through seats in the Senate which the constitution, just put into effect, made 30 percent appointive through the presidency. *The Herald*, January 23–24, 1996, and *L'Expression*, February 9, 1996, tracked these issues.

62. For the anglophone vote analysis, see *Le Messager*, February 1, 1996.

63. *L'Effort Camerounais*, March 9–22, 1996; *Le Quotidien*, March 14, 1996.

64. *L'Expression*, November 17–23, 1992. Success was limited, judging by the 1996 local election results. The SDF won 57 percent of Santa's vote and all its rural council seats, and 44 percent of Nkongsamba's rural council and 60 percent of its urban council votes, translating into sixty-four of seventy-one council seats against the CPDM's four and the MDP's three. The regime's need to appoint urban delegates was clear.

65. *Le Messager*, June 11, 1997 (an edited version).

66. *Cameroon Tribune*, June 9, 1997, is the source.

67. A SDF video is the source here. Ten months later, the SDF took 61 percent of its council votes.

68. Following the SDF victory in Douala's 1996 Urban Council elections, Fru Ndi took a "Villes Propres/Clean Cities" campaign to its streets the last week in February, to demonstrate the party's popular mobilization capacity; *L'Expression*, February 27, 1996.

69. The "Special Duties" in the presidency of a local man, Peter Abety, looked from press reports very much like the arming of villages which were natural SDF and/or SCNC strongholds against each other, on grounds of ancient disputes (we are in Bayart's realm here); dozens were killed and hundreds wounded in mid-1995. *The Herald* covered these stories from Bamenda, with reporters at risk on the spot and, they feared, from regime reprisal after publication. One of the author's travel to a village at the time disclosed men heading for its boundaries with pikes and Dane guns; casualties mounted where newer and heavier arms were frequently used.

70. Attendance at a stormy April 29, 1995, press conference Asanga conducted in Bamenda and a long interview with him July 16, 1995, in Yaounde inform this passage, adding to vigorous press coverage and his own broadsheets the first eight months of 1995. His critique of the SDF is serious, and we return to these issues in Chapter 7; for background, see Ndiva Kofele-Kale, "Class, Status and Power in Postreunification Cameroon: the Rise of an Anglophone Bourgeoisis, 1961–1980," in Irving Markovitz, ed., *Studies in Power and Class in Africa* (New York: Oxford University Press, 1987), pp. 135–169.

71. These are the cases of Basile Kamdoum, Shanda Tonme, Dorothée Kom, Claude Tchepanou and Charly Gabriel Mbock.

72. One could not be in the Bamenda area and disregard its experience and memory since 1990. Two of Fru Ndi's most trusted aides did amputations without anaesthetic on grenade victims, October 2, 1991, and one of them lost a kidney to prison abuse during the state of emergency in 1992. A letter in *Challenge Hebdo*, March 16, 1995, caught the Bamenda perspective on one of the SDF leaders it accused of recent desertion: where, it asked, was Basile Kamdoum drinking champagne when the six youngsters were shot in Bamenda, May 26, 1990?

73. See Julius Ihonvbere, "Elections and Conflicts in Nigeria's Nontransition to Democracy," *Africa Dēmos* III,5 (1996), p. 9 for a harsh judgment on those in Nigeria's political class who have accepted Abacha government offices: "The pro-democracy movement . . . will never recover."

74. Travel by road from Bamenda to Maroua with the SDF's core group and attendance as an observer at the Congress, May 17–29, inform this paragraph, which amplifies previous coverage of the North. The Congress was a triumph of will and work over adversity. A week before departure, there were candid admissions about the dire lack of funds to stage it, and the party's purse at Maroua was down to the cost of a few meals in a greedy seller's market for Fru Ndi's lodging and a "Congress Village" site. But the congress was held.

75. Manga Kuoh, *Cameroun: un nouveau départ* (Paris: L'Harmattan, 1996), p. 95.

76. This version of the NUDP's recurrent theme is from Bello Bouba's General Policy Speech to the Second Ordinary Congress, Ngaoundere, January 4–5, 1997.

77. *West Africa*, March 25–31, 1996, p. 452, and April 15–21, 1996, p. 572, reiterated in the Ngaoundere speech just cited.

78. *Cameroon Post*, December 16, 1996, describes Bello Bouba and Fru Ndi at the SDF's Buea Congress; personal communication from the NUDP's Youth Wing President, February 28, 1997, is the source for their Ngaoundere meeting.

79. A November 13, 1991, manifesto from Douala issued by Fru Ndi and the militant opposition "To The Cameroonian People" over the heads of those still conferring at what it called "the Yaounde mascarade" called for "Ghost Town" to continue until a Sovereign National Conference met. Ndam Njoya took until the Tripartite's third anniversary to respond in full, but an elaborate defence of his role there appeared in a "Special Tripartite" edition of *CDU Voice* 1,4 (1994). The texts are supplemented here by many 1995 interviews with Fru Ndi and one with Ndam Njoya, May 12, 1995.

80. Richard Crook, "Winning Coalitions and Ethno-Regional Politics: The Failure of the Opposition in the 1990 and 1995 Elections in Côte d'Ivoire," *African Affairs* 96, 383 (1997), pp. 215–242.

7

Cameroon and the Prospects for Democratic Transition

Travellers in Douala's airport can hardly miss a wall mural providing the only public view we know of, prior to Youth Day of election year 1997, when Paul Biya looked other than presidential in formal European style. In *Cameroon Tribune*, on postage stamps linking him with football's national team, in photographs studding Cameroon's public space, every other display shows him in a dark suit and tie.[1] But the airport pose uses a leisure suit, a pineapple field background, and the caption "Le bon exemple."

Biya's appearance in this exception to the norm is keyed and calculated to an image which distorts reality. We revert once more to our earlier metaphoric text on food and drink so as to contrast him more directly than before with his principal challenger, for that vocabulary frames Fru Ndi well. It takes a special public occasion to deflect him from Tuesday work on his farm. Even if there is an overtly political guest list and agenda, table talk during meals at his Ntarinkon compound, which one of us shared often during 1995, is frequently about seeds, food supplies and processing, local prices, and the export potential of prime North West Province crops like mangos which remains dormant because the region lacks farm gate to airport transport. Such talk is the essence of Fru Ndi's politics, directed not just to anglophones but to all Cameroonians as their productive capacities and livelihoods dwindle in "la crise/the crisis." Dignitaries on the formal Ntarinkon occasions, if alert, learn quickly why it is wrong to dismiss him as a politician. He knows the prices producers fetch and consumers pay for essentials, and a lot more.

If the story about Fidel Castro and Julius Nyerere is true, that they met and discovered each to be the other's match for agricultural knowledge, shared by no other head of state they knew, Fru Ndi would make a comfortable third. The North West Province and the broadly Cameroonian cultures identify farming and its products as crucial elements in the private and public domains of identity. Fru Ndi quite naturally, without display or pose, satisfies that provider and provisioner role we saw as Michael Schatzberg's focus in the Introduction, planting legitimacy and harvesting public confidence, in words and actions Biya cannot match in image or reality. Known familiarly as "Ni John" but more formally as "The Chairman," Fru Ndi registers an authentic personality which cultivates supporters and attracts followers. Such things matter in Africa's politics, as elsewhere.

Fru Ndi translates his personality into popular speech idioms, in manifold, effortless ways. Here he is in rural Pidgin oratory, transcribed from a videotape of his Furu Awa bush trek, March 1995:

> I no be send helicopter, I walka for road.... I no bring moni, salt, oil, rice, but when we take power, moni whey ee keep for Yaounde will commot (come out).

He *did* bring seed, and promised in this speech that money the CPDM kept in the capital and used to airlift Achidi Achu on political tours, then distributed on the spot, as had recently happened in Furu Awa, would get to the villages for use on essentials with the SDF in power. And from his supporters come remarks in the same idiom about Biya and the CPDM, like this heard in Bamenda a month earlier from a widowed nurse with five children: "We are living from hand to mouth; we work, they eat." Or again, March 20, 1995, in a line of people paying electricity bills: "This country must change.... Yes, from frying pan to fire."

There is potential for such language to spread the politics of opposition if Cameroon's regime continues to lose its capacity to deliver goods in the daily experience Bayart translates into belly politics. In the state-civil society struggle of the 1990s, with all the *tangible* advantages of incumbency favoring the regime and containing Fru Ndi in the conventional politics covered in Chapter 6, it is well to recognize as we introduce this chapter and conclude this study, that Fru Ndi has proved his perseverance and that *intangible* factors more favorable to him, which have claimed some attention above, need again to be reckoned with as we summarize the present and look ahead.

Paul Biya cannot, as is clear by now, be dismissed. Aged sixty-four in 1997, a survivor of one coup attempt, with a young wife, a new child, and the 1996 constitution's gateway to his own presidency until 2010, Biya is not likely to yield power to Fru Ndi's challenge, or any other. Those two, to the public's knowledge and ours, have never met: "ghostly Cameroon" indeed, if that is true. Three brief 1996 texts, easy to obtain overseas, demonstrate the gulf between them. Two were *Jeune Afrique* pieces on Biya's vital political signs, comforting enough for him to have written;

they aged Fru Ndi by a decade past his mid-50s and labelled him a "pidginophone," not just as a mere seller of books who has no capacity in French, the standard critique. The other was a Fru Ndi interview in *West Africa*, closely pitched to *his* fundamental thinking and policy.[2] Anyone reading them will recognize the struggle Cameroon likely faces for many more years, between them, their surrogates, or their successors—who are in neither man's case easy to identify.[3]

Meanwhile, the locus of power in Cameroon's intricate mix of realities and perceptions, its real foundations, knowledge about who will wind up powerful and who powerless, has not yet been determined. Impasse persists at mid-decade, but an end game could develop with all the unpredictability of forces disputing a field where the rules and the roster of players and arbiters surely remain fluid. We have recounted Cameroon's experience with autocracy, 1960–1990, the bullets, ballots, and other features of change, 1990–1997, and the profiles of its two current major protagonists. We seek in closing to delineate the larger forces now in play around them, poised between state and civil society, with choices Robert Fatton among our guides makes clearest, as Cameroon concludes its 1996–1997 cycle of elections. This chapter ranges further than others, away from the documents, behaviors, and institutions of politics which shaped most of the previous text, into the broader spectrum of evidence about social experience and formations. Here, then, is a survey of Cameroon's collective character in the mid-1990s, and an appraisal of its democratic possibilities.

Disarray in the Political Class: Principals and Auxiliaries

Nothing like the solidarity of Ngayap's "classe dirigeante" obtained a decade after his analysis appeared in 1983. We have seen in Chapter 6 how anglophone North West Province massively deserted Yaounde after 1990 even if South West Province was kept closer within the fold, how the erosion of Ahidjo's North-South axis in the 1980s continued into the 1990s, and how the shrinking base of patron-client opportunities in the economy led to a Beti rampart state by 1990. Episodic detail showed the CPDM, as its agent, losing key non-Beti figures like Sengat Kuo among its elders and Ekindi among its cadets by 1991. Defections from the Beti coalition's inner circle followed. A senior Ahidjo official, Victor Ayissi Mvodo, declared himself in late 1996 a candidate within the CPDM against Biya for the next presidential election, and caused embarrassment until his death in mid-1997. From the heart of Biya's own core group, not just his legacy from Ahidjo, Titus Edzoa resigned as minister of health in April 1997, announced *his* presidential candidacy, and was placed under house arrest. As a Beti and intermittently a minister and secretary general at the presidency,

John Fru Ndi with Commonwealth Delegation, Yaounde, 1995 (far left, Tazoacha Asonganyi, General Secretary, Social Democratic Front; near right, François Sengat Kuo). Photo by Kuitché-Dzomaffo, used with permission.

and also Biya's physician and the leader of the Rosicrucian Order which the regime has made its emblematic title society, his announcement and appeal for support had a special flourish for someone very close to Biya for a decade: "Stop being afraid. . . . Do not accept lies."[4]

Less central but still strategic regional and local contributors to the hegemonic mix, stalwarts in Yaounde's political class alliance, were lost as crisis emerged and continued; a sampling of evidence used before, now augmented, is instructive. We saw Ahidjo's anglophone allies Foncha and Muna, for example, defect and then rally with the SCNC at Buea by 1995, following the younger and nationally less prominent East Province governor George Mofor Achu, the prime minister's brother, who resigned and fled Cameroon in 1992. The same commentary on political class and regime slippage applies in the North's politics, if one follows Bello Bouba (who considers himself Ahidjo's heir) from the CNU into exile, then to his own NUDP, the emergence of Ahidjo's own *son* as a force in the NUDP (not the CPDM), and the options Hayatou forged for himself in the SODECOTON affair. Again, for Foumban and the Bamum in the 1990s: the sultan operates a CPDM fiefdom but his kinsman Ndam Njoya avoided the CPDM and leads the CDU. They headed their party slates in the January 1996 election, the sultan was trounced, and Ndam Njoya with four colleagues entered the National Assembly in 1997 with 59 percent of Foumban's votes.[5] These defections from and splits within the ranks constituted breaches in the rampart state, and opened opportunities for its rivals.

The business community was a less public and formal bulwark of the political class, but certainly a principal component. Its profile in the 1990s has been mixed. Recent evidence about the character and dealings of its leaders accompanied domestic press reports of scandal, and requires care, but was no less compelling than documentation printed abroad, for example, by *Jeune Afrique Economie* on the Société Camerounaise de Banque fraud case of 1992, implicating the Biyas.[6] Key CPDM business-industrial backers on its Central Committee like Victor Fotso and Françoise Foning, easy to identify among the Bamileke group based in Douala, faced constant press allegations. Fotso, whose two books of autobiography must be among Africa's best evidence for magnates like himself combining business and politics, articulating a patriot's role in doing so, and who has advocates among progressives, has been linked in the Cameroon press to drug traffic, Foning to huge payment arrears for taxes and both personal and business bills. They appear to span the CPDM experience from politics of the belly to sharper intimidation.[7] Their contested place in Cameroon now is different than the role of bellwether elder statesman, linking private and public sectors more responsibly even if not free of controversy, recognized in obituaries (at least abroad) for an earlier generation's Paul Soppo Priso of Douala, when he died in mid-1996.[8] To utilize another impressionistic measure: the large banks the Ahidjo state created and privileged as the vehicles linking business and politics are now a shadow of what they were a decade ago. But even then, during the oil boom, when Biya and the CPDM inherited a healthier economy, estimates placed the level of informal finance and credit sector (*tontines* and *njangis*) at more than one-third of the nation's private deposits. That level must now be much higher, beyond the state's, party's, or magnate's reaches, as mistrustful Cameroonians evacuate discredited public finance channels if they can (the civil service rank and file can not, for these are their pay stations).[9]

Auxiliaries who constitute an archway of prestige and confer legitimacy on state, party, and the moneyed and managerial bourgeoisie at the heart of the national order and political class, have also cracked or broken ranks, especially on the four province oppositional terrain. Customary rulers, religious leaders, and secular intellectuals provide examples. Perhaps paradigmatic is the North West Province fon (the Grassfields' highest title, a paramount chief hedged by divinity) of Bali-Nyonga, a gifted man with a German doctorate, called to the palace in 1985 and unambiguously a part of anglophone CPDM governance in the early years of both his chieftaincy and the party, rumored as a candidate to be minister of culture in 1991. But the party insisted on an unprecedented move in the local elections of early 1996, placing his name at the top of the CPDM candidate's list, gambling that the tactic would bring them both success. The vote humiliated him, speculation about his removal from the palace followed, and an uncle is the area's SDF leader (the CPDM harasses his family and compound). This continuing episode of a fon made a political commodity, turned into electoral fodder, losing legitimacy, illustrates the precarious grip of the CPDM in local bailiwicks once so secure for Ahidjo.[10] Cyprian Fisiy has shown the fons as collectively beleaguered, caught between the regime's needs and the opposition's popular backing. In fact the province's five fons designated "first class" by the state, with a government salary crucial to their income, include just one with a staunchly CPDM palace, Mankon. Kom's (in elderly hands) wavered in 1995, Bali-Nyonga's (we see here) is disputed, Bafut's discreetly supports the SDF, and Nso's (we will see) is openly an SDF stronghold.[11] It is harder to compare religion's bearings and its leadership's place in Cameroon's political class and public life before and after 1990 than for customary rulers, for there is only scanty pre-1990 evidence. Islam, for instance, provided no pre-1990 challenge to authority, although Nigeria's Maitatsine insurrection in and after 1980 in Nigeria was led by a Cameroonian, and violence during and punishments after the 1984 coup attempt by northerners triggered Muslim outbursts. But tremors, though not eruptions as in 1984, shook Yaounde's northern-dominated Briqueterie quarter early in 1992, against Beti regime members' efforts to buy real estate there, with four deaths reported.[12]

Much more intense and consequential were encounters between the regime and the Catholic Church, with 1990 the turning point, especially because it is the faith which dominates in Beti circles and because, while larger east of the Mungo River, its anglophone-francophone mix is substantial; politics and Catholicism are therefore uniquely interactive. Until 1990, the church's centenary year in Cameroon, it is fair to say that its leaders broadly identified with the Ahidjo-Biya state. Two Betis anchored this entente, Yaounde's Archbishop Jean Zoa as the hierarchy's official voice, the Jesuit priest-scholar Engelbert Mveng as the intellectual and cultural focus for a Catholicism indigenized but kept well within both Rome's views on theology and social gospel, and the Cameroon state's sense of secular authority's primacy.[13] However, the anglophone Archbishop Christian Tumi became Cameroon's first Cardinal in 1985, so that a style of Catholicism shaped by the Dutch Mill Hill Order, historically activist with preferences for the poor like the Maryknoll Order elsewhere, moved from its local and regional roots to a national setting.

The church's newspaper *L'Effort Camerounais* recorded vigorous dialogue, and other sources since 1990 on "la crise/the crisis" revealed an entente under duress. May 17, 1990, a week *before* the SDF-regime clash in Bamenda put secular politics on a new course, the bishops drafted a Pastoral Letter "concerning the economic crisis" and issued it Pentecost Sunday, June 3. It was not a conventional homily. While the government still argued that Cameroon's hardships arose from foreign factors beyond its control, this text specified falling income, failing banks, disappearing jobs, and other domestic calamities as the causes. It referred *inter alia* to "structures of sin" and to church critiques since Vatican II about the failures of Christian and secular conscience as burdens all the world's poor carry, but this bishops' address to Cameroonians went far beyond such standard laments. Recalling the pope's 1985 visit and his praise then for Biya's New Deal, the text immediately jumped to the present: "Why are the banks empty? Who have emptied them? Where are those who emptied them?" Notorious embezzlement of public funds, individual tax arrears of hundreds of billions of CFA francs, and the egoism of the rich and powerful were then cited, working against "the smallest of the small" in Cameroon as elsewhere. Although names weren't specified, they were suspected or known at large, or would soon be through the new journalism of the 1990s, and a collective burden of responsibility concluded the letter:

> the State must create a healthy atmosphere of national solidarity, provide for the security of persons and property, and safeguard the rights of individuals and communities. Democracy must not remain an empty slogan; it must be lived in a responsible way by all citizens.[14]

The charge was clear: Cameroon's 1985–1990 experience violated a fundamental sense of justice, and it was time for accountability.

The bishops here published text the SDF and CDU took another year to match, and continued the critique. During the North West Province state of emergency late in 1992, they collectively rejected government criticism of Bamenda's Archbishop Paul Verdzekov's open sympathy and comfort given those killed, injured and jailed, and praised his "pastoral activity."[15] Three years later, their 1995 year-end communiqué cited alleged ballot rigging prior to January 1996's municipal elections and welcomed reform initiatives to separate state powers and decentralize government, in a setting where "we are all at an impasse" and "impunity should no longer continue to compromise the common good."[16] A postelectoral statement praised the ballot as a step forward for Cameroonians, but criticized the lack of an electoral code and commission devised by all parties.[17] The bishops then called April 20, 1996, for a once monthly "Day of Prayer for all persons who have been assassinated in our country"; those included since 1990 Father Mveng himself a year before, a former archbishop, a former *L'Effort Camerounais* editor, and seven priests or nuns as well as dozens of lay Catholics and others named and unnamed.[18]

Cardinal Tumi and Archbishop Verdzekov, anglophones from the Nso area, which was mentioned earlier and is discussed again below, continued to engage the church in public life, discomfiting the state. Tumi moved from the distant archdiocese of Garoua to Douala's when a vacancy occurred in 1991, adding *his* pastoral words and deeds to the city while constantly being tipped as a potential chairman whenever a national forum on politics or the constitution seemed imminent. Two press interviews in 1994 exemplified his energy in public debate the regime tried to dampen. One favored federalism, citing diocesan autonomy in his own church and Swiss cantons as models. The other revealed that he, perhaps Cameroon's next most notable figure behind Biya and Fru Ndi, had not met the president in three years until the summons to a session canvassing his views on a forthcoming constitutional conference, which Tumi soon walked out of.[19] Verdzekov's demeanor is quiet, but his convictions and his archdiocesan role are likewise direct and therefore contentious. The prayers and sermons he delivers at times of crisis in Bamenda, especially funerals for Catholic activists, are locally renowned. He hosts human rights groups for lay Catholics lodged in "Bishop's House" there. He is articulate beyond obviously pastoral work in, for example, "A Humble and Earnest Appeal" cosigned with his province's bishops, September 25, 1993, urging prime minister Achidi Achu to preserve autonomous anglophone education, a far from innocuous concern.[20] Virtually in the shadow of his cathedral's campanile, Little Mankon parish is in the care of a priest, Father Dinayen Wirba, whose pastoral work, as delivered and as published, is in the liberation theology idiom of justice confronting power. His sermon the week of Father Mveng's killing ranged through teachers' pay arrears and roadway check point extortion by gendarmes ("from the hand of the suppressed to that of the suppressor") before these closing words: "The million dollar question remains. Who killed Fr. Mveng and why? The answer is blowing in the wind."[21]

Not all anglophone Catholics, fewer still francophones, one imagines, approve his language or profess his ways, but it is clear that his church is now far less a comfort to the state, or a sector of the political class, than before 1990. The priests we survey here range from meddlesome to seditious in strict regime terms. Uninvestigated killings, the corruption, poverty, and desperation cited by their words and writings bespeak deeply troubled relations, far removed from the days of *entente*.[22]

Other mainstream confessional Christians, less numerous or strategically placed, have raised critical voices since 1990. The Cameroon Baptist Convention leadership of a largely anglophone communion delivered May 15, 1991, as that year's unrest spread, a message to the Prime Minister. Less rhetorically a challenge than the Catholic letter a year earlier, it nevertheless made clear the New Deal's failed promise and, couched in biblical terms, left no doubt about its advice to heed calls for a Sovereign National Conference. It was quite pointed about a reformed university and a federation among correctives needing attention by "leaders of this nation, past and present, [who] have made mistakes whose awareness incapacitates them in the fact of this present national uprising."[23] Presbyterians (they include Fru Ndi), similarly more anglophone than francophone, have in the 1990s faced inward because of strife between established members and newer fundamentalists over doctrine and practice, but we saw their welcome in 1994 to the AAC when it was prevented by security forces from using Archbishop Verdzekov's cathedral grounds; the Bamenda Proclamation was read from its sanctuary. Here, then, are mainstream Protestant anglophone critics of governance, more measured and less troublesome than Catholics but still a factor in Cameroon's religious experience which denies the regime and its erstwhile political class a full measure of support and consent, an uncontested hegemony. Nothing like an alliance of major denominations such as Kenya's National Council of Churches has placed political criticism, civic education, and pastoral relief work in an ecumenical Christian setting in Cameroon, and politically "quietist" communions thrive (Baha'i, for instance, is very active in Bamenda). But evidence across its spectrum justifies the title of a recent commentary on Cameroon's Christian witness in the 1990s: "Du silence ambigu au réveil politique/From ambiguous silence to political awakening."[24]

The secular intelligentsia has likewise fragmented over the past decade, to the regime's loss. The best comparison point is a publication from 1985 when the ministry of information and culture printed the texts from a week long colloquium, *The Cultural Identity of Cameroon*, in essence a prelude to Biya's *Communal Liberalism* (1987), the regime's flagship statement of progressive political intent and principle. It devoted 500+ pages in "Cultural Identity and_____" chapters to academic disciplines from anthropology through zoology and to most forms of artistic and cultural production. The opening address charged participants to seek "cultural universals" in Cameroon's diversity, with the presumptive goal of a national political culture, very much in Bayart's sense of hegemonic design.[25] Sengat Kuo, the minister, delivered that speech, then (it is rumored) supervised and perhaps wrote Biya's 1987 text; the contrast with his senior advisor's role to Fru Ndi by 1992 is profound evidence of the political class, and political culture, contested.

A smaller regime intelligentsia than in 1985 is still active, staffing ministries, the party apparatus, the state media, and the University of Yaounde; we discussed Jacques Fame Ndongo, its central figure, some of his associates, and their journalistic critics in Chapter 5's analysis of the press. But there is now a broad cultural opposition to the state's "organic intellectuals," the Gramscian term a leading critic Ambroise Kom uses, applying the perspective shaped by his elder colleague (and CPDM defector in the last years before his 1995 death) Thomas Melone, who labelled as a "fonctionnaire" class those who move permanently from the *grande école*/professional school setting into state service. Chapter 1 noted the fierce, formidable criticism by 1980 of Mongo Beti; returned from exile in 1993, he is now the most notable dissenter. The voice emerging from those ranks February 3, 1995, at a Yaounde conference on democratic transition and governance, with a reference to Paul Biya's Cameroon as a "Vichist" (not "Gaullist") state, drew not a murmur of surprise or dissent. The audience included a number of writers in the group, noted in Chapter 5, that since 1990 has published books and presented seminars devoted to the imperative and process of change. These developments measure the gulf now separating many of Cameroon's intellectuals from the state.[26]

Many of those in Melone's "fonctionnaire" class and Ngayap's "classe dirigeante" are permanent or contractual civil servants, who number 130,000 in the mid-1990s and are too diverse to be closely analyzed as we now take their measure, but the sense again is of disarray.[27] The public, unprecedented resignation during the 1992 presidential campaign from Biya's cabinet of his civil service minister, Garoua's Garga Haman Hadji, broke a tradition of government solidarity in a key arena. He remains a figure in opposition with a small party of his own, L'Alliance pour la Démocratie et le Développement, a high approval rating in public opinion, and prime material for any truly new government, conceivably as the opposition's candidate for the presidency in 1997.[28] There is less evidence to cite of middle and lower rank civil service dissent among francophones than the anglophones' Cameroon Public Service Union (CAPSU) and its manifestos provide, but employees in the water and electricity parastatals formed a national federation in 1995 and published a memorandum to Biya stating their case against a host of grievances and the threat of privatization without a workers' voice.[29] Civil service absenteeism and businesses off the work site are tangible signs of what many commentators have for some time cited as generic disdain and opportunism, now fortified by the genuine need to salvage a living. Targets of external austerity demands which have cut their numbers 25 percent in half a decade, civil servants use their urban sites as buffers, and benefit from the state's slow pace of privatization and parastatal reform; they suffer less than those who lack their credentials and protections.[30] But factors already cited like the threats they incurred during 1991, and the steady erosion of pay, benefits, and pensions, have created a drift away from allegiance and performance measures of earlier times, and the drastic fall in government revenues bespeaks a troubled condition.

Cameroon's erstwhile political class has more components than we can cover. There are old schools and their ties still in play like St. Joseph's College, Sasse, for anglophones and Lycée LeClerc, Yaounde, for francophones, more than half a century old and great in influence. They include the neocustomary ethnic associations surveyed in Chapter 6 and newer counterparts like the Rosicrucians and the football clubs which form, compete, and dissolve as the expression of special interests and the channels of access and power.[31] Most such networks have dual potential as rhizomes of the state apparatus and as carriers of civil society's alternatives.

This profile of a once formidable, monolithic political class, splintered and depleted since Ngayap's analysis, concludes with features we can trace only faintly. Not even a sketch of the legal profession is possible from our evidence, a real deficiency, given the importance of judges in cases like the Supreme Court's confirmation of the CPDM's 1992 presidential tally, and the role of lawyers in human rights advocacy and the political opposition since Yondo Black's defiance in 1990. But we can at least survey a second key, if shadowy sector, Cameroon's security force.

Chapter 5 noted evidence of strain during Bamenda's 1991 turbulence among its domestic components. Other tangents are the 1993 demotion of a key Beti general, Benoît Asso'o, for statements critical of political reform rhetoric, and a letter published early in 1995 from a nineteen year army officer veteran, now defected and abroad, about politicization and sheer careerism. Reports of imbalances and tensions persisted, as in 1995 over the isolation of the only senior anglophone army general, James Tataw, and about 1,500 soldiers recruited from the three province regime core area compared with just 1,200 from the other seven.[32] But conflict with Nigeria since 1994 on the coastal Bakassi petroleum border and minor, lesser known disputes near Lake Chad keep up a level of military preparation which France materially assists, so that the armed forces are paid, given raises at sensitive junctures, adequately equipped in the strategic locales, and active enough to be satisfied.[33]

A jolt to the police apparatus occurred with the March 1, 1996, dismissal of Cameroon's internal security chief in various posts over thirty-five years, Jean Fochive. It was notably humiliating, done the day of a Yaounde visit by the French minister of cooperation, whose key brief is Africa, and effected by a gendarme cordon of Fochive's office and a radio announcement. Commentary blamed the fall of "Père Foch" on the CPDM's municipal election failure weeks before in his political fief, Foumban, concern about both the quality of political intelligence and the rise of opportunistic banditry and organized crime in Yaounde just months prior to the OAU summit there, and the need to revamp police operations before 1997's election cycle. This setting includes a rapid growth in the private security industry, employing seasoned or irregular personnel who complicate the policing profile by adding to domestic armaments.[34]

Removing Fochive, the perennial regime bulwark at its security core, fuelled speculation, but nothing in mid-1997 suggested its extension to the ultimate spectre, split loyalties. Still, events which turned Zaire into Congo (Kinshasa), 1996–1997, and the regional spread of unrest in central Africa with all its violence, drew attention. Varieties of action, like coup attempts, or the refusals to act, like his army's for Mobutu, could be consequential for Cameroon, although there is no visible Laurent Kabila there in 1997. What is not in question is the erosion of political class solidarity since 1990. That process will continue if the impasse in politics is not resolved.

An Alternative Political Class?

We have recounted elections with some diligence above, and at the same time warned against using formal electoral politics as a full measure of Cameroonian experience, given the system's flaws and the people's skepticism. The remark of a SDF militant in 1995 is to that point: ostensibly "nonpartisan" local votes are now better indices of Cameroon's politics, for offices in development associations, marketing boards, and the like, where they reflect cumulative opposition at the grass roots since 1990. Achidi Achu, for instance, was defeated in a Bamenda farmer cooperative's leadership vote long before Biya made him prime minister, and the remark serves us as a bridge to discussion of oppositional possibilities as the regime's hegemony wanes, despite its formal electoral edge. The SDF and Fru Ndi become our focus in this effort to discern whether something more favorable than what Fatton thinks lies ahead for Africa is at work in Cameroon. Whereas nothing in the NUDP and CDU, and little in their leaders Bello Bouba and Ndam Njoya, suggests a vanguard, progressive political class option in their ranks, an unmistakably populist mantle identifies the SDF and Fru Ndi. They mix principled refusal to join the politics offered in coalitions, suffering which surely has *not* finished, and underdog anger. They won vote counts and offices in all major cities outside the North, including Yaounde until dislodged, in 1992–1997 elections. With intangible appeals and tangible strengths, they alone now enjoy a growth potential, a capacity to forge electoral alliances across fluid lines of class, ethnicity, and language, and likely less fluid differences of religion, and to mobilize populations on Cameroon's margins.

But does all this constitute the potential for a more just and inclusive political class in Cameroon? Could the SDF counter Fatton's view that opportunism and "disarticulation" are as likely to limit as to promote the growth of civil society's social base and political initiative, pick up the UPC legacy as Mbembe noted was possible, and infiltrate the state? Questions arise and doubts persist. It is not yet clear that SDF appeal stands on its own powers to translate still scattered constituencies into a nationally decisive force, more than other parties' weaknesses. Consider, also, Kofele-Kale's and Asanga's critique of anglophone elites since independence, drawn like moths to Ahidjo's and Biya's candles (they argue) not as innocent dupes or absent-minded fools, but to share spoils. Fru Ndi, we saw, is not just a farmer-bookseller but also a commercial developer in Bamenda. The SDF leadership circle includes both some of the disaffected academics encountered above, and men and women in law and commerce whose grievances and deprivations are in the realms of banks, contracts, and licences; there are potential elements of a Biya-Fotso-Foning parallel, Davidson's piracy, in the SDF. And, perhaps most of all, can the SDF and Fru Ndi continue by and large to skirt the grimmer sides of politics since 1990, the fretful, wasteful, and at times we have noted for Douala, the violent Gramscian "interregnum" symptoms, and pull all Cameroonians beyond such limits?

Intermediate SDF ranks in two areas of North West Province life confirm the scholarship on political class fluidity, not to say ambiguity, and the delicate balance of alignments Fatton emphasizes. Teachers and large scale food traders are among those with influence at the party's Bamenda epicenter and on Fru Ndi's Ntarinkon compound, where SDF policy is decided by the national executive, translated into action, and transmitted to the deeper rank and file. The teachers are salaried professionals who often need now to trade and farm to stay solvent. There are 3,000 anglophones but very few francophones in the Teachers Association of Cameroon, at the cutting edge of upper echelon labor, operating more autonomously of the state each year. They simultaneously press for pay arrears against school officials, work to protect the integrity of anglophone curriculum, exams, and credentials against efforts by the ministry of national education to absorb them, and add to SDF organizing capacities.[35] In the food commodity trade, the core entrepreneurs, many of them women, parallel the teachers' recently militant path, first manifested in their support for 1991's direct action, lately extended to the creation of cooperatives among female farmers; Fru Ndi's wife Rose, trained abroad in catering, is their SDF channel.[36] Both groups defy static readings of their political and economic interests.

A third, more *nationally* active and representative public arena, transport, is as near the roots of the social fabric as schools and food, and as intricate. It is thus another key sector for SDF engagement, providing both a sense of Cameroon's evolving class interests and alignments where capital and labor intersect, and their implications for potentially new political class formation. Long distance trucks and busses attract Bamileke and anglophone venture capitalism, which the SDF courts. Hostility toward and competition with regime Beti are among the reasons their owners pulled trucks off the roads for a week in mid-1995 when the government sought new revenue through increased axle rates, and forced the ministry of transport to bargain a compromise.[37] Smaller scale taxi owners stopped working the streets of southern cities at the same time in 1995, then later in the year, to protest higher petroleum prices. But it was no longer 1991, when a taxi strike was fundamental to "villes mortes." In a motor vehicle industry niche with less capital than there is in bulk cargo hauling, with a driver work force that is outside the trucker's "labor aristocracy" and cannot afford idle time, and a clientele not as tied as years ago to work schedules, certainly with less money for taxis, and now accustomed to walking, their 1995 strike failed. Neither the taxi's owners nor workers are truly organized in trade association or syndicalist terms, and capital-labor differences separate them from each other. The situation and prospects are fluid. Truck and bus owners could gravitate to the SDF, like taxi men in 1991. But work is needed to secure an oppositional politics on the roads.

These arenas, education, food supply, and transport, exemplify the key service and production sectors which Ahidjo's regime took pains to mobilize and discipline across Cameroon's divides of capital and labor, literate and less literate peoples, elites and masses. These were core areas for the national bourgeoisie which his state and party created and relied on for further, downward hegemonic transmission of values and behavior. The CPDM and Biya have lost this capacity. Their politics is rearguard, fighting backfires and brushfires only by dipping deeper into political capital; dwindling interest remains at their disposal, home and abroad. All this is a far cry from Ngayap's 1983 "classe dirigeante" text and from Bayart's 1979 "recherche hégémonique" applied Africa-wide in his 1989 version of power-laden, network-driven feasting. A Sovereign National Conference might have reconstituted, then managed, a modified but still familiar patronage and clientele to include the people, interests, and resources just surveyed, as a recognizable, consensual version of civil society. But the politics of the Yaounde Tripartite of 1991 and the elections which have followed were grudgingly dispensed by the regime's containment policy, leaving the SDF and Fru Ndi to create opportunities and initiatives in a new political market.

Emergent Social Experience, Formations, and Alignments

It remains an open question whether they can assume that vanguard role, turn it to Cameroon's national advantage, and effect a truly democratic transition, in relative peace, fulfilling the optimists' hopes. To do so, they must also provide Cameroonians who are located beyond the people, interests, and resources we have surveyed and calculated thus far with a stake in the process. There are probably a few million such people and their families, less controlled than formerly by the state but not yet decisively mobilized by and committed to an alternative politics. These include the petty marketers known as *sauveteurs* and *buyam-sellam* who fill the large cities' streets, at common risk in the subsistence or "new poverty" pockets of the economy. There is data, 1987–1993, about the state's decision to licence rather than harass street merchants, so as to assert control over and add revenue from this sector, but its political inclination, if any, is a matter of speculation.[38] Our 1995 Bamenda focus group research cited in the Introduction for income data in this population disclosed what one would expect in that time and place, a close to unanimous SDF preference, and the numbers of youth with diplomas, certificates, and degrees forced into this niche by the formal economy's crisis throughout Cameroon implies either oppositional or conscious avoidance (perhaps anomic) tendencies.

More rural populations, or those who ply the roads for a living, seldom active in conventional politics, tending to be dominated by those who are, constitute another untapped civil society possibility. The scholar-priest Jean-Marc Ela's work, using Cameroon as the core of Africa-wide study, identified those drawn into plantation economies as perhaps the largest, most exploited group of subalterns.[39] Plantation life is "quasi militaire" with "un emploi du temps imposé pas les impératifs de la production/a use of time imposed by the imperatives of production." The predominantly migrant, bachelor male population, bringing with it the kind of local affinities which generate tensions, finding release in beer and women, has little access to the new political market and its choices that cities now provide their work forces. The spread of agro-industrial enterprise means that the African state "tend à reprendre un modèle esclavagiste/ tends to resume a slavery model." Working at large sites on others' lands, distant from home, this form of labor remains even farther beyond mobilization for an alternative politics than that tied to northern Cameroon's cotton. The SDF has a formidable task in such areas, which only tea cultivation in one pocket of its North West Province base area approximates. It does less well where this model substantially obtains in South West Province. That it is frontally assaulted when it brings electoral campaigns to the CPDM-Beti southern plantation core zone suggests either contempt for its condition there or fear about its potential. But for now, even though Mongo Beti ran as its candidate for the National Assembly in Mbalmayo, Center Province, in 1997, and despite Ela's belief that the age of the single-party state which channels and profits from these structures is passing, and that "l'ouverture politique/open political space" has become a necessity in Africa, Cameroon's rearguard remains formidable. Ela himself left for Canada not long after Father Mveng's murder in 1995, in peril, according to newspapers and friends.

More promising examples of SDF rural potential in its base area comes from the Furu Awa material we have covered previously, and the case of the Mbororo herders ("cattle Fulani" in more common but imprecise language) north of Bamenda. For many years, and still, subject to the hostility of farmers and predation by a notably demanding CPDM baron, mistrustful of early SDF efforts to establish a link with MBOSCUDA, their community development association, there were in 1995 at least hints if by no means decisive signs of a less defensive, insular mode of Mbororo ethnicity in dealings between the two.[40] But that kind of inclusive, incorporative civil society advance will take long, hard work ahead.

This is tentative analysis, which can only introduce, not satisfy, the need to understand massive numbers of people yet to appear distinctly in politics, and the challenges they present to leaders of any new assemblage of a political class and civil society. We now move to newly emergent social forces which are easier to track, so as to complete our canvass of Cameroon's mid- to late 1990s condition, and to attempt a judgment of the SDF as an agency for resolving the crisis and impasse in this highly fluid national setting.

We start with women and youths, whom the SDF does not, like the CPDM and NUDP, segregate into "wings." Whether it brings them fully into party activity is another issue; scattered ward and street level Bamenda evidence suggests more importance for women than does a count of its national executive officers, where little has happened: three women occupied twenty-three posts in 1991, four occupied forty-five late in 1996. In any event, these are clearly critical populations, hitherto marginal at best in the scheme of Cameroon's public life. How do experiences of women and youths in the 1990s, scanty in the published record thus far except in general "social order" analyses, in aggregate studies like the World Bank's on household, village, and urban economies, or now emerging in monographs like Miriam Goheen's, register in SDF terms?[41]

In short, not decisively, but there are interesting features linked to previous findings, especially among women. One well recorded female presence in Cameroon's public life is, conveniently, from North West Province, dating back forty years to a locally rebellious phenomenon known as *anlu* in the village of Kom.[42] Urban Bamenda and its perimeter now has a counterpart in *takumbeng*, involving female elders. Its roots in the past, its presumptive connection to *anlu* and other Grassfield parallels, with their hints of a distinctively feminist edge in a patriarch's world, need research. But the core of recent *takumbeng* is the public, silent, but eloquent practice which brings post-menopausal women to disputed terrain with sacred grasses and other materials from nature, invoking maternal authority to restore peace, threatening and using bodily exposure against violators, whom it is meant to shame and stop.

Sporadically visible since *anlu* in the region, the practice gained sharper political focus and wider attention in the context of the 1990s. A week after Biya's "sans objet" speech, July 5, 1991, *takumbeng* openly challenged Cameroon's security forces in Bamenda, confronting mainly nonlocal, certainly uncomfortable troops blocking a protest march to the provincial governor's mansion. Since then, what one account calls "les Amazones des SDF" reinforce the oppositional front line there on crisis occasions, most notably as a defence cordon at Fru Ndi's compound when he and 145 others (one-third of them women) were confined there by gendarmes during 1992's state of emergency. Women in the region have translated and adapted the form and spirit of *takumbeng* and its variants to other sites of resistance; cadets joined elders and party placards joined sacred grasses, for instance, when sixty Njinikom women trekked to Bamenda and secured the release from prison of village SDF officers in 1994.[43]

Clearly a localized experience, unmatched elsewhere to our knowledge, *takumbeng* demonstrates a convergence zone within the female sociopolitical field of custom and innovation in the 1990s, and the SDF is both beneficiary and agent for a sense of equity betrayed in public life, needing redress. This is not the sterile setting of the regime's delegations to Beijing for 1995's International Year of the Woman, much publicized by the official press. *Takumbeng* does not, of course, begin to account for the situations of stasis and decline so many Cameroonian women endure, or for their resourceful efforts at amelioration within and beyond family settings, characteristically in their producer cooperatives and credit associations.[44] It does, however, suggest one active forum with an impact in the public realm, mobilized by the one political party with change potential, which the womens' food business network in its core area also utilizes.

The experience of youths, linking them to and political parties, needs a new study like Mbembe's classic *Les jeunes et l'ordre politique en Afrique noire* (1985), written outward from Cameroon about Africa's troubled, perhaps (he feared, not without reason) lost younger generations. He placed the ambience of social and political criticism among *students* in a larger context of unease with family and village legacies of patriarchy, and registered the perceived threat they all posed in terms of precedents like anti-colonial mobilization, Paris in 1968, Soweto in 1976, and a 1982 strike by Yaounde students which drew this response from Ahidjo: "l'ordre régnera à l'Université par tous les moyens/order will rule at the university by all means necessary."[45] In describing the regime and potentially troublesome students as rivals for the campus (with Cameroon's war college adjacent, between it and central Yaounde), and disputes often focussed by carrot and stick tactics over funds and facilities for students, Mbembe anticipated the terrain for the 1991 confrontation there, when the stakes were raised by ethnic tensions and multiparty politics. Turbulence continues there and on some of the campusses since created in order, among other reasons, to disperse the critics. Support for the SDF is widespread in these ranks, but we have no evidence of any organizational form it takes.

Moving from students to youths at large, violence they witnessed and practiced became a special concern during the 1990s, and presents a challenge for those like the SDF who would shape civil society's future. Earlier work from Bayart as well as Mbembe covered the state's effort to displace and divert rebellious youth energies from politics to leisure. The ministry which combined youth and sport in the 1980s is its organizational expression; comprehensive study of music and other features in the popular culture, surely on some writer's agenda, will integrate those themes. But youth violence has been a factor since 1990. Both casual and concerted delinquency and criminalty, of the sort that produced 1991 remarks quoted in Chapter 5 about Rambo and Intifada cultures, and the loss of authority to contain them, was not confined to Douala. Bamenda's "Mile Three Nkwen" for a few weeks in mid-1991 resembled the youth gang milieu Célestin Monga described in Douala, with robberies at gun point, sporadically aimed as never before at the persons and property of foreign missionaries and project workers as well as locals, until neighborhood patrols took root and restored peace. It was hard to know whether these were independent youth ventures, or where the guns came from; some suspected the SDF, some suspected the regime's interest in destabilizing the city in order to justify repression. Near Bamenda, on a smaller rural scale, but of consequence as indices of challenges directed not just to those blamed for the political crisis but also to authority at large, the fon of Bali-Nyonga on his way to arbitrate a land dispute was verbally abused by young men, and the fon of Mendankwe was jostled by youths on the sidelines of a village football pitch. These were shocking provocations to local customary leaders.[46] Another "youth" episode is Burnham's account of the 1991–1992 turbulence in and near Meiganga, Adamawa Province, which led to at least scores and perhaps hundreds of deaths, for it included violence caused in part by older males drawing cadets into past rivalries, settling scores in ways Cameroon elsewhere has surely known in this decade.[47]

Cameroon and the Prospects for Democratic Transition 235

The consequences for youths of these violent features of "villes mortes" and other forms of rebelliousness will bear watching, and the SDF needs as much as all other parties in Cameroon's public life to monitor a situation it does not pretend to master. Will the adolescents in Bamenda who turned their play in mid-1991 to slingshots aimed at the helicopter dropping concussion and tear gas grenades on the city incline more to ballots or bullets as adults, or in more conventional ways pay taxes and rates, or meet other routine obligations they watched their elders refuse? And what of those four years later, who witnessed petrol and tire necklaces for thieves in Bamenda, as vigilante justice took the place of discredited public authority?[48] Dominated subjects witnessing or taking up resistance do not automatically become responsible citizens with the habits of a civil society. Cameroon's future leadership will not be able to take the performance of routine public obligations for granted, as the children of crisis and impasse become adults.

Under- and unemployment are surely factors where youths turn violent, but are not easy to measure. Education is a better known index of their condition. There is some encouraging evidence about it which helps locate the SDF in Cameroon today. The 1994–1995 exam results raised eyebrows.[49] Francophone schools at the prebaccalaureate BEPC level produced a 3–13 percent success level across the range of subjects: "Les résultats confirment la catastrophe . . . dans un environnement académique totalement délabré/The results confirm catastrophe . . . in an academic environment which is in total decay." The context included many pressures, based on lower funding and morale, disputes about favoritism and fraud in the conduct of exams, and any number of other problems the press often reports, with some reference back twenty years to Ndam Njoya as the last minister of education with a concern for honesty and standards. But GCE "A" level exams, the anglophone school system's step above BEPC, produced a 17–59 percent pass range, including 49 percent in the fields most widely tested. This comparative data base has obvious limits, but the context for this measure of anglophone success includes work since 1990 by the teachers in SDF circles discussed above to escape the national ministry's control of curriculum and testing, achieved in 1993 with an independent examinations board in Buea. The almost entirely anglophone Teachers Association of Cameroon is their forum, a loan from the exclusively anglophone Amity Bank secured the funds to pay those who marked 1995 exams, and the teacher group's general secretary speaks of "classroom peace" and a determined reform environment among anglophones as keys to this performance.[50]

It appears, since the debate noted earlier about school boycotts in mid-1991, that activity and morale across an institutional network of anglophones have begun to stabilize, perhaps to improve the region's service in this nationally problematic education sector. There was outrage when Yaounde absorbed anglophone schools in the mid-1980s, and all major anglophone lists of grievances in the 1990s specify schools; indeed, they may be the region's broadest source of concern and resentment.[51] The SDF, beyond electoral and institutional politics, operates as a hub for radial forces like these. It can claim at the very least a deterrent role against the evident pathologies, and perhaps more, for it enables institutional capacity and community where civil society takes shape beyond the state's present and at least short-term future reach. Formulaic arguments that such evidence merely represents exit or withdrawal do not account for the active engagement which this educational dynamic reveals.

Such indirect or direct evidence from civil society is framed not only by the experience of women, youths, and other specific population sectors. Multisectoral action abounds in parts of Cameroon. To complete our review of incipient civil society, we scan two contemporary sites, caught in the contractions and contradictions all accounts of the country convey, yet responsive to the grinding processes of change. This tale of two habitats, a sprawling anglophone North West Province terrain, Kumbo-Nso, and the francophone metropolis of Douala, rural and urban sites which harbor misery and conflict, but also resilience and creativiity, adds substance to previous impressions. It yields some tangible proof of and guarded optimism about democratic practice in the 1990s and the chances it could spread and prevail, against the negative signs which map Cameroon in this and other sources.

First, Kumbo, the core village for 200,000+ people of the traditional Grassfields state of Nso.[52] It is 120 kilometers and three-five taxi hours northeast of Bamenda, subject to the weather variables of the unpaved "Ring Road," the major link for North West Province peoples and commerce and a flash-point grievance against the Cameroon state, from which locals have now "in many ways ... withdrawn."[53] This apparent isolation masks a complex, dynamic history of expansion, with migration surges and war episodes throughout the 19th century, capped by defeat of the powerful rival state of Bamum (Foumban), then stern resistance to Germany until surrender in 1906. Missions, schools, and hospitals began, then British rule from the 1920s both extended their work and left in place a substantial governance role for the fon's governing apparatus. The results included "improvement," then "development," a sturdy village and palace autonomy in the face of formal rule successively radiating from Nigeria, Buea, and Yaounde, and accomplished leaders for Cameroon who maintained their "son of the soil" identity.[54] This was a classic village-capital city "straddle" in Bayart's sense, with village integrity secured by attachment to broader structures which Kumbo-Nso enhanced, providing Cameroon (and our narrative already) a worthy and independent Ahidjo minister (Bernard Fonlon) and two archbishops (Tumi, Verdzakov); one could add professional notables of all kinds, especially scholars.

Why, then, Kumbo-Nso's exit from the state, anticipated above through earlier discussion of the two prelates and their church's distance in the 1990s from the distressed political class? The Biya state, for one, upset the balance of the straddle, most notably in an episode over water, a symbolically charged as well as materially vital resource in a heavily agrarian, partly pastoral economy. Canadian aid in the 1970s, directly to Kumbo-Nso, not via Yaounde, provided a water system and trained locals to manage and maintain it. But the water parastatal, Société Nationale des Eaux du Cameroun (SNEC, now SNAKE in local terms), encroached during the 1980s on the financial and technical sides of the village enterprise. In April 1991, with violent clashes and early SDF work already under way, 300 Kumbo-Nso students came home when the Yaounde campus erupted, then emptied, and brought the national opposition surge with them. They used the water dispute to mobilize their "parliament," adding to the repertoire of resistance already on the ground. The symbolic attribute of "water as life" yielded political capital in "Operation Recover Our Mothers' Farms" and "Operation Water For All" to rally the struggle with SNEC, and videotapes of elders recorded the local, not state-directed history of piped water. Direct confrontations throughout 1991 included a fire which destroyed SNEC's office, with six deaths the cost of that clash.

The violence, as elsewhere, subsided after 1991, but a steady erosion of the state presence and authority was under way. The process led to the choice of a carpenter who helped build and defend the water system to be the new fon in 1993. It is visible in the "Kumbo Water Authority" sign at the entry to his palace, where this agency created by "The General Assembly of the Traditional Council, elites, quarter heads and representatives of consumers" operates since 1994. It takes SNEC's place in a visible, comprehensive demonstration of a successful civil society challenge to Cameroon's state, with a charter and statutes for water's use and management, linking the palace and the entire local community.[55] The previous fon, Goheen notes, "made it clear that he too sided with the people in their discontent with the national government," 1991–1992.[56] Fru Ndi won his third largest margin of victory for the presidency over Biya in 1992 (among Cameroon's fifty-three electoral divisions) in Bui, with Kumbo-Nso at its heart.

Locals vent a litany of complaints about victimization as their challenge to the state continues through the 1990s, moving far beyond water into issues which likely reflect generic frustration and rage wherever opposition surfaces in Cameroon.[57] Work, for example, is complete, under way, or anticipated to pave one road from Bamenda to a smaller, rival village which is a veteran CPDM minister's fief, and another which would tie Kumbo-Nso as a satellite to its historic rival the Bamum-francophone city of Foumban, rather than improving its Ring Road link to Bamenda, its political ally against Yaounde. What are seen as the government's retaliations of other kinds draw fire. It favors lesser neighboring fons and creates new administrative and electoral units, so as to give preference to less intransigent neighbors over Kumbo-Nso. It selects and promotes locals for public office with less education than is available and respected in Kumbo-Nso, and appoints hostile outsiders as local administrators. It advances the interests of large graziers, including the late president of the National Assembly, a local, and some "foreign" landowners and leaseholders, against the region's farmer majority.

All this agiation and action has made Kumbo-Nso notorious in Cameroon as "Baghdad." But this activity and reputation should not be mistaken for disorder. Goheen notes its safety after dark, the brisk commerce by means and methods substituted for the state's, and the palace's discipline by corporal punishment, exercised with a customary legitimacy contrasting what the state, in its decline, has lost.[58] Palace authority has scattered the state's; the latter retains reprisal force but its writ is otherwise severely limited, not just in particulars like taxes refused and rates withheld, but in basic allegiance forfeited. An SDF apparatus is locally entrenched, certainly since 1991 keeps more peace than it disturbs, and can rely on palace initiative and protection.

There is also an orbit outside Kumbo-Nso to consider, with Cardinal-Archbishop Tumi the best example. Its people have significant roles elsewhere in Cameroon, many after study and work abroad, advancing personal, familial, and village interests. Amity Bank is a case in point. A Kumbo-Nso man is its general director in Douala. With no links to the state apparatus beyond its founding charter, and in Cameroon terms thriving while the banks the regime operates and plunders are failing, Amity is reviving Federal Republic era financial autonomy for anglophones. A close kinsman to the fon is its chief legal counsel with chambers in Bamenda.[59] Closer to the SDF, another villager is an executive officer for an American-parent petroleum company in Douala, and on the party's National Executive Committee. The CPDM and parastatals like SNEC lose ground as the SDF, the new water authority at the palace, and new enterprises linked by personnel or interest to the SDF on some combination of political and business grounds create new networks of choice and opportunity. They integrate private and public sectors, the palace and the alternative political party, with some claim now and chance in future to be a better (not in Fatton's terms predatory, in Davidson's terms piratical, or in Bayart's terms belly politicking) leadership.

In important ways, though not in full reality, this is more than "exit." Kumbo-Nso *is* now largely self-governing, by people no more to be dismissed than Fru Ndi. If it is democratic neither in the way its palace remains patriarchal (its women, Goheen's intense focus, and its youths we have briefly described *do* have voices but not yet any decisive role), nor in the way it seeks to maintain control of its own hinterland, a potentially effective local counterhegemony to Cameroon's state is emerging. And women and youths will likely find channels opened in this transitional time and space for their voices to move from Kumbo-Nso's margins toward its center, in an alliance forged by an old palace and a new political party with more, and more vital, people, resources, and interests in common than a faltering regime has recently mustered and served.

Douala, more complex and important (its population far exceeds 1,000,000), remains more within the state's reach in 1997 than does Kumbo-Nso. The latter was reclaiming its ground in 1991, but recall Biya's message during that year's trip to Douala, the buckle of the regime's lifeline, joining Limbe with its petroleum and Yaounde with its administration to its own industry, commerce, and port: "Me voici donc à Douala/So here I am in Douala" to start a speech which matched the city's own defiance. Kumbo-Nso and places like it mark the regime's retreat, but there can be no withdrawal of the state from Douala, its irreducible center.

It is contested *and* disorderly. Observation confirms press reports of lights out on the Wouri Bridge, wrecks in potholes, floods from clogged drains, garbage uncollected.[60] We noted the 1991 accounts of youth beyond discipline. There are also all the adults foraging for a toehold of decency like a thirty-eight year old migrant from the Grassfields in 1992 to Douala, Felix Chia. He described himself in 1995, as a "redundant" teacher, "lucky to be employed" with a shop job in New Bell, a family of eight in a shack near a malarial swamp six kilometers from both a bus stop and a primary school, needing medicine which a World Bank project withheld because of concern that funds would wind up in a ministry rather than a designated NGO local health center two kilometers from his house.[61] About the time he arrived in 1992, "Les Dockers du Port" voiced a more settled population sector's grievances in an open letter to Paul Biya, citing swollen numbers and low competence in dock management, fraud and corruption in security, filth and danger at the work site, and calling *him* "inaccessible au petit peuple."[62] Harbor conditions and management remained a point of contention three years later, 1995, joined by reports of extortionate, politicized direction of Douala's public markets.[63]

Throughout Douala, there is organized and casual violence producing local headlines like "Peur sur la Ville/Fear over the City" in 1991 and "La barbarie organisée" in 1996.[64] They were reproduced abroad in 1997 by a Paris article on private security, "un secteur plein d'avenir/an enterprise with a future" in a city of burglaries, gates with padlocks, drug trafficking connected with Nigeria, and an apparatus and mentality of citizen self-defence replacing an overwhelmed public security force.[65] Such is the local context, close to African norms, for a much compromised urban environment and a civic politics which now leans toward the SDF. In 1992 the titular head of the city's core Duala ethnic elite and soon-to-be president of Ngondo, its council of customary leaders, Prince René Manga Douala Bell, widened the wedge between local patriarchs and the CPDM which Sengat Kuo had created a year before, by making a very public appearance at the SDF's inaugural Bamenda Congress. His interviews and speeches criticized government policy on a broad front, including a remark from this veteran of French military service in Indo-China that it was time for anglophones to lead Cameroon.[66] That same year, other highest echelon leaders including Cardinal Tumi, the former foreign minister and OAU secretary general William Eteki Mboumoua, and Douala's most prominent indigenous businessman Paul Soppo Priso, called on the CPDM to postpone all 1992 elections ("seeds of conflict, disorder and chaos") until fundamental electoral reforms were in place.[67]

Douala is a good example of what passes for Cameroon's electoral politics in the 1990s. Wouri division's nine National Assembly seats, which Douala dominates, were split in 1991 between UPC (five), CPDM and NUDP (two each), with the CPDM list receiving just 28 percent of the votes cast. The regime deficit grew worse; the 1992 presidential vote count there gave Fru Ndi 69 percent and Biya 15 percent, and the 1996 municipal election gave the SDF five of the city's six urban councils (victories then negated by the regime appointments of urban delegates). The consequences for the 1997 legislative poll? Wouri division was split from one into four electoral districts, and the CPDM announced victories in five seats to the SDF's four, against a SDF popular vote of 41 percent to the CPDM's 34 percent. Deep skepticism is aroused about such electoral politics in this much contested political site.

The regime does retain far more control in less disciplined Douala than in well disciplined Kumbo-Nso. But there is, in both the village and the metropolis, an unmistakable, one is inclined to say irreversible, SDF growth as the CPDM contracts, unless *force majeure* is applied, and it needs to be phased into any plausible version of politics at the national level. The CPDM retains power but cannot close the gulf, demonstrated in the passage on the CPDM's 10th anniversary starting Chapter 5, between the two Cameroons voiced in the radio version from Yaounde and the taxicab version from Bamenda. If it is to be closed, if any optimism is appropriate, if civil society and the state are to converge and promote real citizenship and civility, the SDF is the source for that effort, with its potential for mobilization across customary and newer affinities, rural and urban constituencies, and class, age, gender, and ethnic lines.

Failure in that effort will consign the politics of promise during the 1990s to irrelevance. There is also the risk made clear by Peter Geschiere's recent study of witchcraft, drawing largely on Cameroon, as the alternative domain of accusation between rich and poor, old and young, kin and familiars of all kinds. It sharpens hostility between accumulators and levellers within local societies, outside public channels of authority, and creates still higher tension if the state needs to adjudicate.[68] Much is at risk in these years of social and economic insecurity, if actors in political society cannot break Cameroon's impasse, and make public institutions nonviolent, effective and accountable sites of conflict resolution, surely a prime objective and necessity for any democratic civil society.

Foreign Factors

This fundamentally domestic study has at times noted France's large stake in and impact on Cameroon, and we cannot neglect a look beyond the borders as the book concludes. From Biya's 1991 nod to his democratic tutor, Mitterrand, through his shuttle trips between Paris and Yaounde which surely prime Cameroon policy and enrage many Cameroonians beside Mongo Beti, there is evidence of the community of French heritage, *la francophonie*, at work. Conventional wisdom regards this as a reciprocal bond, with France guarding a resource base, language, culture, and rationale for United Nations Security Council status in its black African redoubt, and African heads of state in turn securing regime survival. But events since 1989 challenge these ways. France's and Africa's politicians are less personally attached since the deaths of Houphouët-Boigny in 1993 and France's *éminence grise* for Africa since the 1950s, Jacques Foccart, in 1997. The French patron's advice and the African clients' consent to devalue the CFA franc in 1994 loosened basic material ties in the region Cameroon shares, and Mobutu's fall in 1997 implicated France, which backed him too long, in ways that could lead to fundamental strategic shifts there. Signs of French interest in Nigerian and South African ventures displacing older axes had already surfaced. *Will France consistently prefer Cameroon over Nigeria in their current disputes, given the politics of petroleum?*[69] And, even if his French presidential campaign in 1995 signified no change in African policy, will Lionel Jospin as prime minister from 1997, with European elections generally tilting to the left, make a difference in controversial business and environment issues like the Chad-Cameroon oil pipeline to Kribi, South Province, which the Biya regime favors over the domestic opposition we noted in Chapter 6? Three 1995 "sphere of influence" studies of francophone Africa disclosed pressures on France's *status quo* protection of historic links to autocratic African regimes, created by the force of global fiscal and economic structures which, variously, favor Japan, the U.S.A., and Germany in competition with France over the region and in the European arena which Africa most closely services.[70]

There is no doubt about France as still *the* singular influence in Cameroon. GATT 1993 data in one basic category, merchandise imports and exports, showed imports from France at 45 percent of Cameroon's total, four times the amount from any other region or country, and exports to France at 23 percent, the same as to a recently growing client, Spain, but well above the level to all others. The import-export imbalance with France and a very level graph for most indices since 1980 are worth noting. One altered sector of the trade, and its alleged character, are most striking. The biggest change in any import-export item in the years after 1985 was the rise in raw wood exports, which became the joint target of legislation in the first multiparty National Assembly and of fierce press assaults, protesting sale of Cameroon's patrimony and profits by the French, including Mitterrand's son, Jean-Christophe, and his clients.[71]

Commentary on this issue and others leads to speculation about French patience with Biya and the CPDM, facing such criticism, which a SDF presence in the National Assembly from June 1997 is likely to maintain. The SDF, however, muted its francophobic voices by mid-decade. Judging the regime to be vulnerable, it made an uneasy but reciprocated approach to France when Gilles Vidal took Yves Omnes' place as ambassador to Cameroon in 1993 (we noted a cartoon about it in Chapter 5). Fru Ndi and Vidal met in both Yaounde and Bamenda, Fru Ndi met France's consul at Garoua on the way to the Maroua Congress in 1995, and then called off a SDF boycott of French goods. Lower level meetings and technical assistance initiated French links to both anglophone provinces; Bamenda by 1997 had a new abattoir and market site with French assistance, and L'Alliance Française planned a bureau there.[72] All this disturbed those anglophone sovereigntists who perceive a French string and Cameroonian puppet in every high place, and an enemy in all francophones. But the SDF initiatives reflected a good deal of opinion, in Bamenda at least, that time is on its side. France, it is argued, must soon reckon the long term costs of keeping Biya in power, and open channels to his rivals, or it will soon face the global movements, noted above, favoring Japan and the U.S.A. in Africa and Germany in Europe, in ways bound to help all Cameroonians and to hurt France if it does not bend.

There is hardly a more stubborn force globally than *la francophonie*, and it is common to hear from all sides of domestic opinion in Cameroon, with complacency or rage, that France will never permit an anglophone in its presidency. But French policy in sub-Saharan Africa is surely mutable. CFA devaluation marks a permanent policy change from protectionism, which can hardly please those with a stake in Cameroon's political impasse, and may encourage those favoring democratic transition who argue a connection between more open market forces and pluralist politics.

Conclusion

Regime creation and consolidation in Cameroon, over twenty-five years from 1960, and contraction since, are obvious. So, now, is opposition growth. The *core* opposition is the SDF. With its populist history and the appeals it projects beyond the ranks it already commands or attracts, it is more open textured, less opportunistic, ethnocentric, and "class closed" than any of its rivals. It has some serious competitors for Cameroonians' political party and communal group allegiances: sovereigntist anglophones; the North's Fulbe-Muslim elite; regime Betis; restless Bamilekes wherever they are; South West Province skeptics. The SDF four state federation constitutional proposal, perhaps mediated by the CDU's, offers a way to attract those peoples, and to alter the framework of the state immediately and society over time, based on a shared allocation of political initiatives and material resources and a sound base in electoral practices and human rights. Since a miracle is unlikely to grace the millenial turn we approach, it is Cameroon's best chance to break its impasse. It could satisfy Schatzberg's sense of legitimacy, Mbembe's sense of historical process and the need to escape a half century of Aujoulat, Ahidjo, and Biya as Cameroon's political spinal cord, and a sufficient number of democratic though not utopian sensibilities, as they are understood by both the global and African sources our Preface canvassed. Whether the unitary state can be peacefully dissolved remains problematic, but the SDF has nearly a decade's survival experience, clean hands, and no significant political debts outstanding. It understands the need for, and practices within prudent limits, what Julius Ihonvbere has recently called attention to as a weakness in prodemocracy efforts in Africa: confrontation with the state.[73] It is a prime example among exceptions to the pervasive Afro-pessimism of the 1990s.

Cameroon's national recovery under any auspices will be lengthy. Some combination of exhaustion and cynicism is a factor, unmistakable on some of the ground described in this book. Achille Mbembe and Janet Roitman in their 1995 study of the extreme improvisation life in Cameroon now imposes on its people. "We are not describing havoc in Cameroon": despite this disclaimer, the surreal becomes normal in the ways space is contested, transactions are negotiated, services have collapsed, and resources are cannibalized on the Yaounde and Maroua terrains they describe.[74] "Normalcy" there goes beyond the uncertainty which anglophones understand as their lot. But we have shown Kumbo-Nso and Bamenda creating disciplines of their own, away from most state domination, and conducting a SDF politics worthy of the term. The challenge, though, is formidable. For Cameroon at large, a 1994 opinion poll found 39.6 percent responding "Aucun/None" when asked about political party preference, a category up 50 percent from the year before and far ahead of the SDF at 15.5 percent and the CPDM at 11.5 percent.[75] Fru Ndi's own reputation failed to translate, or no longer translated, into national confidence in his party as an institution. Without an anchored opposition or regime, undisciplined parties and unruly contests in elections axiomatically risk the Pandora's box of particularistic rivalry. Nor have the national conference variants proved to be a panacea. Cautious use of both approaches makes sense, along lines resembling South Africa's, which has thus far negotiated a path between federalism and centralism. The SDF has long advocated a parallel: transition (not coalition) government for two years, giving time (but not too much) to write a constitution and other texts to be put to referendum, and to test the parties' "cohabitation" possibilities, before any vote to determine executive and legislative powers and put them to work. One more South African factor is needed: limits to the politics of vengeance.

It challenges the optimist's faith to believe that Cameroon's institutional fabric is now resilient enough for such processes to work. The most favorable scenario for the 1990s would have been initiatives combining, first, national conference consultative processes, then an electoral cycle which was adequately free and fair, internationally prepared and monitored. None of this happened. Wishful thinking in mid-1997 suggests *la francophonie* and the Commonwealth, given Cameroon's joint membership, as post-1997 election cosponsors of a "time certain" deliberative process. If nothing like *this* happens, and impasse continues, with the prospects of a Biya presidency until 2010, the morbid, dystopic setting Mbembe and Roitman describe could trigger worse scenarios. Action in the streets and arms brought from the barracks could resume. Either could happen first, and would activate the other.

In that event, Cameroon might well fall from the place it occupied with eleven other African polities in William Zartman's 1995 study, on the "serious danger" edge of collapse which faced eight others in a "maximum danger" zone.[76] He specified five criteria for state collapse in Africa: loss of power from a conflicted center to the peripheries; the retreat of the central power from "its social bases" to "its innermost trusted circle"; government's avoidance of difficult choices and hard decisions; a defensive incumbency which balances repression and concession, postpones elections and abandons a coherent sense of program; "the ultimate danger sign... when the center loses control over its own state agents, who begin to operate on their own account." The Cameroonian experience of the 1990s matches or approximates the second, third and fourth of these conditions, and there are signs of the first and fifth.

Failing the will and skill to turn governance back from that brink, Cameroon could join not just Congo (Kinshasa), a point of comparison throughout this text, but two others in Zartman's deepest cavity, Rwanda and Burundi, which many in Cameroon say they fear, Fru Ndi among them. As Burnham in 1996 closed his study of the North, so must we, in 1997, conclude ours for Cameroon at large: "the prospects for peace look bleak."[77]

Notes

1. See the cover of *West Africa*, July 22–28, 1996.
2. *Jeune Afrique*, March 20–26, 1996, pp. 15–18, followed up by the same author François Soudan's text on Biya's tactical skill and shrewd team, *Jeune Afrique*, July 3–9, 1996, pp. 70–71; *West Africa*, July 15–21, 1996, pp. 1102–1103.
3. Succession is always contested. In mid-1997, Edouard Akame Mfoumou, Biya's kinsman, with experience running key ministries, was the CPDM's current favorite in that vulnerable situation. Nfor Susungi, following an African Development Bank career of the kind that now creates politicians in the region, took over *Cameroon Post* and surfaced in high SDF ranks; he could succeed Fru Ndi at need.
4. *Le Messager*, June 18, 1997.
5. *Le Messager*, January 29, 1996, for the details, and *L'Expression* the same day for the (translated) headline: "Was [Sultan] Mbombo Njoya Booby-Trapped?"
6. *Jeune Afrique Economie*, May 1992, pp. 106–126, with research and writing by Célestin Monga for the Paris magazine which, for this issue, was denied Cameroon entry.
7. From Fotso (b.1926), see *Tout Pour la Gloire de Mon Pays* (Yaounde: CEPER, n.d. but likely the late 1970s) and *Le Chemin de Hiala* (Paris: Editions de Septembre, 1994). For allegations about him and Foning: *Le Quotidien*, October 8, 1992; *Challenge Nouveau*, January 4, 1996; *La Nouvelle Expression*, November 18–21, 1994 and July 7–11, 1995; *Dikalo*, August 7, 1995. Fotso relocated his enterprises in the mid-1990s to his natal Bandjoun, West Province, adding a bank and the mayoralty to the technical institute which bears his name. Of all the research questions we would like to have evidence for, to turn rumor into data, but (despite some effort) do not, SDF funding is the most elemental: do the big entrepreneurs hedge their political bets?
8. *Jeune Afrique*, June 5–11, 1996, pp. 32–33; *West Africa*, July 1–7 1996, p. 1029.
9. David Blandford, et al., "Oil Boom and Bust: The Harsh Realities of Adjustment in Cameroon," in David Sahn, ed., *Adjusting to Policy Failure in African Economies* (Ithaca: Cornell University Press, 1994), p. 147.

10. *The Herald*, January 23–24, 1996, on the fon's precarious post-election situation; interview in Bali with his uncle, Prince Fonyonga, May 6, 1995.

11. This canvass draws on newspaper accounts, frequent talks with observers of, and visits to, four of the five palaces in 1995, and a February 20, 1995, interview with the fon of Bali-Nyonga. He remarked that the Central Committee of the CPDM (he is a member) has created local problems for him, which he has no authority or resources to solve: "They've taken all power from fons." Cyprian Fisiy, "Chieftaincy in the Modern State: An Institution at the Crossroads of Democratic Change." *Paideuma* 41 (1995), pp. 49–63, elaborates points made here, as do other papers in that *Paideuma* special issue honoring E. M. (Sally) Chilver.

12. *Le Messager*, January 23, 1992, summarized press coverage of incidents which triggered language about sacred lands which read like Israel and Palestine disputing Jerusalem.

13. Richard Bjornson, *The African Quest for Freedom and Identity: Cameroonian Writing and the National Experience* (Bloomington: Indiana University Press, 1991) pp. 53, 135, 180–183.

14. *Pastoral Letter of the Episcopal Conference of Cameroon Concerning the Economic Crisis which the Country is Undergoing*, June 3, 1990, pp. 16–21, 28.

15. *Statement of the Standing Committee of the Bishops of Cameroon*, December 11, 1992.

16. *Meeting of the Standing Committee of the National Episcopal Conference of Cameroon, Yaounde: Final Communiqué*, December 15, 1995.

17. *L'Expression*, February 9, 1996.

18. Archbishop Verdzekov's Bamenda prologue to the bishop's declaration, June 23, 1996, noted in the absence of any official investigation "no longer any doubt that [Mveng] was assassinated."

19. *The Herald*, October 31–November, 2 1994; *Cameroon Post*, February 24–March 3, 1995. We have a photocopy of the rationale explaining his withdrawal from Biya's constitutional talks; evidently in his own hand, it is forthright and eloquent.

20. *Newslink*, 1,1 (December 27, 1993) prints a typical sermon; he hosts the Catholic Human Rights Association, which he also chairs, and the Action by Christians for the Abolition of Torture; *A Humble and Earnest Appeal by the Bishops of the Bamenda Ecclesiastical Province to the Right Honorable Simon Achidi Achu, M.P., Prime Minister*, September 25, 1993, is the text on schools.

21. The Bob Dylan vocabulary in the sermon matches the Brazilian Catholic sources in his extensive library, "Power to the People" T-shirt, and preparation of counselling, then communion, for francophone gendarmes troubled by their Bamenda role, observed in his parish church, March 30, 1995. The text on Father Mveng is from the *Sunday Report* series of his published sermons, available in Bamenda, key sources for any reconstruction of that area's history since 1991.

22. Luc Sindjoun, "Le champ social camerounais," *Politique Africaine* 62 (1996), pp. 59–60: "La contestation semble depuis 1990 bénéficier ... de l'onction religieuse de l'épiscopat catholique."

23. *The Position of the Leadership of the Cameroon Baptist Convention on the Issues Currently Rocking Cameroon Peace and Stability of the Years*, Bamenda, May 1991, pp.

18, 28, 30. The Executive Secretary who led the delegation and issued the text soon became fon of Oku, and left the religious front line and any other activity likely to be construed as oppositional.

24. Galia Sabar-Friedman, "Church and State in Kenya, 1986–1992: The Churches' Involvement in the 'Game of Change,'" *African Affairs* 96, 382 (1997), pp. 25–52 for Kenya; *L'Expression*, November 7, 1995, for Cameroon.

25. Ministry of Information and Culture. *The Cultural Identity of Cameroon* (Yaounde, 1985), p. 14.

26. See Ambroise Kom, "Writing Under A Monocracy," *Research in African Literatures* 22,1 (1991), pp. 83–92, and Milton Krieger, "Building the Republic through Letters: *Abbia: Cameroon Cultural Review*, and Its Legacy," *Research in African Literatures* 27,2 (1996), pp. 155–177. Collectif, *Changer le Cameroun: Pourquoi pas?* (Yaounde: C3, 1990) is the key primary source.

27. The estimate of numbers, 1 percent of all Cameroonians, is from Manga Kuoh, *Cameroun: un nouveau départ* (Paris: L'Harmattan, 1996), p. 126.

28. For a synopsis of him and his party, see *The Star Headlines*, March 29, 1996. *Challenge Mensuel*, September and November, 1993, published opinion polls from SYNAPSE (Douala) based on 1,000 responses which ranked Garga Haman the fourth most favored figure in Cameroon; Fru Ndi ranked first.

29. For CAPSU, see, e.g., *The Messenger*, February 22, 1994: "We Follow CAPSU Not Biya"; for the largely francophone parastatal labor manifesto, see *Dikalo*, August 16, 1995.

30. Material on public finance cited in the Introduction from Sahn, ed., *Adjusting to Policy Failure*, applies here.

31. Sankie Maimo, *Sasse Symphony* (Limbe: Nooremac Press, 1989) for the college's golden jubilee is a good example of the school tie celebration genre. Football, of course, needs more attention than we can give it here. Anyone who left Cameroon in 1991 and returned in 1995 found the Baham, West Province, team a new power, financed at his village in the interim by the minister of communications, Augustin Kontchou Kouomegni, reflecting a swift rise to favor, as does his vast estate near the convergence of roads from Bamenda, Douala and Yaounde. This is the most conspicuous sign of the recent Biya regime's splendors we know of.

32. *Newschampion*, January 25, 1995; *Le Messager*, August 5, 1995.

33. *Jeune Afrique*, May 15–21, 1996.

34. See *Jeune Afrique Economie*, March 18, 1996, pp. 64–67, for the well informed David Ndachi Tagne's story on Fochive, and a survey on "Les ravages du grand banditisme" referring to roughly twenty private security firms now operating, foreign and domestic, some of them the subject of investigative press reports in mid-1995 about perils *they* posed. Armed Yaounde episodes in 1995 included the shooting death of the inspector general of the ministry of culture and robbery at the house of the retired French ambassador.

35. An interview with and materials from TAC general secretary Sammy Besong Arrey-Mbi, July 31, 1995, in Bamenda, supply this detail. Seven months there in both 1991 and 1995 inform this entire section, and the next. Anonymity for some sources is required.

36. There were tangles, *The Messenger* claimed, May 18, 1995, in a story about rivalries among these women and SDF favoritism to "its own."

37. 37. This account combines mutually supportive press reports, July-October 1995, Bamenda and Yaounde conversations with owners and drivers, and direct observations.

38. Kengne Fodouop, "Le secteur informel dans le contexte des ajustements au Cameroun: L'exemple de la "vente à la sauvette," *Labour, Capital and Society* 26,1 (1993), pp. 42–61.

39. Passages below are from Jean-Marc Ela, *Quand l'Etat pénètre en brousse.... Les ripostes paysannes à la crise* (Paris: Karthala, 1990), pp. 234, 246, and *Afrique: L'irruption des pauvres: Société contre ingérence, pouvoir et argent* (Paris: L'Harmattan, 1994), p. 182. The entire texts are germane, as is Piet Konings, *Labour Resistance in Cameroon: managerial strategies and labour resistance in the agro-industrial plantations of the Cameroon Development Corporation* (London: James Currey; Portsmouth, N.H.: Heinemann, 1993).

40. Lucy Davis, "Opening Political Space in Cameroon: the Ambiguous Response of the Mbororo," *Review of African Political Economy* 64 (1995), pp. 213–228, and further observation by one of the authors in 1995.

41. Miriam Goheen, *Men Own the Fields, Women Own the Crops: Gender and Power in the Cameroon Grassfields* (Madison: University of Wisconsin Press, 1996).

42. For the broad regional setting, see Paul Nkwi, "Traditional female militancy in a modern context," in Jean-Claude Barbier, ed., *Femmes du Cameroun* (Paris: Karthala-ORSTOM, 1985), pp. 181–191, and Eugenia Shanklin, "Anlu Remembered: the Kom Women's Rebellion of 1958–61," *Dialectical Anthropology* 15 (1990), pp. 159–181. Emmanuel Fru Doh, "Women, Events and the Revitalization of Culture: Takumbeng as a Mankon Cultural Phenomenon in the Bamenda Grassfields," University of Buea Conference on Cameroon Writing, 30 November–4 December, 1994, cited above, Chapter 5, gives *takumbeng* details.

43. Personal witness is the source for July 5, 1991; other details come from *Génération*, May 24–31, 1995 ("les Amazones"), *Jeune Afrique Economie*, December 1992, p. 116, *L'Expression Nouvelle*, December 18, 1992 (state of emergency), and *The Herald*, October 31–November 2, 1994 (Njinikom and variants).

44. Goheen, *Men Own The Fields*, throughout.

45. Achille Mbembe, *Les jeunes et l'ordre politique en Afrique noire* (Paris: L'Harmattan, 1985), p. 111 for Ahidjo, pp. 110–121, 133–160 respectively for students and youth at large in this context (there are others, including a "sex, drugs, rock and roll" analysis, which are equally telling). The film *Le Quartier Mozart* augments Mbembe.

46. For analysis of youth in a similar but more channelled mode in that region, see Nicolas Argenti, "Mbaya: New Masquerades, Violence, and the Nation State in Cameroon," a paper delivered at the 39th Annual Meeting of the African Studies Association, San Francisco, November 23–26, 1996. Details from the Bamenda area in this section are from our sight or reliable, corroborated secondary sources.

47. Philip Burnham, *The Politics of Cultural Difference in Northern Cameroon* (Washington: Smithsonian Institution Press, 1996), pp. 137–138.

48. See *Sunday Report*, 16,10 of March 5, 1995, taken from a sermon by Father Ndinayen Wirba of Little Mankon Parish, Bamenda, describing and reflecting on an Ash Wednesday necklacing.

49. *La Nouvelle Expression*, August 1–4, 1995, for francophones, August 25–28, 1995, for anglophones.

50. Interviews with the TAC's Sammy Besong Arrey-Mbi, July 31, 1995, and Lawrence Tasha, Amity Bank's managing director, August 14, 1995.

51. An "Open letter to all English-Speaking Parents" from their student-children, October 20, 1985, printed in Albert Mukong, ed., *The Case for the Southern Cameroons* (CAMFECO, 1990), pp. 107–119, and still well known to the public, remains a comprehensive indictment of the unitary Cameroon state's work among anglophones, lacking only the later constitutional program for redress.

52. See Goheen, *Men Own The Fields*, for both the text and its bibliographical leads to an excellent print legacy, much of it from locals. This section also draws on frequent, though brief, visits in 1991 and 1995, and interviews which in some cases require confidentiality.

53. Goheen, *Men Own The Fields*, p. xii, cited earlier and now again because of its salience for our Congo (Kinshasa) comparison and the emerging autonomy of East Kasai Province there, which the regime change may not deter; Howard French, *New York Times*, September 18, 1996, p. A4.

54. Goheen, *Men Own The Fields*, pp. 52–56.

55. Statutes of the Kumbo Water Authority, Article 1.1. The same SNEC-village conflict in Bali-Nyonga, covered in *West Africa*, September 12–18, 1994, pp. 1592–1593, with a text slanted more in SNEC's favor, adds to the fon's problems there.

56. Goheen, *Men Own The Fields*, p. 39, for this and the next passage on palace preferences in the 1990s.

57. This paragraph utilizes obvious sights as well as semi-sensitive talks, and a range of press reports too frequent to cite, touching local administrative decisions and personnel. It is one passage which gossip and rumor would truly dramatize.

58. Goheen, *Men Own The Fields*, pp. 15, 39 on palace floggings, which are of interest because many locals praise them, and they in part assert elders over youths in a setting full of changes. One wonders if less popular fons get away with flogging.

59. See *Cameroon Post*, June 26–July 3, 1995, for one episode ("20 Thugs Disrupt Amity Bank Meeting") in a recurrent 1995 attempt by Yaounde authorities to bring Amity under their wing. It stays independent, and goes elsewhere than France for offshore capital.

60. For one example, *Le Messager*, April 20, 1992, reported 700 tons of garbage mounting each day, and troops sent to clean the rich suburbs.

61. *The New Expression*, January 26, 1995.

62. *Galaxie*, February 3, 1992, in a document comparable to the 1995 letter from water and electricity workers to Biya, cited earlier in this chapter.

63. On the harbor, see *La Nouvelle Expression*, April 25–28, 1995, and *The Herald*, May 8–10, 1995. On markets, see *Le Messager*, July 24, 1995, *La Nouvelle Expression*, August 16–18, 1995, and *Galaxie*, August 21, 1995.

64. *La Détente*, December 23, 1991; *Le Quotidien*, January 25, 1996.
65. *Jeune Afrique*, March 12–18, 1997, pp. 34–35.
66. *Cameroon Post*, June 1–8, 1992; *Challenge Hebdo*, June 10, 1992.
67. "Call by Certain Independendent Personalities and Church Leaders of Wouri Division in Reaction to the Upcoming Elections," (n.d.).
68. Peter Geschiere, *Sorcellerie et politique en Afrique—La viande des autres* (Paris: Karthala, 1995), translated by the author and Janet Roitman as *The Modernity of Witchcraft: Politics and the Occult in Postcolonial Africa* (Charlottesville: University Press of Virginia, 1997).
69. For a review of this dispute not dominated by petroleum, current to 1995, see Thomas Weiss, "Migrations et conflits frontaliers: une rélation Nigeria-Cameroun contrariée," *Afrique Contemporaine* Numéro spécial (1996), pp. 39–51.
70. Guy Martin, "Continuity and Change in Franco-African Relations," *The Journal of Modern African Studies* 33,1 (1995), pp. 1–20; Gordon Cumming, "French Development Assistance to Africa: Towards a New Agenda?" *African Affairs* 94 (1995), pp. 383–398; Peter Schraeder, "From Berlin 1884 to 1989: Foreign Assistance and French, American and Japanese Competition in Francophone Africa," *The Journal of Modern African Studies*, 33,4 (1995), pp. 539–567.
71. General Agreement on Tariffs and Trade, *Trade Policy Review: Cameroon, 1995* (Geneva, May 1995), pp. 11–13; Samuel Efoua Mbozo'o, *L'Assemblée Nationale du Cameroun* (Yaounde: Hérodote, 1994), pp. 125–126; *The Messenger*, May 18, 1995; *Challenge Hebdo*, June 29, 1995; *La Nouvelle Expression*, March 22, 1996.
72. Standard press reports of France-anglophone Cameroonian meetings and SDF briefings about quieter, more routine contacts are used here, as well as *The Herald*, June 29–July 2, 1995, for data on French aid for South West Province education. One guesses that the Cameroon government's consent for these contacts is not a factor.
73. Julius Ihonvbere, "On the Threshold of Another False Start? A Critical Evaluation of Prodemocracy Movements in Africa," *Journal of Asian and African Studies* XXXI,1–2 (1996), pp. 125–142, esp. pp. 130–131.
74. Achille Mbembe and Janet Roitman, "Figures of the Subject in Times of Crisis," *Public Culture* 7 (1995), pp. 323–352, p. 327 for the quoted passage.
75. Fondation Friedrich Ebert, "Les Sensibilités Politiques des Camerounais," October 1994, polling 1,000 people.
76. William Zartman, ed., *Collapsed States: The Disintegration and Restoration of Legitimate Authority* (Boulder: Lynne Rienner, 1995), pp. 2–3, 10–11.
77. Burnham, *The Politics of Cultural Difference*, p. 171.

Appendix 1

First Draft of the Letter by the National Coordination for Democracy and Multi-party System (NCDM)

Men and Women of Cameroon!

After a quarter of a century of the AHIDJO era, characterized on the political front by a drift towards monarchy in the so-called "Lawful State" and on the economic front by a state of perpetual take-off, the Cameroonians—all the Cameroonians—took it for granted that the man of renewal, master of Cameroon's destiny, would turn the painful page of this drift towards monarchy once and for all, and make possible the progress we have waited for so long.

Today it is to be seen that the disappointment of the Cameroonians, of all Cameroonians who care about their nation's fate, and the rising anger of the masses are in proportion to the hopes that had been raised by the coming of a renewal.

We must not again accept, for a third time, the unhappy experience of an anti-people regime. A new wind is rising. The wind of freedom and democracy is blowing throughout the world. We cannot remain indifferent to it.

Men and Women of Cameroon,

This is a grave moment. We must choose between taking action and resigning ourselves to our fate.

The damaging manifestations of crisis which are now the Cameroonians' daily fare are coming to be seen even as fatal. A general drop in income, particularly for peasants and public sector employees; massive dismissals and cutbacks in employment; a marked deterioration in the educational standards of our young people; an increasingly disturbing health situation, etc, etc. The difficulties and sufferings of today are no more than the direct result of a policy essentially designed for a tiny minority who have little concern for the interests of the people.

Moving beyond such a historical anachronism, the legitimate aspirations of the Cameroonian people must put aside and relegate to the archives of history the negative legacy of the past and the one-party system: UC, UNC, RDPC, etc.

Throughout its history, it cannot be ignored that Cameroon has taken up the widest range of challenges. Far from retreating behind an ideologically monolithic wall, the national liberation movement developed based on the premises of ideological pluralism, a constructive, responsible multi-party system and the rotation of political power. The heroic struggle of the founding fathers of the UPC, Ruben UM NYOBE, Felix-Roland MOUMIE and Ernest OUANDIE, was always carried on in a climate of dialogue, tolerance, equal rights for all individuals, respect for the values on which our political system is based and the outstretched hand. This policy has nothing in common with the so-called "moralization" and/or "democratization" of Mr. Paul BIYA'S regime.

Men and Women of Cameroon,

We must give a new drive to our economy. We must restore the confidence of our people.

This can only become possible in a climate of genuine freedom, democracy and progress.

Cameroon, our beautiful country, must be reborn with a multi-party system and a responsible National Assembly, completely independent of the executive. It is there, at the summit of the most fundamental principles, principles around which the social majority of Cameroonians and the political majority of the avant garde may unite in solidarity, that today's fight, which is the fight of all Cameroonians, must take place.

Appeal from the National Coordination for
Democracy and a Multi-party System.

Second Draft of the Appeal of the National Coordination for Democracy and a Multi-party System (NCDM)

Men and Women of Cameroon!

After a quarter of a century of the AHIDJO era, characterized on the political front by a drift towards monarchy and on the economic front by a state of perpetual take-off which never gained altitude, the Cameroonians—all the Cameroonians—took it for granted that the man of rigour, of order, the new master of Cameroon's destiny, would turn this painful page in our country's history once and for all, and make possible the progress we have waited for so long.

Today we face the disappointment of the Cameroonians, which is constantly, unceasingly increased by the demands of policies whose direction bears no relation to everyday problems.

We must not again accept, for a third time, the unhappy experience of a regime which oppresses the people.

At a time when the winds of freedom and its corollary, democracy, are blowing throughout the world, we have no right to fold our arms and do nothing.

Men and Women of Cameroon,

This is a grave moment. Brandishing the spectre of crisis, those in power keep demanding ever greater sacrifice of us.

Throughout its history, it cannot be ignored that Cameroon has taken up the widest range of challenges.

Far from retreating behind a wall of monolithic dogmatism, the national liberation movement developed based on the premises of ideological pluralism, a constructive, responsible multi-party system and the rotation of political power.

Running a country within a democratic system requires a balance of power.

The heroic struggle of our patriots was always carried on in a climate of dialogue, tolerance, equal rights for all individuals and respect for the values on which our political system is based, rejecting tribalism or regionalism of any kind. Such policies have nothing in common with those of the RDPC.

Therefore we call upon you, with all our force, so that together we may demand of those now in power to recognize our civil rights and fundamental freedoms.

Cameroon, our beautiful country, must be reborn with a multi-party system and a responsible National Assembly, completely independent of the executive.

It is there, at the summit of the most fundamental principles, principles around which the social majority of Cameroonians and the political majority of the avant-garde may unite in solidarity, that today's fight, which is the fight of all Cameroonians, must take place.

You may pledge us your participation in this noble and sublime battle by signing and returning this appeal: appeal from the National Coordination for Democracy and a Multi-party System.

Source: Amnesty International Report, July 23, 1990. Translated from the French version by Amnesty International.

Selected Bibliography

Abada, Marcelin Nguele. "Ruptures et continuités constitutionelles en république du Cameroun, "*Revue Juridique et Politique, Indépendance et Coopération* 50,3 (1996), 272–293.
Achebe, Chinua. *A Man of the People*. New York: Anchor, 1967.
Africa Contemporary Record. New York: Africana Publishing, 1968/69–.
Africa Research Bulletin. Exeter: Africa Research, 1964–.
Africa South of the Sahara. London: Europa Publications, 1971–.
Ahmadou, Ahidjo. *Contribution to National Construction*. Paris: Présence Africaine, 1964.
———. *As Told by Ahmadou Ahidjo*. Monaco: Paul Bory, 1968.
———. *The Political Philosophy of Ahmadou Ahidjo*. Monaco: Paul Bory, 1968.
———.*Nation and Development in Unity and Justice*. Paris: Présence Africaine, 1969.
Ake, Claude. *A Political Economy of Africa*. Harlow: Longman, 1981.
Alima, Joseph-Blaise. *Les Chemins de l'Unité*. Yaounde: Afrique Biblio Club, 1977.
All Anglophone Conference. *The Buea Declaration*, 1993.
———. *The Bamenda Proclamation*, 1994.
Amin, Julius A. *The Peace Corps in Cameroon*. Kent: The Kent State University Press, 1992.
Amin, Samir. *Neo-Colonialism in West Africa*. Translated from French by Francis McDonagh. New York: Monthly Review Press, 1974.
Amnesty International Reports. London, 1974/75–.
Apter, David, and Carl Rosberg, Jr., eds., *Political Development and the New Realism in Sub-Saharan Africa*. Charlottesville: University Press of Virginia, 1994.
Ardener, Edwin. "The Nature of the Reunification of Cameroon." In *African Integration and Disintegration*, ed. Arthur Hazlewood, 285–337. London: Oxford University Press, 1967.
Argenti, Nicolas. "Mbaya: New Masquerades, Violence and the Nation State in Cameroon." Paper presented at the 39th Annual Meeting of the African Studies Association, San Francisco, November 23–26, 1996.
Austen, Ralph. "Tradition, Invention and History: The Case of the Ngondo (Cameroon)," *Cahiers d'Etudes Africaines* XXXII,2 (1992), 285–309.
———. *The Elusive Epic: Performance, Text and History in the Oral Narrative of Jeki la Njambè (Cameroon Coast)*. Atlanta: African Studies Association Press, 1995.
Austin, Dennis. *Politics in Africa*. Hanover: University Press of New England, 1984.
Ayina, Egbomi, "Pagnes et politique," *Politique Africaine* 27 (1987), 47–54.

Azevedo, Mario. "The Post-Ahidjo Era in Cameroon." *Current History* 86, 520 (1987), 217–220, 229–230.

———. "Ethnicity and Democratization: Cameroon and Gabon." In *Ethnic Conflict and Democratization in Africa*, ed. Harvey Glickman, 255–288. Atlanta: African Studies Association Press, 1995.

Bandolo, Henri. *La Flamme et la Fumée*. Yaoundé: Sopecam, 1985.

Bayart, Jean-Francois."One Party Government and Political Development in Cameroon." *African Affairs* 72, 287 (1973), 125–144.

———. "The Birth of the Ahidjo Regime." In *Gaullist Africa: Cameroon Under Ahmadu Ahidjo*, ed. Richard Joseph, 45–65. Enugu, Nigeria: Fourth Dimension Publishers, 1978.

———. *L'Etat au Cameroun*. Paris: Presses de la Fondation Nationale des Sciences Politiques, 1979.

———. "La Rentrée Politique du President Biya," *Marchés Tropicaux et Méditerranéens* (January 11, 1985), 69–71.

———. "Civil Society in Africa." In *Political Dominance in Africa*, ed. Patrick Chabal, 109–125. Cambridge: Cambridge University Press, 1986.

———. "Cameroon." In *Contemporary West African States*, ed. Donal Cruise O'Brien, John Dunn, and Richard Rathbone, 31–48. Cambridge: Cambridge University Press, 1989.

———. *The State in Africa: The Politics of the Belly*. Translated by Mary Harper, Christopher and Elizabeth Harrison. London and New York: Longman, 1993. Originally published as *L'Etat en Afrique: La politique du ventre*. Paris: Fayard, 1989.

Bederman, Sanford. *The Cameroons Development Corporation: Partner in National Growth*. Bota, West Cameroon: Cameroons Development Corporation, 1968.

———. "The Demise of the Commercial Banana Industry in West Cameroon," *Journal of Geography*, 70,4 (April 1971), 230–234.

Bello Bouba, Maïgari. *General Policy Speech* to the Second Ordinary Congress of the National Union for Democracy and Progress, Ngaoundere, 1997.

Benjamin, Jacques. "Le fédéralisme camerounais: l'influence des structures fédérales sur l'activité économique ouest-camerounaise," *Canadian Journal of African Studies*, 5,3 (1971), 281–306.

———. "Dix ans de fédéralisme camerounais 1961–1971." In *Fédéralisme et Nations*, ed. Roman Seryn, 279–290. Montreal: Les Presses de l'Université du Québec, 1971.

———. *Les camerounais occidentaux: la minorité dans un état bicommunautaire*. Montreal: Les Presses de l'Université de Montréal. 1972.

———. "The Impact of Federal Institutions on West Cameroon's Economic Activity." In *An African Experiment in Nation Building: The Bilingual Cameroon Republic Since Reunification*, ed. Ndiva Kofele-Kale, Boulder: Westview Press, 1980.

Benjamin, Nancy, and Shantayanan Devarajan. "Oil Revenues and the Cameroonian Economy." In *The Political Economy of Cameroon*, ed. Michael Schatzberg and William Zartman, 161–188. New York: Praeger, 1986.

Beti, Mongo. *Main basse sur le Cameroun: autopsie d'une décolonisation*. Paris: François Maspero, 1977.

Bickford, Kathleen. "The A.B.C.s of Cloth and Politics in Côte d'Ivoire," *Africa Today* 4,2 (1994), 5–19.
Bishops of the Bamenda Ecclesiatical Province (Catholic). *A Humble and Earnest Appeal . . . to Simon Achidi Achu.* September 25, 1993.
Biya, Paul. *Le message du renouveau: discours et interviews du President Paul Biya.* Yaounde: Editions Sopecam, 1984.
———.*Communal Liberalism.* London and Reading: Macmillan Publishers, 1987
Biyiti bi Essam, J-P. *Cameroun: complot et bruits de bottes.* Paris: L'Harmattan, 1984.
Bjornson, Richard. *The African Quest For Freedom and Identity: Cameroonian Writing and the National Experience.* Bloomington: Indiana University Press, 1991.
Blandford, David, et al. "Oil Boom and Bust: The Harsh Realities of Adjustment in Cameroon." In *Adjusting to Policy Failure in African Economies,* ed. David Sahn, 131–163. Ithaca: Cornell University Press, 1994.
Blassel, Frederic. "Les Habits Neufs du Renouveau." *Marchés Tropicaux et Méditerranéens* (April 5, 1985), 799–802.
Boulaga, Fabien Eboussi. *Les Conférences Nationales en Afrique Noire.* Paris: Karthala, 1993.
Bratton, Michael. "Civil Society and Political Transitions in Africa." In *Civil Society and the State in Africa,* ed. John Harbeson, Donald Rothchild, and Naomi Chazan, 51–81. Boulder: Lynne Rienner Publishers, 1993.
Bratton, Michael and Nicolas van de Walle, "Popular Protests and Political Reform in Africa," *Comparative Politics,* 24,4 (1992), 419–442.
Brian, Murphy. "Cameroon-Africa's Happy Exception." *International Perspectives* (March/April 1983), 21–23.
Burnham, Philip. *The Politics of Cultural Difference in Northern Cameroon.* Washington: Smithsonian Institution Press, 1996.
Callaghy, Thomas. "Civil Society, Democracy, and Economic Change in Africa: A Dissenting Opinion." In *Civil Society and the State in Africa,* ed. John Harbeson, Donald Rothchild, and Naomi Chazan, 231–253. Boulder: Lynne Rienner Publishers, 1993.
Cameroon Baptist Convention. *The Position of the Leadership . . . on Issues Currently Rocking Cameroon.* May 1991.
Cameroon Democratic Union. *Manifesto.* 1991.
———. *Avant-Projet de Constitution du Cameroun.* October 29,1993.
Cameroon National Union. *As Told by Ahmadou Ahidjo.* Monaco: Paul Bory, 1968.
———. *Ahmadou Ahidjo: Ten Years of Service to the Nation.* Monaco: Paul Bory, 1968.
———. *The New Deal Message.* Yaounde: Editions Sopecam, 1983.
———. *The New Deal: Two Years After.* Yaounde: n.p., 1985.
———. *Paul Biya: le Président de tous les Camerounais.* Yaounde: Union Nationale Camerounaise, 1985.
Carter, Gwendolen, ed. *National Unity and Regionalism in Eight African States.* Ithaca: Cornell University Press, 1966.
Chabal, Patrick. *Political Domination in Africa: Reflections on the Limits of Power.* Cambridge: Cambridge University Press, 1986.

———. *Power in Africa: An Essay in Political Interpretation.* Basingstoke, Hampshire: Macmillan, 1992.

Chafer, Tony. "French African Policy: Towards Change," *African Affairs* 91,362 (1992), 37–51.

Chazan Naomi, "Africa's Democratic Challenge." *World Policy Review* 9,2 (1992), 279–307.

———. "Between Liberalism and Statism: African Political Cultures and Democracy." In *Political Culture and Democracy in Developing Countries,* ed. Larry Diamond, 67–105. Boulder: Lynne Rienner Publishers, 1993.

Chem-Langhee, Bongfen. "Southern Cameroons Traditional Authorities and the Nationalist Movement, 1953–1961," *Afrika Zamani* 14/15 (1984), 147–163.

Chikeka, Charles O. *Britain, France, and the New Africa State.* New York: Edwin Mellen Press, 1990.

Chinje, Eric. "The Media in Emerging African Democracies: Power, Politics and the Role of the Press," *The Fletcher Forum of World Affairs* 17,1 (1993), 49–65.

Chipman, John. *French Power in Africa.* Oxford: Basil Blackwell, 1989.

Clark, John. "The Constraints on Democracy in sub-Saharan Africa: The Case for Limited Democracy" *SAIS Review* (Summer-Fall 1994), 91–108.

Collectif. *Changer le Cameroun: Pourquoi pas?.* Yaounde: C3, 1990.

———. *Le Cameroun Eclaté?* Yaounde: C3, 1992.

———. *Le 11 Octobre, 1992.* Yaounde: C3, 1993.

Collier, Ruth. "Political Change and Authoritarian Rule." In *Africa,* ed. Phyllis Martin and Patrick O'Meara, 295–309. Bloomington: Indiana University Press, 1977.

———. *Regimes in Tropical Africa: Changing Forms of Supremacy, 1945–1975.* Berkeley and Los Angeles: University of California Press, 1982.

Corbett, Edward. *The French Presence in Black Africa.* Washington, D.C.: Black Orpheus, 1972.

Crook, Richard. "Winning coalitions and ethno-regional politics: the failure of the opposition in the 1990 and 1995 elections in Côte d'Ivoire," *African Affairs* 96,383 (1997), 215–242.

Cumming, Gordon. "French Development Assistance to Africa: Towards a New Agenda?" *African Affairs* 94,376 (1995), 383–398.

Davidson, Basil. *Let Freedom Come: Africa in Modern History.* Boston: Little, Brown, 1978.

———. *The Black Man's Burden: Africa and the Curse of the Nation-State.* New York: Random House, 1992.

Davis, Lucy. "Opening Political Space in Cameroon: the Ambiguous Response of the Mbororo," *Review of African Political Economy* 64 (1995), 213–228.

Decalo, Samuel. *Psychoses of Power, African Personal Dictatorships.* Boulder: Westview Press, 1989.

———. "The Process, Prospects and Constraints of Democratization in Africa," *African Affairs,* 91,362 (1992), 7–35.

Decraene, Philippe. "Cameroun: L'Irresistible Ascension de M. Biya," *L'Afrique et l'Asie Modernes* 138 (1983), 3–11.

DeLancey, Mark. "Credit for the Common Man in Cameroon," *Journal of Modern African Studies,* 15,2 (1977), 316–322.

———. "The Construction of the Cameroon Political System: The Ahidjo Years, 1958–1982," *Journal of Contemporary African Studies*, 6,1–2 (1987), 3–24.
———. *Cameroon: Dependence and Independence*. Boulder: Westview Press, 1989.
Diamond, Larry. "Introduction: Political Culture and Democracy." In *Political Culture and Democracy in Developing Countries*, ed. Larry Diamond, 1–33. Boulder: Lynne Rienner Publishers, 1993.
———. *Prospects for Democratic Development in Africa*. Palo Alto: Stanford University, Hoover Institution, 1997.
Diamond, Larry, Juan Linz and Seymour Lipset, eds., *Democracy in Developing Countries*, Volume 2. Boulder: Lynne Rienner Publishers, 1984.
Dinka, Fongum Gorji. "Pour un nouveau contrat social," *Peuples Noirs-Peuples Africains* 50 (1986): 50–65.
Dodge, Dorothy. *African Politics in Perspective*. Princeton: Van Nostrand, 1966.
Dongmo, Jean-Louis. *Le dynamisme bamileke*. Yaounde: CEPER, 1981.
Duignan, Peter and Robert Jackson, eds. *Politics and Government in African States*. Palo Alto: Stanford University, Hoover Institution Press, 1986.
Dumont, René. *L'Afrique noir est mal partie*. Paris: Editions du Seuil, 1962.
Dunn, John, ed. *West African States: Failure and Promise*. Cambridge: Cambridge University Press, 1978.
Ecole Supérieure des Sciences et Techniques et de l'Information (ESSTI). *Paul Biya, 5 Ans Après... Les Camerounais Jugent Leur President*. Yaounde: Impression ESSTI, 1987.
Ejedepang-Koge, N. S. *Change in Cameroon*. Alexandria, Virginia: ARC Publications, 1985.
Ekeh, Peter. "Colonialism and the Two Publics in Africa: A Theoretical Statement," *Comparative Studies in Society and History*, 17,1 (1975), 91–112.
Ela, Jean-Marc. *Quand l'Etat pénètre en brousse... Les ripostes paysannes à la crise*. Paris: Karthala, 1990.
———. *Afrique: L'irruption des pauvres: Société contre ingérence, pouvoir et argent*. Paris: L'Harmattan, 1994.
Emerson, Rupert and Martin Kilson, eds. *The Political Awakening of Africa*. Englewood Cliffs, New Jersey: Prentice-Hall, 1965.
Emily, Florent, et al. *Paul Biya ou l'incarnations de la rigueur*. Yaounde: Editions Sopecam, 1983.
Epale, Simon-Joseph. *Plantations and Development in Western Cameroon 1885–1975: A Study in Agrarian Capitalism*. New York: Vantage Press, 1985.
Etonga, Mbu. "An Imperial Presidency: A Study of Presidential Power in Cameroon." In *An African Experiment in Nation Building: The Bilingual Cameroon Republic Since Reunification*, ed. Ndiva Kofele-Kale, 133–157. Boulder: Westview Press, 1980.
Eyinga, Abel. *Introduction à la politique camerounaise*. Paris: L'Harmattan, 1984.
Eyoh, Dickson. "From Economic Crisis to Political Liberalization: Pitfalls of the New Political Sociology for Africa," *African Studies Review* 39, 3 (1996), 43–80.
Eyongetah, Tambi, Robert Brain and Robin Palmer. *A History of the Cameroon*. London: Longman, 1987.
Falola, Toyin, and Julius Ihonvbere. *The Rise & Fall of Nigeria's Second Republic: 1979–84*. London: Zed Books, 1985.

Fatton, Jr., Robert. *Predatory Rule: State and Civil Society in Africa*. Boulder: Lynne Rienner Publishing, 1992.
———. "Africa in the Age of Democratization: The Civic Limitations of Civil Society," *African Studies Review*, 38,2 (1995), 67–99.
Femia, Joseph V. *Gramsci's Political Thought: Hegemony, Consciousness, and the Revolutionary Process*. Oxford: Clarendon Press, 1981.
Fisiy, Cyprian. "Chieftaincy in the Modern State: An Institution at the Crossroads of Democratic Change," *Paideuma* 41 (1995), 49–63.
Fodouop, Kengne. "Le secteur informel dans le contexte des ajustements au Cameroun: L'exemple de la vente à la sauvette," *Labour, Capital and Society* 26,1 (1993), 42–61.
Fombad, Charles. "Freedom of Expression in the Cameroonian Democratic Transition," *The Journal of Modern African Studies* 33,2 (1995), 211–226.
Fondation Friedrich Ebert. "Les Sensibilités Politiques des Camerounais," October 1994.
Forje, John W. *The One and Indivisible Cameroon: Political Integration and Socio-Economic Development in a Fragmented Society*. Lund: University of Lund, 1981.
Fotso, Victor. *Tout pour la Gloire de mon Pays*. Yaoundé: CEPER, n.d.
———. *Le Chemin de Hiala*. Paris: Editions de Septembre, 1994.
Friedland, William H. and Carl G. Rosberg, Jr., eds. *African Socialism*. Palo Alto: Stanford University Press, 1964.
Frimpong-Ansah, Jonathan. *The Vampire State in Africa: The Political Economy of Decline*. Trenton: Africa World Press, 1992.
Fru Doh, Emmanuel. "Women, Events and the Revitalization of Culture: Takumbeng as a Mankon Cultural Phenomenon in the Bamenda Grassfields." Paper presented at the University of Buea Conference on Cameroonian Writing, November 30–December 4, 1994.
Fru Ndi, John. *General Policy Address* to the Second Ordinary Convention of the Social Democratic Front, Bafoussam, 1993.
Gaillard, Philippe. *Ahmadou Ahidjo (1922–1989)*. Paris: Groupe Jeune Afrique, 1994.
Gardinier, David E. *Cameroon: United Nations Challenge To French Policy*. London: Oxford University Press, 1963.
General Agreement on Tariffs and Trade. *Trade Policy Review: Cameroon, 1995*. Geneva, May 1995.
Geschiere, Peter. "Hegemonic Regimes and Popular Protest-Bayart, Gramsci and the State in Cameroon (1)." In *State and Local Community in Africa*, ed. Wim van Binsbergen, Filip Reyntjens, and Gerti Hesseling, 309–347. Brussels: Centre d'Etude et de Documentation Africaines, Cahier 2–4, 1986.
———. *The Modernity of Witchcraft: Politics and the Occult in Postcolonial Africa*. Translated by Peter Geschiere and Janet Roitman. Charlottesville: The University Press of Virginia, 1997. Originally published as *Sorcellerie et Politique en Afrique—La viande des autres*. Paris: Karthala, 1995.
Geschiere Peter and Piet Konings, eds. *Pathways to Accumulation in Cameroon*. Paris: Karthala; Leiden: Afrika-Studiecentrum, 1993.
Gobata, Rotcod. *The Past Tense of Shit*. Limbe: Nooremac Press, 1993.

Goheen, Miriam. *Men Own the Fields, Women Own the Crops: Gender and Power in the Cameroon Grassfields.* Madison: The University of Wisconsin Press, 1996.
Gros, Jean-Germain "The Hard Lessons of Cameroon," *Journal of Democracy* 6,3 (1995), 112–127.
Guie, Honoré, "Organizing Africa's Democrats," *Journal of Democracy*, 4,2 (1993), 119–123.
Gutkind, Peter, and Immanuel Wallerstein, eds. *The Political Economy of Contemporary Africa.* Beverley Hills: Sage Publications, 1976.
Hall, John, ed. *Civil Society: Theory, History, Comparison.* Cambridge: Polity Press, 1995.
Hayward, Fred, ed. *Elections in Independent Africa.* Boulder: Westview Press, 1987.
Heilbrunn, John. "Social Origins of the National Conferences in Benin and Togo," *The Journal of Modern African Studies,* 31,2 (1993), 277–299.
Hermet, Guy, Richard Rose, and Alain Rouquie, eds. *Elections Without Choice.* New York: John Wiley and Sons, 1978.
Hughes, Arnold and Roy May. "The Politics of Succession in Black Africa," *Third World Quarterly* 10,1 (January 1988), 1–22.
Hugon, Philippe. *Analyse du Sous-Développement en Afrique Noir.* Paris: Presses Universitaires de France, 1968.
Hyden, Goran. *Beyond Ujamaa in Tanzania: Underdevelopment and an Uncaptured Peasantry.* Berkeley and Los Angeles: University of California Press, 1980.
———. *No Shortcuts to Progress: African Development Management in Perspective.* Berkeley and Los Angeles: University of California Press, 1983.
Hyden, Goran, and Colin Leys. "Elections and Politics in Single-party Systems: The Case of Kenya and Tanzania," *British Journal of Political Science* 2, 4 (1972), 389–420.
Ihonvbere, Julius. "The Third World and the New World Order in the 1990s," *Futures* (December 1992), 987–1002.
———. "The State, Human Rights and Democratization in Africa," *International Issues* 37,4 (1994), 59–80.
———. "Elections and Conflicts in Nigeria's Nontransition to Democracy," *Africa Dēmos* III,5 (1996), 8–11.
———. "On the Threshold of Another False Start? A Critical Evaluation of Prodemocracy Movements in Africa," *Journal of Asian and African Studies* XXXI, 1–2 (1996), 125–142.
Jackson, Robert, and Carl Rosberg, Jr., eds. *Personal Rule in Black Africa: Prince, Autocrat, Prophet, Tyrant.* Berkeley and Los Angeles: University of California Press, 1982.
Johnson, Willard. "The Cameroon Federation: Political Union between English and French-speaking Africa." In *French-Speaking Africa: the Search for Identity,* ed. William Louis, 205–220. New York: Walker, 1965.
———. "The Unions des Populations du Cameroun in Rebellion: The Integrative Backlash of Insurgency." In *Protest and Power in Black Africa 1886–1966,* ed. Robert Rotberg and Ali Mazrui, 671–692. New York and London: Oxford University Press, 1970.

———. *The Cameroon Federation: Political Integration in a Fragmented Society*. Princeton: Princeton University Press, 1970.
Joseph, Richard. "Ruben Um Nyobe and the 'Kamerun' Rebellion," *African Affairs* 73,293 (1974), 428–444.
———. "National Politics in Post-war Cameroon: The Difficult Birth of the U.P.C.," *Journal of African Studies* 2,2 (1977), 201–239.
———. *Radical Nationalism in Cameroun: Social Origins of the U.P.C. Rebellion*. Oxford: Oxford University Press, 1977.
———. "Radical Nationalism in French Africa: The Case of Cameroon." In *Decolonization and African Independence: The Transfer of Power*, ed. Prosser Gifford and William Louis, 321–345. New Haven: Yale University Press, 1988.
Joseph, Richard, ed. *Gaullist Africa: Cameroon Under Ahmadu Ahidjo*. Enugu: Fourth Dimension, 1978.
Jua, Nantang. "UDEAC: Dream, Reality or the Making of Sub-imperial States," *Afrika Spectrum* 21, 2 (1986): 211–223.
———. "Cameroon: Jump-starting an Economic Crisis," *Africa Insight* 21,3 (1991), 162–170.
———. "State, Oil and Accumulation." In *Pathways to Accumulation in Cameroon*, ed. Peter Geschiere and Piet Konings, 131–159. Paris: Karthala; Leiden: Afrika Studiecentrum, 1993.
Kale, Peter. *Political Evolution in the Cameroons*. Buea: n.p., 1967.
Kamga, Victor. *Duel Camerounais: Démocratie ou Barbarie*. Paris: L'Harmattan, 1979.
Kofele-Kale, Ndiva. "Cameroon and its Foreign Relations," *African Affairs* 80,319 (1981), 197–217.
———. *Tribesmen and Patriots: Political Culture in a Poly-ethnic African State*. Washington, D.C.: University Press of America, 1981.
———. "Ethnicity, Regionalism, and Political Power: A Post-Mortem of Ahidjo's Cameroon." In *The Political Economy of Cameroon*, ed. Michael Schatzberg and William Zartman, 53–82. New York: Praeger, 1986.
———. "Class, Status, and Power in Postreunification Cameroon: the Rise of an Anglophone Bourgeoisie 1961–1980." In *Studies in Power and Class in Africa*, ed. Irving Markovitz, 135–169. New York and Oxford: Oxford University Press, 1987.
Kofele-Kale, Ndiva, ed. *An African Experiment in Nation Building: The Bilingual Cameroon Republic Since Reunification*. Boulder, Colorado: Westview Press, 1980.
Kom, Ambroise. "Writing Under A Monocracy," *Research in African Literatures* 22,1 (1991), 83–92.
Kom, David. *Le Cameroun: essai d'analyse économique et politique*. Paris: Editions Sociales, 1971.
Konings, Piet. "L'Etat, agro-industrie et la paysannerie au Cameroun," *Politique Africaine* 22 (1986), 120–137.
———. *Labour Resistance in Cameroon: managerial strategies and labour resistance in the agro-industrial plantations of the Cameroon Development Corporation*. London: James Currey; Portsmouth, N.H.: Heinemann, 1993.
Kraus, Jon. "Building Democracy in Africa," *Current History* 90,556 (1991), 209–218.

Krieger, Judith. "Women, Men, and Household Food in Cameroon." Ph.D Dissertation, Department of Anthropology, University of Kentucky, 1994.

Krieger, Milton. "Cameroon's Democratic Crossroads, 1990–1994," *The Journal of Modern African Studies* 32,4 (1994), 605–628.

———. "Building the Republic through Letters: *Abbia: Cameroon Cultural Review*, and Its Legacy," *Research in African Literatures* 27,2 (1996), 155–177.

Kuissu, Siméon. "Le Kamerun à la Croisée des Chemins," *Journal of African Marxists*, 9 (1986), 24–40.

Kuoh, Manga. *Cameroun: un nouveau départ*. Paris: L'Harmattan, 1996.

Lancaster, Carol. "Democracy in Africa," *Foreign Policy*, 85 (1991/92), 148–165.

Lemarchand, René. "Uncivil States and Civil Societies: How Illusion Became Reality," *The Journal of Modern African Studies* 30,2 (1992), 177–191.

Le Vine, Victor. "The Cameroon Federal Republic." In *Five African States*, ed. Gwendolen Carter, 263–360. Ithaca: Cornell University Press, 1963.

———. *The Cameroons: From Mandate to Independence*. Berkeley and Los Angeles: University of California Press, 1964

———. "The Politics of Partition in Africa: The Cameroons and the Myth of Unification," *Journal of International Affairs* XVIII,2 (1964), 198–210.

———. "Cameroon Political Parties." In *Political Parties and National Integration in Tropical Africa*, ed. James Coleman and Carl Rosberg, Jr., 132–184. Berkeley and Los Angeles: University of California Press, 1964.

———. *The Cameroon Federal Republic*. Ithaca: Cornell University Press, 1971.

———. "The Politics of Presidential Succession," *Africa Report* 28, 3 (1983), 22–26.

———. "Leadership and Regime Changes in Perspective." In *The Political Economy of Cameroon*, ed. Michael Schatzberg and William Zartman, 20–52. New York: Praeger, 1986.

Liebenow, Gus. *African Politics: Crisis and Challenges*. Bloomington: Indiana University Press, 1986.

Linz, Juan, and Alfred Stepan, *Problems of Democratic Transition and Consolidation: Southern Europe, South America, and Post-Communist Europe*. Baltimore: Johns Hopkins University Press, 1996.

Lippens, Philippe, and Richard Joseph. "The Power and the People." In *Gaullist Africa: Cameroon Under Ahmadu Ahidjo*, ed. Richard Joseph, 111–126. Enugu, Nigeria: Fourth Dimension, 1978.

Logo, Patrice Bigombe, and Hèlène-Laure Menthong. "Crise de légitimité et évidence de la continuité politique," *Politique Africaine* 62 (1996), 15–24.

Lokanga, Mujos Lokula. "Precipitated Elections: The Biya Drug," *Le Messager*, August 14, 1991.

Lusignan, Guy de. *French-Speaking Africa Since Independence*. New York: Praeger Publishers, 1969.

MacGaffey, Janet. *The Real Economy of Zaire: The Contribution of Smuggling and Other Unofficial Activities to National Wealth*. Philadelphia: University of Pennsylvania Press, 1991.

Manor, James, ed. *Rethinking Third World Politics*. London: Longman, 1991.

Martin, Guy. "Continuity and Change in Franco-African Relations," *The Journal of Modern African Studies* 33,1 (1995), 1–20.

Mbaku, John Mukum. "Effective Constitutional Discourse as an Important First Step to Democratization in Africa," *Journal of Asian and African Studies* XXXI, 1–2 (1996), 39–51.

Mbarga, Emile. *Les Institutions Politiques du Cameroun*. Yaounde: Presses de l'Imprimerie Nationale, 1982.

Mbembe, Achille. *Les jeunes et l'ordre politique en Afrique noire*. Paris: L'Harmattan, 1985.

———. "La Palabre de l'indépendance: les ordres du discours nationaliste au Cameroun (1948–1958)," *Revue Française de Science Politique* 35, 3 (1985), 459–487.

———. "Pouvoir des morts et langage des vivants: les errances de la mémoire nationaliste au Cameroun," *Politique Africaine*, 22 (1986), 37–72.

———. "Pouvoir, violence et accumulation," *Politique Africaine*, 32 (1990), 7–25.

———. "Provisional Notes on the Postcolony," *Africa* 61, 1 (1992), 3–37.

———. "La violence derrière le multipartisme," *Afrique Magazine* 97 (November 1992), 56–60.

———. "Crise de légitimité, restauration autoritaire et déliquescence de l'Etat." In *Pathways to Accumulation in Cameroon*, ed. Peter Geschiere and Piet Konings, 345–374. Paris: Karthala; Leiden: Afrika-Studiecentrum, 1993.

———. "La 'chose' et ses doubles dans la caricature camerounaise, *Cahiers d'Etudes Africaines* 36 141/142 (1996), 143–70.

Mbembe, Achille, ed. Ruben um Nyobe, *Le Problème National Kamerunais*. Paris: L'Harmattan, 1984.

Mbembe, Achille and Janet Roitman. "Figures of the Subject in Times of Crisis," *Public Culture* 7 (1995), 323–352.

Mbuyinga, Elenga. *Tribalisme et Problème National en Afrique Noire: Le cas du Kamerun*. Paris: L'Harmattan, 1989.

Mendo Ze, Gervais. *Pour un Multipartisme Réfléchi en Afrique Noire: Le cas du Cameroun*. Yaounde: Editions Gaps/Gideppe, 1990.

Mentang, Tatah. "New Deal Electoral Politics in Cameroon: A Political Analysis," *ESSTI* 3 (1986).

Migdal, Joel S. *Strong Societies and Weak States: State-Society Relations and State Capabilities in the Third World*. Princeton: Princeton University Press, 1988.

Monga, Célestin. *The Anthropology of Anger: Civil Society and Democracy in Africa*. Translated by Linda Fleck and Célestin Monga. Boulder: Lynne Rienner Publishers, 1996. Originally published as *Anthropologie de là Colère*. Paris: L'Harmattan, 1994.

Moutard, G. "Quelles chances pour la politique du President Biya," *Afrique Contemporaine* 25,139 (1986), 20–35.

Muigai, Githu. "Kenya's Opposition and the Crisis of Governance," *Issue: A Journal of Opinion* 21,1–2 (1993), 26–43.

Mukong, Albert. *The Problem with the New Deal*. Bamenda, Cameroon: n.p., 1984.

———. *What is to be Done?*. Bamenda, Cameroon: n.p., 1985.

———. *Prisoner Without a Crime*. Bamenda, Cameroon: Alfresco, 1985.

———. *My Stewardship in the Cameroon Struggle*. Enugu, Nigeria: Chuka, 1992.

Bibliography

———. "Where Things Went Wrong." Paper presented to the Buea All Anglophone Conference, April 2–3, 1993.

———."Let's Keep Cameroon One." Paper presented to the Second Ordinary Convention of the Social Democratic Front, Bafoussam, 1993.

Mukong, Albert, ed. *The Case for the Southern Cameroons*. USA: CAMFECO, 1990.

Muna, Bernard. *Cameroon and the Challenges of the 21st Century*. Cameroon: Tama Books, 1993.

Mveng, Engelbert. *Histoire du Cameroun*. Paris: Présence Africaine, 1963.

Nabudere, Dani. "The One-Party State in Africa and its Assumed Philosophical Roots." In *Democracy and the One-Party State in Africa*, ed. Peter Meyns and Dani Wadada, 1–24. Hamburg: Institut Für Afrika-Kunde, 1989.

National Democratic Institute for International Affairs. *Déclaration Post-Electorale Préliminaire*. October 14, 1992.

———. *An Assessment of the October 11, 1992 Election in Cameroon*. Washington, D.C., 1993.

National Episcopal Conference of Cameroon (Catholic). *Pastoral Letter . . . Concerning the Economic Crisis*. June 3, 1990.

———. *Statement*. December 11, 1992.

———. *Final Communiqué*. December 15, 1995

National Union for Progress and Democracy. *Proposition de Loi Portant Revision de la Constitution*. 1995.

Ndifor, Akwanka. "A Political Turning Point," *Africa Report* 36, 4 (1991): 17–19.

Ndongko, Wilfred. "A Comparative Analysis of French and British Investment Policies in Cameroon," *Pan-African Journal* 7,2 (1974), 101–110.

———. *Planning for Economic Development in a Federal State: The Case of Cameroon, 1960–1971*. Munchen: Weltforum Verlag, 1975.

———. "The Financing of Economic Development in Cameroon," *Africa Development* 2,3 (1977), 59–76.

———."The Political Economy of Development in Cameroon: Relations Between the State, Indigenous Business, and Foreign Investors." In *The Political Economy of Cameroon*, ed. Michael Schatzberg and William Zartman, 83–110. New York: Praeger, 1986.

Ndongo, Jacques Fame. *Le Prince et le Scribe*. Paris: Berger-Levrault, 1988.

Nelson, Harold, et al. *Area Handbook For the United Republic of Cameroon*. Washington, D.C.: U.S. Government Printing Office, 1974.

Ngayap, Pierre Flambeau. *Cameroun: Qui Gouverne? De Ahidjo à Biya, l'héritage et l'enjeu*. Paris: L'Harmattan, 1983.

Ngoh, Victor. *Cameroon 1884–1985: A Hundred Years Of History*. Limbe: Navi Group Publication, 1988

———. *Constitutional Developments in Southern Cameroons, 1946–1961*. Yaounde: CEPER, 1990.

Ngu, Manfred. "Anglophone, Francophone: False and Real Matters About the Need of Democracy and Development in Kamerun," *Peuples Noirs-Peuples Africains* (January–August 1985), 150–173.

Ngugi wa Thiong'o. *Petals of Blood*. New York: Dutton, 1978.

Ngwana, A. S. "Cameroon Democratic Party," *Peuples Noirs-Peuples Africains* (January–August 1985): 174–190.
Nkwi, Paul. "Traditional female militancy in a modern context." In *Femmes du Cameroun*, ed. Jean-Claude Barbier, 181–191. Paris: Karthala-ORSTOM, 1985.
Nyamnjoh, Francis Beng. "Contrôle de l'information au Cameroun: Implication pour les recherches en communication," *Afrika Spectrum* 28,1 (1993), 93–115.
Nyang'oro, Julius E. "Reform Politics and the Democratization Process in Africa," *African Studies Review* 37,1 (1994), 133–149.
Nyerere, Julius K. *Freedom and Unity*. London: Oxford University Press, 1967.
Nzouankeu, Jacques. "The Role of the National Conference in the Transition to Democracy in Africa: The Cases of Benin and Mali," *Issue: A Journal of Opinion* 21,1–2 (1993), 44–50.
Ododa, Harry. "Voluntary Retirement by Presidents in Africa: Lessons From Sierra Leone, Tanzania, Cameroon, and Senegal," *Journal of African Studies* 15, 3/4 (1988), 94–99.
Osaghae, Eghosa. "The Study of Political Transitions In Africa," *Review of African Political Economy* 22,64 (1995), 183–197.
Oyono, Dieudonné. "Introduction à la politique africaine du Cameroun," *Mois en Afrique* 18,207–208 (1983), 21–30.
———. "Le coup d'état manqué du 6 avril 1984 et les engagements du politiques etrangère du Cameroun," *Mois en Afrique* 20,223–224 (1984/85), 48–56.
Pagès, Monique. "L'explosion de la presse en Afrique francophone au sud du Sahara," *Afrique Contemporaine* 159 (1992), 77–82.
Péan, Pierre. *L'argent noir: corruption st sous-développement*. Paris: Fayard, 1988.
Phillipson, Sir Sydney. *Report on the Financial, Economic and Administrative Consequences to Southern Cameroons of Separation from the Federation of Nigeria*. Prime Minister's Office, Southern Cameroons, 1959.
Republic of Cameroon. Ministry of Information and Culture. *The Cultural Identity of Cameroon*. Yaounde, 1985.
———. *Rights and Freedoms: Collection of Recent Texts*. Yaounde: SOPECAM, 1991.
———. Technical Committee on Constitutional Matters. *Preliminary Draft Constitution* ("The Owona Constitution"). May 17, 1993.
———. Consultative Committee on Constitutional Reform. *Proposals of the President of the Republic for Constitutional Reform* ("The Biya Constitution"). December 15, 1994.
Robinson, Pearl. "The National Conference Phenomenon in Francophone Africa," *Comparative Studies in Society and History* 36,3 (1994), 575–610.
Rodney, Walter. *How Europe Underdeveloped Africa*. Dar es Salaam: Tanzania Publishing House, 1974.
Rosberg, Carl, Jr., and Thomas Callaghy, eds. *Socialism in sub-Saharan Africa: A New Assessment*. Berkeley and Los Angeles: University of California Press, 1979.
Rowlands, Michael. "Accumulation and the Cultural Politics of Identity in the Grassfields." In *Pathways to Accumulation in Cameroon*, ed. Peter Geschiere and Piet Konings, 71–97. Paris: Karthala; Leiden: Afrika-Studiecentrum, 1993.
Rubin, Neville. *Cameroon: An African Federation*. London: Pall Mall Press, 1971.
Rudin, Harry. *Germans in the Cameroons*. New Haven: Archon, 1968.

Sabar-Friedman, Galia. "Church and State in Kenya, 1986–1992: The Churches' Involvement in the 'Game of Change'," *African Affairs* 96,382 (1997), 25–52.

Sam-Kubam, Patrick, and Richard Ngwa-Nyamboli, *Paul Biya and the Quest for Democracy in Cameroon*. Yaounde: Editions CLE, 1985.

Sandbrook, Richard. "Hobbled Leviathans: Constraints on State Formation in Africa," *International Journal* XLI,4 (1986), 707–733.

Schatzberg, Michael. "The Metaphors of Father and Family." In *The Political Economy of Cameroon*, ed. Michael Schatzberg and William Zartman, 1–19. New York: Praeger, 1986.

———. *The Dialectics of Oppression in Zaire*. Bloomington: Indiana University Press, 1988.

———. "Power, Legitimacy and 'Democratization' in Africa," *Africa* 63,4 (1993), 445–461.

Schatzberg, Michael, and William Zartman, eds. *The Political Economy of Cameroon*. New York: Praeger, 1986.

Schilder, Kees. "La Démocratie aux champs: les présidentielles d'octobre 1992 au Nord-Cameroun," *Politique Africaine* 50 (1993), 115–122.

———. *Quest for self-esteem: State, Islam and Mundang ethnicity in northern Cameroon*. Aldershot: Avebury, 1994.

Schissel, Howard. "Cameroon's Economy: Myth or Reality," *West Africa* (September 12–18, 1983), 2107–2108.

Schraeder, Peter. "From Berlin 1884 to 1989: Foreign Assistance and French, American and Japanese Competition in Francophone Africa," *The Journal of Modern African Studies* 33,4 (1995), 539–567.

Scott, Frederick. "Biya's New Deal," *Africa Report* 30, 4 (1985), 58–61.

Shanklin, Eugenia. "*Anlu* Remembered: the Kom Women's Rebellion of 1958–1961," *Dialectical Anthropology* 15 (1990), 159–181.

Sindjoun, Luc. "Cameroun: Le système politique face aux enjeux de la transition démocratique (1990–1993)." In *L'Afrique Politique 1994*. Paris: L'Harmattan, 1994.

———. "Le champ social camerounais," *Politique Africaine* 62 (1996), 57–67.

Sklar, Richard. "Democracy in Africa," *African Studies Review* 26,3/4 (1983), 11–24.

Social Democratic Front. *Resolutions* of the Second Ordinary Convention, Bafoussam, 1993.

———. *Resolutions* of the National Executive Committee, August 22, 1994.

———. *Proposed Constitution for a New Social Order in Cameroon*. December 16, 1994.

———. *Resolutions* of the Third Ordinary Convention, Maroua, 1995.

Stark, Frank. "Persuasion and Power in Cameroon," *Canadian Journal of African Studies* 14,2 (1980), 273–293.

———. "Federalism in Cameroon: The Shadow and the Reality." In *An African Experiment in Nation Building: The Bilingual Cameroon Republic Since Reunification*, ed. Ndiva Kofele-Kale, 101–132. Boulder: Westview Press, 1980.

Suret-Canale, Jean. *French Colonialism in Tropical Africa*. Translated by Till Gottheiner. London: C. Hurst & Company, 1964.

Syndicat National des Enseignants du Supérieur (SYNES). *The University in Cameroon: an institution in disarray*. Yaounde, 1992.

Takougang, Joseph. "The Post-Ahidjo Era in Cameroon: Continuity and Change," *Journal of Third World Studies* X,2 (1993), 268–302.

———. "The Demise of Biya's New Deal in Cameroon, 1982–1992," *Africa Insight* 23,2 (1993), 91–101.
Tamar, Golan. "A Certain Mystery: How can France do Everything that it does in Africa and get away with it?" *African Affairs* 80,318 (1981), 3–11.
Tangri, Roger. *Politics in Sub-Saharan Africa*. London: James Currey, 1985.
Tunteng. P-Kevin. "External Influences and Subimperialism in Francophone West Africa." In *The Political Economy of Contemporary Africa*, ed. Peter Gutkind and Immanuel Wallerstein, 212–231. Beverly Hills: Sage Publications, 1976.
U.S.A. Department of State. *Country Reports on Human Rights*.
van de Walle, Nicolas. "The Politics of Public Enterprise Reform in Cameroon." In *State-Owned Enterprises in Africa*, ed. Barbara Grosh and Rwekaza Mukandala, 151–174. Boulder: Lynne Rienner, 1994.
———. "Neopatrimonialism and Democracy in Africa, with an illustration from Cameroon." In *Economic Change and Political Liberation in sub-Saharan Africa*, ed. Jennifer Widner, 129–157. Baltimore: John Hopkins University Press, 1994.
Vansina, Jean. "Mwasi's Trials," *Daedalus* 111, 2 (1982), 49–72.
Vengroff, Richard. "Governance and the Transition to Democracy: Political Parties and the Party System in Mali," *The Journal of Modern African Studies* 31,4 (1993), 541–562.
Wallerstein, Immanuel. *The Modern World System*. New York: Academic Press, 1974.
Warnier, Jean-Pierre. *L'Esprit d'entreprise au Cameroun*. Paris: Karthala, 1993.
Weiss, Thomas. "Migrations et conflits frontaliers: une rélation Nigeria-Cameroun contrariée," *Afrique Contemporaine* Numéro spécial (1996), 39–51.
Welch, Claude, Jr. *Dream of Unity: Pan-Africanism and Political Unification in West Africa*. Ithaca: Cornell University Press, 1966.
Wells, F. A. and W. A. Warmington, *Studies in Industralization: Nigeria and the Cameroons*. London: Oxford University Press, 1962.
Wiseman, John A. *Democracy in Black Africa*. New York: Paragon House Publishers, 1990.
Wongibe, Edwin. *The Social Democratic Front and the Thorny Road to Social Justice*. Bamenda: n.p., 1991.
Wonyu, Eugène. *De L'U.P.C. à L'U.C.: Témoignage à l'aube de l'independance (1953–1961)*. Paris: L'Harmattan, 1985.
World Bank. *Cameroon: Diversity, Growth, and Poverty Reduction*. Report No. 13167-CM, April 4, 1995.
World Press Review. New York: Stanley Foundation, 1980–.
Woungly Massaga, Ngouo. *Où va le Kamerun?*. Paris: L'Harmattan, 1984.
———. *Combat Pour La Démocratie*. Yaounde: Imprimerie Saint-Paul, 1993.
Zartman, William, ed. *Collapsed States: The Disintegration and Restoration of Legitimate Authority*. Boulder: Lynne Rienner, 1995.
Zolberg, Aristide. *One-Party Government in the Ivory Coast*. Princeton: Princeton University Press, 1969.

Newspapers and Magazines

Africa
Africa Concord

Bibliography

Africa Confidential
Africa Dēmos
Africa News
African Recorder
Africa Report
Africa Watch
Afrique Magazine
L'Aurore
Cameroon Analysis
Cameroon Express
Cameroon Life
Cameroon Monitor
CamerooNow
Cameroon Post
Cameroon Tribune
La Caravane
CDU Voice
Challenge Hebdo
Challenge Mensuel
Challenge Nouveau
Le Combattant
Le Courrier de la Semaine
La Détente
Dikalo
L'Effort Camerounais
Galaxie
La Gazette
Génération
The Herald
Jeune Afrique
Jeune Afrique Economie
Marchés Tropicaux et Méditerranéens
Le Messager
Le Monde
New African
The New Nation
Newschampion
Newslink
The New Standard
New York Times
La Nouvelle Expression
NUDP News
OCALIP Info
Le Progrès
Le Quotidien
SDF News
The Sketch

The Star Headlines
Sunday Report
Time Magazine
La Vision Hebdo
Wall Street Journal
Washington Post
Washington Times
West Africa

Index

AAC-SCNC, 168–169
Abada, Marcelin Nguele, 194
Abasi, Awal, 74
Abdoulaye, Maikano, 65, 70–71
Achebe, Chinua, 1, 2, 19, 21
Achu, George Mofor, 151, 220
Achu, Simon Achidi, 145–146, 168, 191, 198–201, 218, 228
Adamawa Province, 72, 135, 137–138, 150, 169, 176, 200, 234
Advanced Democracy, 109, 115–116
Advanced School of Mass Communication, 118
African Studies, 14
Afro-narcotism, 16
Ahidjo, Ahmadou, 3, 4, 10, 22, 23, 24, 35–58, 60 (n 64), 80, 90, 97–98, 102, 104–105, 119, 135, 145, 168, 198–207, 243
    administration, 94, 107
    alliance, 107
    and Biya, 140
    and elite alliance, 28
    and France, 3
    and maquis, 22
    and patronage, 52
    appointed prime minister, 37
    appointed vice-prime minister, 37
    as French puppet, 65
    authoritarian rule, 115
    awarded Dag Hammarsjoëld Peace Price, 69
    critics, 23
    defeat of the UPC, 22
    effigies, 128
    Father of the Nation, 67, 71
    French puppet, 65
    Fulbe, 35
    legacy, 173
    loyalists, 4, 75
    Northern Muslim, 35
    presidency, 132
    regime, 230
    relinquishes power, 63–76
    replaces Jua, 48
    resignation as chairman of CNU, 73
Akame, Fouman, 83
Ake, Claude, 15
Akwa, Dika, 132
All Anglophone Conference (AAC), 163–164, 167, 183, 189, 203, 225
Alliance for the Reconstruction of Cameroon-Sovereign National Conference (ARC-SNC), 142–143, 147–148
Algeria, 35
ALUCAM. See Société Aluminium du Cameroun,
Amin, Idi, 14, 55
Amin, Samir, 15
Amity Bank, 235, 238
Amnesty International, 105, 139
Anglophone, 6, 49, 50, 52, 75, 91, 94–96, 107, 132, 137, 144, 163–165, 169, 189–190, 199, 202–203, 224–225, 242–244
    and petroleum, 4
    cameroonians, 94
    council, 166
    critics, 225
    elites, 229
    federalist movement, 161

frustrations, 143
leaders, 168
private press, 121
province, 95, 108
Anglo-Saxon, 166, 187
Anlu, 232
Anta Diop, Cheikh, 127
Anyang, Luc, 71
Anyangwe, Carlson, 167, 183
Asanga, Siga, 202–203
Asonganyi, Tazoacha, 204, 220
Assale Charles, 43, 54
Association of African Jurists, 92
Ateba, Koko, 92
Aujoulat, Louis-Paul, 26, 243
Aurillac, Michel, 141
Aurore Plus, 127
Austin, Ralph, 182
Avant-Projet de Constitution du Cameroun, 186
Awa, Furu, 201, 218, 232
Ayaba Ward, 116

Badjika, Mohamadou Ahidjo, 177, 198
Bafoussam, 116, 128, 130, 137–138, 144, 150, 164–165, 202
Bafut, 222
Baghdad, 238
Baha'i, 225
Bakassi peninsula, 206, 227
Bakweri, 94
Balandier, Georges, 18
Bali, 78
Bali-Nyonga, 221–222, 234
Bamenda, 6, 7, 9, 28, 45, 58, 78, 80, 105–107, 115–116, 124–125, 128–130, 134–136, 138–139, 142–144, 147, 168, 196–204, 238, 242, 244
  capital of North West Province, 5, 28
  congress, 79, 83, 107, 240
  New Deal congress, 108
  proclamation, 166–167, 225
  rally, 144
  resistance, 152
  six, 105
Bamileke, 22, 57, 72, 94, 96, 119, 135–136, 139, 171, 180–182, 203, 221, 243
  businessmen, 100
  ethnicity, 136
  region, 38, 39
Bamoun Division, 42
Bandolo, Henri, 68, 86 (n 23)
Bank of Credit and Commerce, 112 (n 47)
Banque Camerounaise de Développment (BCD), 100
Banque des Etats d'Afrique Centrale, 172
Barkindo, Haddabi, 171
Basil, Emah, 94, 96
Bassa, 94, 139
Batanga, 182
Bayart, Jean-François, 1, 16–29, 50, 66, 77, 82, 127, 136, 174, 182 218, 236
  and binary thought, 17
  and ethnofascism, 94
  and pathological, 17
  and politics of the belly, 1, 19, 22, 26
  and recherche hégémonique, 18, 25, 41, 79, 84, 115, 230
Bayas, 94
BBC Television, 139
Bebey-Eyidi, Marcel, 43
Bedzigui, Célestin, 182
Bejem, Julienne, 112 (n 65)
Bell, Luc René, 139, 168
Benin, 5, 20, 123, 127
  and Sovereign National Conference, 20
Benjamin, Jacques, 23, 169
BEPC, 235
Betchel, William, 40
Beti, 65, 71, 94–96, 100, 107, 124–125, 127, 139, 145, 171, 181–182, 227, 243
  irredentism, 181

Beti, Mongo, 21, 23, 24, 32 (n 34), 77, 232, 241
Biafra, 54, 206
Bill of Rights, 190, 192
Biya, Jean-Irène, 111 (n 18)
Biya, Paul, 4–10, 24, 26, 63–85, 90–95, 97–98, 103, 107–108, 115, 118, 121, 124, 127, 131, 133, 137–138, 141, 145, 148–151, 171, 184, 195, 241–243
   Ahidjo's choice for president, 65
   and Beti, 119
   and Communal Liberalism, 76, 225
   and Grand Débat, 202
   and New Deal, 84, 95, 99, 100–101
   appointed prime minister, 65
   bestowed as Fon of Fons, 78
   critics, 102
   early reforms, 89
   elected CNU Chairman, 73
   election as vice president of the CNU, 66
   Mitterrand's best pupils, 109
   southern Christian, 65
   struggle with Ahidjo, 69–71
   successor to Ahidjo, 63–66
Biyidi, Alexandre, 23. See also Beti, Mongo
Bjornson, Richard, 24
Black, Yondo. See Black, Yondo Mandengue
Black, Yondo Mandengue, 103–104, 113 (n 70), 116, 124, 162, 202, 227
   and SDF, 113 (n 73)
Bloc Démocratique Camerounais, 26
Boh, Herbert, 120
Bokassa, 14
Bongo, Omar, 73
Boston Bank, 100
Botswana, 3, 4, 8
Bouba, Bello Maïgari, 66, 67, 71, 107, 135, 143, 145, 147–148, 150, 160–161, 173, 174, 177, 184, 192, 199, 206–207, 220, 228
   as Biya's first prime minister, 132
   exile in Nigeria, 132

Boulou, 94
Brasseries du Cameroun, 57, 129
Bratton, Michael, 20, 76
Brazil, 7
Bretton Woods, 199
Brigades Mixtes Mobiles (BMM), 54, 55, 90
Briqueterie, 222
Britain, 83
British, 16
   administration, 35
   Cameroon, 41
   rule, 236
   Southern Cameroons, 35, 162
   trust territory, 4
British Broadcasting Corporation, 92
Buea, 55, 58, 163–165, 168, 183, 189, 236
   declaration, 164, 166
Burkina Faso, 89
Burnham, Philip, 8, 173, 176, 234, 245
Burundi, 245
Bwanga, Rudolph, 112 (n 65)

CAM-AAC meetings, 163
Cameroon Airlines (CAMAIR), 75, 174
Cameroon Anglophone Movement (CAM), 143, 162
   first national convention, 163
Cameroon Bank, 99–100
Cameroon Baptist Convention, 225
Cameroon Bar Association (CBA), 103
Cameroon Calling, 195
Cameroon Democratic Front (CDF), 107
Cameroon Democratic Party, 83
Cameroon Democratic Union (CDU), 132, 135–136, 141, 143, 145, 184, 192–193, 200, 207, 220, 223, 243,
   first Bafoussam rally, 135
   manifesto, 134
   rally in Bamenda, 135
Cameroon Development Corporation (CDC), 79, 175, 199
Cameroon Federalist Committee, 162
Cameroon Life, 82

Cameroonian, 3–9, 28, 36, 50, 54, 76, 78, 82–85, 90, 93–94, 102, 110, 121, 130, 140, 199, 245
  authorities, 35
  disenchanted, 89
  economy, 56
  misery for, 3
  politicians, 35
Cameroon National Union (CNU), 3, 4, 24, 26, 46, 58, 65, 66, 68, 69, 71, 79–80, 94, 108, 135, 168, 220
  central committee, 50, 65–71, 75–76, 78, 80
  change to CPDM, 79
  congress, 64, 78
  creation of, 3
  extraordinary session of, 73
  political bureau, 50, 52, 66, 75, 78
  veterans, 93
Cameroon People's Democratic Movement (CPDM), 4, 8, 26, 80, 89, 105–106, 115–116, 120–121, 123, 125, 131, 135–137, 138, 142–143, 145, 147, 150–151, 161–162, 180–182, 194–208, 240, 242, 244
  central committee, 79–83, 93, 108, 221
  congress, 133
  extraordinary congress, 195
  first national congress, 108
  fortress, 140
  governance, 116
  increase in membership, 108
  membership, 81
  militants, 104
  political bureau, 81, 84, 93, 113 (n 82)
  strongholds, 150
Cameroon People's National Convention (CPNC), 44, 46
Cameroon Post, 117, 143, 149, 208
Cameroon Press and Publishing Company (SOPECAM), 91, 122
Cameroon Public Service Union (CAPSU), 226

Cameroon Radio and Television (CRTV), 115, 117, 127, 137, 140–141, 146, 149, 183
Cameroon Report, 91
Cameroon's Flag, 144
Cameroon's Intelligence Service, 93
Cameroon Times, 91
Cameroon Tribune, 7, 68, 77, 83, 106, 117–119, 123, 137–138, 217, 141–142, 144, 149, 167
Cameroon United Congress, 45
Canada, 232
Castro, Fidel, 217
Catholic Bishops, 106
Catholic Church, 40, 113 (n 81), 117, 222
Center-South, 72
Central Africa, 7
Central African Republic, 7
Centre National de Documentation (CND), 54, 90
Centre National des Etudes et des Recherches (CENER), 93, 104
Cercle de Réflexion et d'Action pour le Triomphe du Renouveau (CRATRE), 181
Césaire, Aimé, 127
CFA franc, 6, 7, 44, 54, 56, 57, 97–101, 123, 131, 241
  devaluation, 122, 169, 243
Chad, 7, 181
Chad-Cameroon, 241
Chaillot Declaration, 103
Chaillot Palace, 103
Challenge Hebdo, 122, 127, 141
Chase Manhattan, 100
Chazan, Naomi, 20
Cheysson, Claude, 68
Chia, Felix, 239
China, 93
Chirac, Jacques, 99
Christian, 94, 107
Clinton, 209 (n 2)
Cold War, 85, 90, 102
Commercial Avenue, 105, 138, 148
Communal Liberalism, 76, 89, 115

Communism, 102
Compagnie Française pour le Développement des Fibres Textiles (CFDT), 169, 174
Congo (Kinshasa), 3, 8, 19, 27, 208, 228, 245
  clandestine economies, 19
  patron-client networks, 19
Congrès de la Maturité, 64
Congress of Freedom and Democracy, 108
Congress of the People, 189
Congressional Black Caucus, 161
Constitutional Council, 186, 193
Consultative Constitutional Committee, 195
Côte d'Ivoire, 4, 36, 68, 208
Courant des Forces Progressistes, 133
Croatia, 164
Crook, Richard, 208
CRTV. See Cameroon Radio and Television
Custom's Economic Union of Central Africa, 67

Daïssala, Dakolé, 200
Daouda, Youssoufa, 70
Daoudou, Sadou, 53, 65, 67, 70, 171
Davidson, Basil, 15, 16, 17, 31 (n 12), 229
  and inheritance elites, 15
Debré, Bernard, 172
De Gaulle, Charles, 23, 64
DeLancey, Mark, 26, 52
Démocrates Camerounais (DC), 42–44
D'Estaing, Giscard, 91
Dien Bien Phu, 35
Dikalo, 122, 172
Dinka, Fongum Gorji, 91, 162
Diouf, Abdou, 73, 89
Direction Général d'Etudes et de la Documentation (DIRDOC), 90
Djam-Yaya, Oumarou, 72
Djibril, Cavaye Yegue, 171
Djon Djon, Charles René, 112 (n 65), 113 (n 70)

Doe, 17
Donga and Mantung, 106
Douala, 5, 7, 49, 58, 64, 79, 81, 119, 126–128, 133, 135, 138–141, 144, 145, 150, 182, 201, 240
  seaport, 55
Douala-Mbanga railway, 49
Douala Ten, 104–105, 108
Douala-Yaounde railroad, 55
Doumba, Joseph Charles, 75–76, 78–80
Draft Constitution, 186
Dschang, 55
  University of, 116
Dutch Mill Hill Order, 222

East Cameroon, 44–54, 57
Eastern Europe, 85, 90, 102, 108
Ebogo, 113 (n 82)
Ebolowa, 44, 197
Eboua, Samuel, 67, 120, 135, 138, 143, 160, 198
Edea, 57
Edzoa, Titus, 111, (n 18), 113 (n 82), 181, 219
Egbe, Emmanuel Tabi, 75
Ekah-Nghaky, Nzo, 45
Ekane, Anicet, 112 (n 65), 113 (n 70)
Ekane, Denis, 79
Ekindi, Jean-Jacques, 81, 113 (n 82), 133, 138, 148, 219
Ekwe, Henriette, 112 (n 65)
Elad, Sam Ekontang, 167–168, 183, 192
Ella, Jean-Marc, 231–232
Emane, Benoît Asso'o, 110 (n 18), 113 (n 82), 227
Endeley, Emmanuel, 44, 45
English-speaking Cameroonians, 78, 94–95
  as second class citizens, 96
  continued marginalization of, 95
  students, 95, 96
Equatorial Guinea, 181
Eritrea, 169
Essingan, 181
Etats-Généraux, 123
Eton, 182

Etonga, Mbu, 51
Etoudi. See Etoudi Palace
Etoudi Palace, 118, 131, 174, 196, 204
Europe, 136, 208, 242
European and Caribbean banks, 99
Ewondo, 94, 182
Expression, 121
Expression Nouvelle, 121
Eyinga, Abel, 23, 24, 25, 32 (n 34)

Fako. See Fako Division
Fako Division, 79, 106, 150
Fang, 181
Fatton, Robert Jr., 16, 17, 19, 20, 27, 28, 33 (n 54), 151, 228, 238
 and civil society, 21
Federal Constitution, 41
Federal Coordinating Committee, 41
Federal Republic of Cameroon, 4, 22, 48, 238.
Federalism and Unitarism, 143
Federation and Separation, 143
Feko, Vinvent, 112 (n 65)
Fifth Development Plan, 97
Fisiy, Cyprian, 222
Foccart, Jacques, 40
Fochive, Jean, 93, 227–228
Fofe, Joseph, 79, 113 (n 82)
Fon, 78
Foncha, John Ngu, 45–53, 60 (n 43), 105, 167–168, 220
Foning, François, 181, 221
Fonlon, Bernard, 53, 237
Fotso, Victor, 181, 221
Foumban, 132, 145, 150, 160, 200, 208, 221, 236, 238
 conference, 163
 constitution, 50
 constitutional convention, 41
 Sultan of, 71
 Sultanate of, 132
Fourth Estate, 120
Franc zone, 97
France, 56, 57, 64–65, 68, 71–73, 83, 102, 103, 175, 183, 186, 194, 196, 242
 military intervention in Cameroon, 39
France d'Outre-mer, 36
Franco-African Summit, 103
Franco-Cameroonian Agreements, 38
Francophone, 96, 107, 137, 163–164, 189, 192, 202–203, 224, 238, 242
 Africa, 123
Free West Cameroon Movement, 162
Freedom of Mass Communication, 109, 117, 121
French, 35, 39, 199, 241
 administration, 35
 aid, 57
 assistance, 112 (n 49)
 authorities, 35–37
 Cameroon, 35
 heritage, 241
 high commissioner, 36
 influence, 37, 174
 investments, 38
 Ministry of Corporation, 227
 revolution, 24
 rule, 35
 speaking-Cameroonians, 94–95
 support, 35, 40
 territory, 96
 troops, 39
 zone, 35
Frenchness, 96
Front de Libération du Québec (FLQ), 23
Front of Allied for Change (FAC), 206
Front Populaire pour L'Unité et de la Paix (FPUP), 42–43
Fulbe, 35, 72, 145, 150, 173–174
 and Gbaya, 138
Fulbeization, 174, 179
Fulbe-Muslim, 135, 145, 180, 243
Fundamentalists, 225

Gabon, 7, 73, 97, 181
Galaxie, 121
Gang of Four, 76
Gardinier, David, 42

Index 277

Garoua, 68, 72, 93, 107, 117, 174, 177, 198, 226, 242
  congress, 184
Gassagay, Antar, 160
Gaullist, 226
  Africa, 27
  state, 41
  strategy, 44
Gbaya, 179
GCE. See General Certificate of Education
Gendarmarie, 71–73, 168, 190
General Agreement on Tariffs and Trade (GATT), 6, 9, 242
General Assembly of the Traditional Council, 237
General Certificate of Education (GCE), 95, 235
Génération, 119, 122, 174–178
Germany, 236, 242
Geschiere, Peter, 15, 241
Ghost Towns. See Villes Mortes
Goheen, Miriam, 232, 239
Golden Pen Award, 121
Gorbachev, Mikhail, 82
Gramsci, 14, 17
Gramscian, 226, 229
Grand Débat, 183, 191, 202
Grand North, 145, 169, 197, 200, 206
Grassfield, 128, 136, 221, 233, 236, 239
Great National Party, 43
Great North. See Grand North
Great West, 197, 205
Group Parlementaire, 185
Guinness, 57

Hadji, Garga Haman, 148, 207, 226
Hamani, Gabriel, 112 (n 65)
Haut-Nkam, 144
Hayatou, Sadou, 113 (n 82), 125, 141, 145, 171–177, 180, 220
Hirschman, Albert, 21
Hôtel des Députés, 132
Houphouët-Boigny, Félix, 4, 73, 241
House of Chiefs, 49
House of Commons, 103

Hurd, Douglas, 103
Hyden, Goran, 14, 17, 18, 19

Ibadan, 195
Ibrahim, Saleh, 73
Ihonvbere, Julius, 243
Indo-China, 240
Indomitable Lions, 7
Ingram, Joseph, 172
International Monetary Fund (IMF), 99, 101, 103, 194–195
  debt payment, 194
  debt renegotiation, 142
  structural adjustment program, 101
Intifada, 234
Islam, 150

Japan, 242
Jaspin, Lionel, 241
Jehovah's Witnesses, 54, 92
Jeune Afrique, 77, 218
Jeune Afrique Economie, 100, 221
Jeune Afrique Plus, 119
Jeunesse Démocratique du Cameroun, 39
Johnson, Willard, 22, 23, 41, 45, 47, 48
Joseph, Richard, 3, 24, 26, 27, 29, 39, 42, 77
Jua, Augustin, 45, 48, 53
Jua, Nantang, 97, 98

Kabila, Laurent, 228
Kamerun National Democratic Party (KNDP), 45
Kamga, Paul Fokam, 53
Kamga, Thomas, 61 (n 70)
Kamga, Victor, 43
Kenya, 2, 150, 208
Kenyans, 1
Kenya National Council of Churches, 225
Keutcha, Jean, 53
Kirdi, 174, 179
Kodock, Frédéric, 143, 145, 147, 160, 192, 198–200

Kofele-Kale, Ndiva, 32 (n 34), 46, 49, 93, 229
Kom, 78, 232
Kom, Ambroise, 140, 226
Kondo, Samuel, 79
Kouomegni, Augustin Kontchou, 124
Kribi, 181, 241
Kumba, 49, 168
Kumbo-Nso, 124, 127, 236–239, 242, 244
Kumbo Water Authority, 237
Kuo, François Sengat, 77–80, 113 (n 82), 133, 142, 148, 219, 225, 240
Kuo, Manga, 206
Kwa-Moutome, Francis, 112 (n 63)
Kwayeb, Enoch, 53

Laakam, 180–181
Laakam and Essingan, 182
L'Alliance Français, 242
L'Alliance pour la Démocratie et le Développement, 226
Labarang, Mohamad, 78
La Baule, 103
La Caravane, 121
La Fédération Internationale des Editeurs de Journaux, 121
Lake Chad, 227
Lamarchand, René, 20, 21
Lamido of Rey Bouba, 171
La Nouvelle Expression, 3, 9, 120, 121, 170, 177, 195
La Prince et le Scribe, 118
La Vision Hebdo, 138
Le Canard Enchaîné, 120
Lecco, Felix Sabal, 75
Le Combattant, 118
Le Courrier, 121
L'Effort Camerounais, 54, 117, 223–224
Le Jeune Observateur, 138
Le Messager, 5, 8, 77, 82, 117–123, 127, 171, 197–198
Le Monde, 195
Le Patriote, 121
Leviathan, 1
Le Vine, Victor, 22, 23, 64–65, 93

Liberal Democratic Alliance, 167
Liberty Laws, 109, 115
Liberty Square, 128
Limbe, 49, 79, 150, 181, 198
L'Organisation Camerounaise pour la Liberté de la Presse (OCALIP), 122–123
Littoral. See Littoral Province
Littoral Province, 35, 79, 126–127, 135, 150, 160, 166, 200
Louis XV, 195
Loum, 57
Lycée LeClerc, 227

Maidadi, Seidou, 179
Makas, 94
Mali, 5, 123, 125
Mandela, Nelson, 5, 20, 124
Manifeste du Front National Unifié, 43
Mankon, 78
Maquisard, 128, 130
Maroua, 116, 137, 179, 197, 201, 204
  congress, 204–205, 242
  SDF national congress at, 116, 166, 179, 204–205
Marxist, 13–14
Maryknoll Order, 222
Mass Tea Party, 2
Massaga, Woungly, 132, 198
Massock, Mboua, 127
Mayi-Matip, Théodore, 38, 43
Mbaku, John Mukum, 182
Mbalmayo, 232
Mbam, 171
Mbappe, Robert Mbella, 198
Mbembe, Achille, 3, 24, 25, 26, 27, 29, 127, 136, 182, 189, 206, 229, 243
Mbere Division, 179
Mbida, André-Marie, 36, 37, 43, 71
Mbo, 198
Mbouda, 128, 130
Mbororo, 232
MBOSCUDA, 232
Mboumoua, William Eteki, 61 (n 70), 95, 240
Mecca, 180

Index

Meiganga, 138, 234
Meillassoux, Claude, 8
MEKONGO, 119
Meloné, Thomas, 81, 113 (n 82), 226
Mendankwe, 6, 7, 116, 234
Mengueme, Jean-Marcel, 78
Messmer, Pierre, 141
Mfoundi, 150
Mifi Division, 150
Military Operational Command, 139
Ministry of Finance, 123
Ministry of Industrial and Commercial Development, 180
Ministry of Public Function, 148
Ministry of Territorial Administration (MINAT), 127, 123, 139, 147, 149, 151, 160, 196, 199, 205, 208
Minute by Minute, 91
Mitterrand, François, 64, 70–71, 73, 103, 109, 124, 241
Mitterrand, Jean-Christophe, 172, 242
Mobutu. See Mobutu Sese Seko
Mobutu, Sese Seko, 17, 73, 228, 241
Monga, Celestin, 5, 21, 27, 121, 127, 234
  open letter to Biya, 117
  trial, 117
Monsieur Sans Objet, 126
Mouelle, Ebenezer Njoh, 80
Moumié, Félix-Roland, 40
Mouvement d'Action Nationale Camerounaise (MANC), 38, 43
Movement for Democracy and Progress, 160, 198
Mouvement for the Defence of the Republic (MDR), 145, 192, 179, 200
Mouvement Progressiste (MP), 133
Mukong, Albert, 92, 162–165
Multipartisme administratif, 189
Muna, Ben, 148, 202–203
Muna, Solomon Tandeng, 45, 46, 48, 60 (n 43), 75, 88 (n 76), 167–168, 220
Mundang, 179
Mungo Division, 150

Municipal Stadium, 135, 142, 168
Munzu, Simon, 167–168, 183
Muslim, 65, 72, 94, 97, 132, 145, 173–174, 222
Muslim-Christian rivalry, 74
Muslin-Fulbe, 72
Musonge, Peter Mafany, 199
Mustapha, Hamadou, 161, 171
Mveng, Engelbert, 195, 222, 224, 232
Mvodo, Ayissi Victor, 52, 219
Mvomeka'a, 100, 131

Nairobi, 1, 2
Namibia, 5
National Alliance for Democracy, 199
National Assembly, 38, 50, 81, 83, 108, 125, 142, 145–150, 171, 179–186, 189, 194–195, 198, 206–208, 221, 240, 242
  and balance of power, 115
  and presidency, 115
  opposition, 161
National Charter of Freedom, 76–77, 92
National Charter of Liberties, 185
National Coordinating Committee, 45
National Coordination for Democracy and Multiparty System (NCDM), 103
National Coordination of Opposition Parties and Associations 206
National Democratic Institute for International Affairs (NDI), 148–149, 199
National Electoral Commission, 187
National Electric Company (SONEL). See Société Nationale d'Electricité
National Gendarmarie. See gendarmerie
National Liberation Front, 35
National Party for Progress, 160
National Petroleum Company (SNH), 75
National Produce Marketing Board (NPMB), 75, 99

National Union for Democracy and
    Progress (NUDP), 107, 121, 132,
    135, 141, 143, 145, 147–148, 150,
    171–177, 184–192, 197–207, 220,
    240
Nchankou, Adamou, 104
Ndando, Victoria Tomedi, 133
Ndi, John Fru, 5, 29, 105, 120, 124, 128,
    137, 139, 143, 147–148, 150–151,
    196–207, 237–245
  among anglophones, 134
  and Social Democratic Front, 5, 28,
    29
  CPDM legislative candidate, 28
  meeting with the Clintons, 161
  populist, 135
  SDF phenomenon, 28
  The Chairman, 218
Ndioro, Justin, 171–172, 174, 177, 180
Ndongmo, Albert, 104
Ndongo, Jacques Fame, 118, 121, 226
Neville, Rubin, 57
New Bell, 239
New Deal, 67, 75, 77, 81, 82, 84, 93, 95,
    115, 225
  administration, 99
  congress, 78, 80
  programs, 98
New Plan of Action, 142
New Social Order, 92, 189
Ngafor, James Chi, 104
Ngango, Georges, 78–80
Ngaoundere, 55, 72
Ngayap, Pierre Flambeau, 28, 39, 227
  and la classe dirigeante, 41, 52, 84,
    93, 107, 115, 219
NGO, 239
Ngondo, 78, 182, 240
Nguema, Macias, 14, 55
Ngugi wa Thiong'o, 1, 2, 9, 19, 21
Nigeria, 3, 7, 68, 97, 162, 181, 189, 206,
    222, 236, 240–241
Nigerian, 2, 6, 122, 129
Ninyim, Pierre Kamdem, 42–43
Njana, Mono, 110 (n 16)
Njangi, 6, 116, 130, 221

Njawe-Monga trial, 124
Njawe, Pius, 82, 117, 121–122
Njinikom, 233
Njoya, Adamou Ndam, 132, 134–135,
    137, 140, 143, 148–150, 160, 186,
    192, 200–208, 220, 235
Njoya, Ibrahim Mbombo, 71, 106
Nkemayong, Paul, 90
Nkongsamba, 104, 144, 150, 160, 198
Nlend, Henri Hogbe, 147
Norris, Chuck, 127
North America, 136
North-South axis, 26, 38, 145, 219
Northern alliance, 174
Northern Christian, 71
Northern Muslim, 65–68, 74, 96–97
Noun Division, 70, 145, 150
Nouvelles du Cameroun, 181
Nso, 78, 224
Ntarinkon, 217, 229. Also see
    Ntarinkon Park
Ntarinkon Palace, 201
Ntarinkon Park, 105, 196, 204
Ntumazah, Ndeh, 132
NUDP News, 184
Nujoma, Sam, 5, 20
Nyerere, Julius, 217
Nyong-et-Sanaga, 38

Ojuku, 54
Okala, Charles, 43
Omgba, Damase Omgba, 110 (n 18)
Omnes, Yves, 242
Onana, Antoine, 104
Onobiono, James, 79
Operation Antelop, 98
Operational Command, 125, 142
Operational Command Structure. See
    Operational Command
Organization of African Unity (OAU),
    95, 195, 228, 240
Ouandié, Ernest, 57, 104, 162
Owana, Charles Onana, 53
Owona Constitution, 184
Owona, Joseph, 181–194
Oyono, Ferdinand, 23

Index

Paribas Cameroun, 100
Parti de la Solidarité du Peuple, 132
Parti des Démocrates Camerounais, 36
Parti des Travailliste Camerounais, 43
Parti Progressistes, 42
Parti Socialistes Camerounais, 43
Paysans Indépendants, 38
Péchiney-Ugine, 57
Pefok, Jomia, 138
Perspectives Hebdo, 171–172, 174
Pidgin, 128
Pidginophone, 219
Pokam, Sindjoun, 110 (n 16)
Pondi, Paul, 209 (n 2)
Popoli, 120
Portuguese, 16
Preliminary Draft Constitution, 183
Presbyterian, 225
Presidential Guard, 73
Presidential Monarchist, 118
Presidential Monarchy, 189
Priso, Paul Soppo, 221, 240
Produce Marketing Board (ONCPB). See National Produce Marketing Board.
Proposition de Loi Portant Revision de la Constitution, 185
Protestant, 225

Québec, 23, 169, 182

Radio France International, 73
Rallié Upecistes, 38–39
Ramadier, Jean, 36–37
Rambo, 234
Reagan, Ronald, 99
Republic of Cameroon, 4, 40, 83, 92, 96, 163
Republican Guard, 69, 74
Ring Road, 236, 238
Rodney, Walter, 15
Roitman, Janet, 243–244
Rosicrucians, 227. Also see Rosicrucian Order.
Rosicrucian Order, 220

Rotoprint, 122
Rubin, Neville, 57
Rural Development Bank (FONADER), 75
Rwanda, 3, 245

Sack, Joseph, 79
Salatou, Adamou, 71
Sanaga-Maritime Division, 38, 81
Sankara, Thomas, 89
Sans Objet, 126
Saro-Wiwa, Ken, 21
Sauveteurs, 231
Schatzberg, Michael, 8, 218, 243
Schilder, Kees, 173, 174, 179
Schwarzenegger, Arnold, 124, 127
Semengue. See Semengue Pierre
Semengue, Pierre, 72, 74
Sende, Joseph, 91
Senegal, 36, 68, 73, 89
Senegalese, 63
Senghor, Léopold, 4, 36, 63
Senior Divisional Officer (SDO), 139
Sentinelle, 122
Service des Etudes et de la Documentation (SEDOC), 39, 54
Showa Arabs, 180
Sierra Leone, 63
Social Democratic Forum, 199
Social Democratic Front (SDF), 5, 27–29, 105–106, 108, 116, 120–121, 127–128, 131, 133, 135, 138, 141–143, 149–150, 162, 164, 168, 179–180, 189, 193–208, 237–244
  among Anglophones, 136
  Bafoussam congress, 189
  Bafoussam convention, 166
  bloody birth, 117
  CDU split, 148
  coalitions, 167
  credo, 139
  draft constitution of the Federal Republic of cameroon, 189
  first national convention, 147

led-coalition, 160
national executive committee, 203
SDF Echo, 131
Socialist Party, 64
Société Aluminium du Cameroun (ALUCAM), 57
Société Camerounaise de Banque (SCB), 99–100, 221
Société de Développment du Coton (SODECOTON), 169–172, 174–180, 220
Société des Transports Urbains du Cameroun (SOTUC), 119–120, 170
Société Mobilière d'Investissement du Cameroun (SMIC), 170–174, 176, 179, 180, 182
Société Nationale d'Electricité (SONEL), 49, 75
Société Nationale de Raffinage (SONARA), 91
Société Nationale des Eaux du Cameroun (SNEC), 237–238
Somaliland, 164
Sonac Street, 134, 165, 184, 188
South Africa, 3, 5, 92, 244
Southern Cameroons, 35, 40, 41, 96, 166. Also see British Southern Cameroons
Southern Cameroons National Council (SCNC), 167–169, 189, 192, 220
Southern Cameroons Peoples' Conference (SCPC), 167
Southern Christian, 65–66
Sovereign National Conference (SNC), 121, 123, 128, 133, 138, 143, 161, 183, 202, 225, 230
rejected by Biya, 131
Soviet Union, 54, 82, 102
Soviet. See Soviet Union
Soweto, 234
Soyinka, Wole, 21
Stalin, 45
Stark, Frank, 47
Stevens, Siaka, 63

St. Joseph's College, 227
Structural Adjustment Program (SAP), 101
Sunji, Nganso, 72
Supreme Court, 91, 149–151, 227
Switzerland, 40

Tabi, Emmanuel Egbe, 45, 46, 52
Takumbeng, 128, 151, 233
Takumbeng AMAZONS, 151
Tanzania, 4, 63
Tataw, James, 72, 125, 227
Tchangue, Pierre, 79
Tchinaye, Vroumsia, 53
Tchiroma, Issa, 161
Tchoungui, Simon, 46
Teachers Association of Cameroon, 227, 235
Tekam, Jean-Michel, 107, 112 (n 65), 113 (n 70)
The Herald, 119, 121, 123, 171, 192
The Messenger, 122
The New Social Order, 92
Third World, 97
Tiko, 49
Tonkam, Senfo, 124
Tontine, 6, 100, 116, 221
Torré, Xavier, 37
Tripartite Conference, 121, 141, 143, 183
Tripartite Talks. See Tripartite Conference
Tripartite Technical Committee, 183
Tsoungui, Gilbert Andze, 71, 74, 111
Tumi, Cardinal, 107, 113 (n 81), 192, 222–224, 238, 240
Tutu, Desmond, 152

Um Nyobe, Ruben, 24, 26, 38, 40
UNESCO, 123
Unified Front, 43
Unified Party, 43
Unilateral Declaration, 169
Union Camerounaise (UC), 38, 41, 42, 44
Union Démocratiques des Femmes Camerounaises, 39

Union des Populations du Cameroun
(UPC), 3, 24, 26–27, 35–44, 58,
80, 83, 91, 131, 135, 141, 143,
147–148, 150, 198–200, 206, 240
  armed resistance by, 35
  armed struggle, 109
  fragment, 145
  insurection, 36
  Kodock, 161
  MANIDEM, 160
  radical, 37
  rebels, 38, 39
  rebellion, 39–40
  reinstated, 123
  resistance, 38, 104
  terrorists, 39
  uprising, 73
United Kingdom, 167
United National Front, 43
United Nations, 162–164, 167–168
  Charter, 190
  flag, 168
  security council, 241
  trust territory, 35
United Republic, 4, 50, 53. Also see
United Republic of Cameroon
United Republic of Cameroon, 69, 83, 92, 96
United States of America (U.S.A), 102, 167, 189, 199, 242
Universal Declaration of Human Rights, 190
UPCistes, 3
Urban Transport Company (SOTUC), 75
U.S.A., See United States of America

van de Walle, Nicolas, 5, 20, 33 (n 46), 76, 195
Vatican II, 223
Verdzakov, Paul, 223–225, 237
Vichist, 226
Victoria. See Limbe
Vidal, Gilles, 242
Vietnam, 35

Villes Mortes, 116, 126, 128, 130, 136, 139–140, 142, 230, 235

Wadjiri, Ibrahim, 71
Wakai, Nyo, 151
Wallerstein, Immanuel, 15
Washington, D.C., 148, 161
WCPDM, 81
West Africa, 77, 219
West Cameroon, 44–54
West Cameroon Electric Power (POWERCAM), 49
Wirba, Dinayen, 224
World Bank, 7, 99, 101, 103, 172, 175, 239
  report, 6
  structural adjustment program, 101
World Cup, 7
Wouri Division, 81, 150, 240

Yaounde, 7, 29, 41, 43, 48, 51, 55, 67, 69, 70, 73–74, 91–94, 96, 104, 108, 116, 119, 124–125, 127–128, 130, 132, 138–140, 145, 150–151, 180, 190–192, 194, 201, 203, 236–237, 242
  military garrison, 110 (18)
  tripartite, 207, 230. Also see Tripartite Conference
  declaration, 141–143, 147, 206–207
  university of, 55–56, 77, 96, 105, 113 (n 75), 202, 204
Yaya, Moussa, 67
YCPDM, 81
Yondo, Black, 5

Zaire, 8, 27, 73, 228. Also see Congo (kinshasa)
Zairois, 8
Zambia, 5, 20, 125, 167
Zartman, William, 244–245
Zebaze, Benjamin, 122
Zéro Mort, 124, 126
Zero Option, 169
Zoa, Jean, 113 (n 81), 222

DATE DUE

DEMCO 38-296